In Search of Fame and Fortune:

The Leahy Family of Engineers 1780-1888

Brendan O Donoghue

This publication has received support from the Heritage Council under the
2006 Publications Grant Scheme

SUPPORTED BY THE HERITAGE COUNCIL

LE CUIDIÚ AN CHOMHAIRLE OIDHREACHTA

Published in Ireland by
Geography Publications,
Kennington Road,
Templeogue, Dublin 6W.

© Brendan O Donoghue

ISBN 0 906602 920

Design and typesetting by Keystrokes Digital, Brunswick Place, Dublin 2.
Printed by The Leinster Leader, Naas, Co. Kildare.

For Bernie

Contents

List of illustrations	vi
Introduction and acknowledgements	ix
Abbreviations	xiv

THE LEAHYS IN IRELAND, 1796-1846

I	Patrick Leahy – Country Surveyor	3
II	A Family Business Develops, 1830-34	33
III	First Cork County Surveyors, 1834	51
IV	Road Building in Cork, 1834-46	73
V	Public Buildings in Cork	97
VI	Private Practice, 1834-46	105
VII	Railway Engineering, 1836-46	125
VIII	Loss of County Surveyor posts	143

THE LEAHYS ABROAD, 1846-88

IX	At the Cape of Good Hope, 1849-51	159
X	Edmund and Matthew in Turkey, 1852-58	175
XI	Family Matters, 1852-60	197
XII	Matthew and Denis in Trinidad, 1860-63	207
XIII	Edmund Leahy in Jamaica, 1858-63	221
XIV	The Jamaica Tramway Scandal, 1862-65	233
XV	Edmund Leahy and the Cashel Election, 1868	257
XVI	Edmund Leahy, 1864-88	273
XVII	The Last Years of the Leahy Family	283
XVIII	Conclusion	293

Appendix I: Family correspondence in the papers of Archbishop Leahy	303
Appendix II: Memoranda of a Visit to the Site of the Ruins of the Ancient City of Sizicus in Asiatic Turkey, by E Leahy CE, 1857	311
Bibliography	315
Index	329

List of illustrations

FIGURES

Roads in West Cork laid out by Edmund Leahy, 1834-46	79
Railways promoted by the Leahys, 1844-46	131
Cape Town and adjoining areas in the 1860s	163
Turkey and adjoining countries, ca 1850	179
Trinidad in 1844	209
Jamaica in the 1860s	226
The Jamaica railway system in the 1880s	249

PLATES

Land surveyors at work	6
Advertisement in *Clonmel Advertiser*, 1818	16
Map of the town and lands of Thurles, 1819	18
Killenaule Estate Rental Reference Book, 1828	20
Patrick Leahy's *Tables of Weights & Measures*, 1826	24
Anne Street, Clonmel	36
St Patrick's College, Thurles, 1850	37
Engraved map of the City of Waterford and its Environs, 1834	40
Advertisement in *Tipperary Free Press*, 1833	42
Edmund Leahy's plans for a suspension bridge, 1833	44
Documents relating to Patrick Leahy's admission as an Associate of the Institution of Civil Engineers (London), 1834	46
Grand Jury (Ireland) Act, 1833	53
Patrick Leahy's first report to the grand jury of County Cork	59
Plans for proposed Allihies – Castletownbere road, 1836	81
Sketch of tunnel on proposed Allihies – Castletownbere road	81
Fly-over bridge at Douglas, Cork	85
Bridge at Bandon, County Cork	87
Plaque on Bandon Bridge	88
Donemark Bridge, Bantry	89
Title page of Edmund Leahy's *Treatise on Making and Repairing Roads*	91
Leahy's offices at South Mall, Cork	107

Lismore workhouse	113
Swivel bridge at Blennerville, Tralee	116
Report on Bandon Navigation, 1842	119
Cork & Bandon Railway plans, 1844	128
Cork & Bandon Railway Act, 1845	130
Turning the first sod for the Cork & Bandon Railway	133
Convict agitation at the Cape Colony	166
Patrick Leahy's application for a position at the Cape Colony	167
Edmund Leahy's Statement of Claim on the Ottoman Government	187
Archbishop Patrick Leahy	198
Denis Leahy's plans for a new building at the Colonial Hospital in Port of Spain, Trinidad	216
Letter of 10 January 1859 from Edmund Leahy in Jamaica	225
Prospectus of the Jamaica Tramroad, 1862	236
Jamaica election returns, 1863	240
Plans of the Jamaica Tramroad	241
Reports on the Jamaica Tramway presented to the House of Lords, 1865	251
Cashel Election Commissioners Report, 1869	266
Cathedral of the Assumption, Thurles	285
Edmund Leahy's death certificate	288
Edmund Leahy's grave	289

COLOUR PLATES
MAPS PREPARED BY THE LEAHYS, 1811-34
(Between pages 16 and 17)

Part of the map of the Lough Neagh District submitted to the Bogs Commissioners in 1813
Map of the bogs in Derry and Antrim, north of Lough Neagh
Map of the Barony of Slievardagh, 1818
Table of Mineralogical Reference and Geological Description in Slievardagh map, 1818
Killenaule, New Birmingham and the proposed extension of the Grand Canal as shown on the Barony of Slievardagh map, 1818
Mullinahone and district, as shown on the Barony of Slievardagh map, 1818
Kilcooly, as shown on the Barony of Slievardagh map, 1818
Map of the Coal District near Killenaule, 1824
Book of 51 watercolour maps of Waterford city and corporation lands, 1831-32
Section of one of the 51 watercolour maps of Waterford, 1831-32
Waterford city centre, 1831-32

Section of the manuscript map of Waterford showing the Artillery Barracks, the Infantry Barracks and the Jails, 1831-32

Detail of house and grounds at Woodstown, Waterford, 1831-32

Map of Waterford Corporation land in the townland of Skibbereen showing numbered fields

Engraved Map of the City of Waterford and its Environs, 1834

Table of Explanation and Reference from the engraved map of the city of Waterford and its Environs

Part of Waterford city centre as shown on the engraved map of the city of Waterford and its Environs

Book of sections of the sewers of the city of Waterford, 1833

Comments and recommendations by Patrick Leahy on the gradients of the Waterford sewers

Sections of two of the Waterford city sewers, 1833

Introduction and acknowledgements

The career of Patrick Leahy (1806 -1875), Catholic Archbishop of Cashel and Emly from 1857 to 1875, is well documented.[1] But why should anyone attempt to write a biography of the archbishop's father, also Patrick Leahy, a land surveyor and engineer who practiced mainly in North Tipperary in the early decades of the nineteenth century, or of the archbishop's younger brothers, Edmund, Denis and Matthew Leahy, who followed their father into the emerging civil engineering profession in the 1830s? Why should these four engineers be rescued from the historical obscurity reserved for the hundreds of their contemporaries who are not ranked among those who made a substantial or lasting contribution to society? And why, of the many engineers who worked in Ireland in the first half of the nineteenth century, should one set out to document the lives of men who did not complete a single major engineering or architectural project for which they would deserve to be remembered.

My interest in the Leahy family began more than twenty-five years ago when, long after it had been closed down and dismantled, I set out to trace the events that led to the establishment of the Cork & Bandon Railway which I had known and travelled in my youth. The readily available sources identified Edmund and Patrick Leahy as the engineers who had promoted the company and designed the line in 1844, but provided little or no information about them, except the fact that they were employed as county surveyors by the County Cork Grand Jury. This immediately whetted my interest: as a senior official in the then Department of Local Government, and applying twentieth-century standards in relation to conflicts of interests and double-jobbing, I was struck by the fact that two senior public officials had apparently been able to act as railway promoters on a large scale (and to carry on a substantial private practice in engineering and architecture as well) while continuing to hold their public offices. Subsequent research led me to pursue three interlinked lines of enquiry:

- the origins and early years of the county surveyor system itself and the backgrounds and achievements of the men who became the first county surveyors in 1834, holding public offices which were the first in Ireland or in England to be filled on a merit basis instead of by patronage;

- the circumstances which led to the promotion and establishment of the Cork & Bandon Railway, the engineering and financial issues involved, and the railway mania which developed in West Cork in 1845-46; and

Introduction

- the lives of the Leahy family themselves, their careers in Ireland up to 1847, and their subsequent activities abroad.

Instead of attempting to include the results of these three strands of research in one unwieldy volume, this book focuses primarily on the careers of the Leahys while, at the same time, outlining the origins of the county surveyor system and the frenetic but short-lived rush to build a network of railways in West Cork in the 1840s.

Patrick Leahy senior was in practice as a land surveyor from 1796 onwards and his three sons had joined him in his surveying and engineering practice by 1834. Although their combined careers involved work in Ireland, England, Europe, Africa and the Americas between 1796 and 1888, there are few references to the Leahys in published work on the history of engineering in Ireland or elsewhere. Patrick's activities in the period up to 1830 are briefly noted in *Plantation Acres* - the definitive work by J H Andrews on the hundreds of land surveyors who practiced in Ireland - with the apt comment that these surveyors' ambitions after about 1815 could no longer be expected to leave much trace for posterity.[2] In Leahy's case, however, and almost by way of exception to the general rule, it emerges that a substantial body of his work as surveyor/cartographer in the 1820s and early 1830s has survived.

Between 1834 and 1846, Patrick and Edmund Leahy served as the first county surveyors for the East and West Ridings of County Cork, respectively, aided by Denis and Matthew Leahy, but their road building and other activities in these important positions are rarely mentioned in writings on the nineteenth-century development of the county. Colin Rynne's magnificent book *The Industrial Archaeology of Cork City and its Environs* does, however, note the Leahys' "injudicious" appointments as county surveyors for adjoining areas, their willingness to court controversy, their neglect of their grand jury duties in favour of private work and their eventual dismissal.[3] Moreover, Frederick O'Dwyer's study of *The Architecture of Deane and Woodward*[4] notes the difficulty experienced by other engineers in obtaining grand jury work in Cork in the 1840s in competition with the Leahys. David M Nolan, in the course of a comprehensive but unpublished study of the activities of the County Cork Grand Jury between 1836 and 1899,[5] comments on these issues in greater detail, describing the Leahys' as not being entirely honest in their dealings with the grand jury, and noting the jury's lack of decisiveness in dealing with Edmund even when he had publicly demonstrated his contempt for his employers.

With the exception of Edmund Leahy's activities in relation to the Cork & Bandon Railway, which were described as "blundering buffoonery" in a paper written by Joseph Lee in 1967,[6] the extensive railway promotion and engineering work undertaken by

In search of fame and fortune: the Leahy family of engineers, 1780-1888

Edmund and other family members in Ireland during the 1844-46 period, when the railway mania was at its peak, has seldom been noted in the extensive literature on the history of railways in Ireland. And finally, while the three sons of Patrick Leahy pursued careers in engineering and engaged in a variety of other activities outside Ireland from 1848 to 1888, no notice of these activities has ever been published, notwithstanding the fact that the files of the British Colonial Office, Foreign Office and War Office, and the volumes of the British Parliamentary Papers, probably contain more material generated by or relating to the Leahys – particularly to Edmund - than to any other Irish engineers of the period.

Against this background, and while acknowledging that the actual achievements of the Leahys would not qualify them for inclusion in the top rank of Irish nineteenth century engineers, I hope that this study will be of interest under a number of headings. In a limited way, it complements the magisterial overview by J H Andrews of the Irish country surveyors, by presenting a picture of how Patrick Leahy, one of these surveyors, attempted to survive and to adapt to changing circumstances in the early decades of the nineteenth century, when the establishment of the Ordnance Survey in 1824 seriously affected the employment prospects of the traditional surveyor. It sketches the establishment of the county surveyor system in 1834, the activities of the Leahys in this new role and their impact on infrastructural development in Cork in the following years. It outlines the scale of the railway mania in West Cork in the 1840s and describes some of the abortive schemes promoted by the Leahys in the area during those years. And finally, it documents how readily the members of one family of engineers were able to gain access to Secretaries of State and senior officials in the Foreign and Colonial Offices in the years after 1847, and to obtain public employment in various colonies and further afield; this may help to promote interest in the activities generally of Irish engineers and other professionals in the administration of the British Empire in the middle years of the nineteenth century, a subject which has yet to be comprehensively studied.

Although it was never my objective to trace the Leahy family history, I found myself irresistibly drawn into doing so on discovering that some family correspondence had survived in the papers of Archbishop Patrick Leahy in the Cashel Diocesan Archives.[7] It is, of course, notoriously difficult, without prior inside or family information, to trace in any detail the history of a relatively "ordinary" landless Irish Catholic family originating in the early nineteenth century. However, the fact that, after 1847, most members of the Leahy family lived at least periodically in England (for which records of births, deaths and marriages from 1837 onwards, and census records from 1851 to 1901, are readily available on the internet), made it possible to piece together much of the life history of the eight children of Patrick Leahy and his wife, Margaret, including their four daughters, and to

Introduction

identify some grandchildren who were alive in the early decades of the twentieth century. This would not have been possible if the family had remained in Ireland where Census records for the years before 1901 have been destroyed, and where civil records of births, marriages and deaths began only in 1867, are still not available in electronic searchable format, and can be searched in their original form only with difficulty.

My research on the Leahy family was carried out in two phases, separated by almost twenty years, because of the demands of my public service positions. I am grateful to many people for their advice and assistance at various stages. Among these are Professor J H Andrews, the late Professor Sean de Courcy, Dr R C Cox, Dr Arnold Horner, Dr Freddie O'Dwyer, Peter J O'Keeffe, Dr Edward McParland, the late Professor Simon H Perry, and Dom Mark Tierney OSB. Others who helped in particular ways were Michael Doody, former Waterford City Manager; Jerry Guerin, Superintendent, St Mary's Catholic Cemetery, Kensal Green, London; the late Rev John J Lambe PP, Gortnahoe and Glengoole; Rev Vincent Leahy OP, Newbridge College, Co Kildare; Sheila Macauley, Boston; Professor Donal P McCracken, University of Durban-Westville, Durban, South Africa; Donal and Nancy Murphy, Nenagh; Monsignor Christy O'Dwyer, formerly Cashel and Emly Diocesan Archivist; Michael O'Malley, Senior Engineer, South Tipperary County Council; and Glenroy Taitt, Archdiocesan Pastoral Centre, St Augustine, Trinidad. Helen Kelly, professional genealogist, Dublin, was enormously helpful in tracing the early history of the Leahy family in Ireland and I valued her advice on Leahy family matters generally.

I am indebted also to a large number of archivists and librarians: Julia Barrett, Architecture Library, University College, Dublin; Tim Cadogan, Reference Librarian, Cork County Library; John Callanan, Institution of Engineers of Ireland; Joan de Beer, Deputy National Librarian, National Library of South Africa; Siobhan Fitzpatrick, Royal Irish Academy; Stewart Gillies, Reference Specialist, and Ed King, British Library Newspaper Library, Colindale, London; Seamus Helferty, Principal Archivist, University College, Dublin; Mary Higgins, Trinity College, Dublin; Bridget Howlett, Senior Archivist, London Metropolitan Archives; Rachel MacGregor, Archivist, Birmingham City Archives; Donal Moore, Waterford City Archivist; Mrs M K Murphy, Archivist, Institution of Civil Engineers, London; Mary O' Doherty, Royal College of Surgeons in Ireland; John O' Gorman, Local Studies, Tipperary Libraries, Thurles; Colm O' Riordan and Ann-Martha Rowan, Irish Architectural Archive; Dr Matthew Parkes and Petra Coffey, Geological Survey of Ireland; David Sheehy, Dublin Diocesan Archives; Sarah Strong, Archives Officer, Royal Geographical Society, London; Martin C Tupper, Islington Local History Centre, London; and Julia Walsh, Documentation Officer, South Tipperary County Museum, Clonmel.

In search of fame and fortune: the Leahy family of engineers, 1780-1888

Research for the later chapters of this book would have yielded limited results but for the magnificent service provided by The National Archives (formerly the Public Record Office), London. The on-line catalogue, the efficient document ordering and delivery systems and the self-service copying facilities available at Kew make it a pleasure to work there. At the National Archives of Ireland, I much appreciated the help of the Director, Dr David Craig, Catriona Crowe, Aideen Ireland and John Delaney. My greatest debt, however, is to the National Library of Ireland where, for many years before I had any official link with the institution, front-line staff in the Reading Rooms were invariably friendly and helpful and willing to offer sound advice to a novice researcher. I am grateful, in particular, to Jim O'Shea who helped me in the 1980s to understand how best to make use of the British Parliamentary Papers. In more recent years, Elizabeth Kirwan and Tom Desmond have provided invaluable assistance with manuscript collections and manuscript maps, while Joanna Finnegan has been of enormous help in accessing the Library's prints and drawings collection and the Ordnance Survey map collection. I am very grateful also to other members of the staff of the Library for assistance of various kinds, especially Kevin Browne, Brian McKenna, Gerry Lyne, Sandra McDermott, Bernard Devaney and Sara Smyth, and to David Monahan for his photographic work.

This book would never have been written but for Dr William Nolan who, because of his own personal interest in all aspects of Tipperary history, has encouraged me for several years to complete my research on the Leahys. Dr Nolan gave generously of his time in reading and commenting on successive drafts and offered valuable suggestions on the content. I am indebted also to him and to his wife, Teresa, for undertaking publication of the work under the Geography Publications imprint.

Finally, my long-suffering wife, Bernie, has not only had to live with my Leahy obsession for over twenty years but has also been induced to undertake archival and other research in Ireland and in London during those years. In recognition of her support in this and in so many other matters, the book is dedicated to her.

REFERENCES
1. Christopher O'Dwyer, *The Life of Dr Leahy, 1806-1875*, MA thesis, St Patrick's College, Maynooth, 1970; CE 28 January 1875; Dom Mark Tierney, OSB, Calendar of the papers of Dr Leahy, Archbishop of Cashel, 1857-1875, NLI Special List 171; aspects of Dr Leahy's career are extensively dealt with by Emmet Larkin in some of the seven published volumes of his history of The Roman Catholic Church in Ireland between 1850 and 1891.
2. J H Andrews, *Plantation Acres: An Historical Study of the Irish Land Surveyor and His Maps*, Ulster Historical Foundation, 1985, page 354
3. Colin Rynne, *The Industrial Archaeology of Cork City and its Environs*, Stationery Office, Dublin, 1999, page 179
4. Frederick O'Dwyer, *The Architecture of Deane and Woodward*, Cork University Press, 1997
5. David M Nolan, *The County Cork Grand Jury, 1836 – 1899*, A Thesis for the Degree of Master of Arts, UCC, 1974
6. Joseph Lee, "The Construction Costs of Early Irish Railways 1830-1853", in *Business History*, Vol 9, 1967
7. These letters are reproduced in Appendix I.

Abbreviations

BC Minutes	Minute Book of the Bogs Commissioners 1809-13 (NAI 1137/77)
BL	British Library
CC	*Cork Constitution*
CDA	Cashel Diocesan Archives
CE	*Cork Examiner*
CO	Colonial Office
CSO	Chief Secretary's Office
CSORP	Chief Secretary's Office Registered Papers (NAI)
DDA	Dublin Diocesan Archives
DGIN	Directors General of Inland Navigation
FO	Foreign Office
HC	House of Commons
ICEI	Institution of Civil Engineers of Ireland
ILN	*Illustrated London News*
IRG	*Irish Railway Gazette*
NAI	National Archives of Ireland
NLI	National Library of Ireland
ODNB	Oxford Dictionary of National Biography
OP	Official Papers (NAI)
OPW	Office of Public Works
OS	Ordnance Survey
TICEI	*Transactions of the Institution of Civil Engineers of Ireland*
TNA	The National Archives, London (formerly Public Records Office)
PRICE	*Proceedings of the Institution of Civil Engineers*
VHD	Diary of Sir Vere Hunt

In search of fame and fortune: the Leahy family of engineers, 1780-1888

DISTANCES

The Irish mile measured 2240 yards as against the English or statute mile of 1760 yards. The yard was divided into 3 feet or 36 inches. The Irish perch measured 7 yards, whereas the English perch measured $5\frac{1}{2}$ yards; 4 English perches (22 yards) made a chain.

CURRENCY

The pound sterling was divided into 20 shillings or 240 pence. Cash amounts are written as £ s d.

MONEY VALUES

The Composite Price Index 1750 to 2003, published by the UK Office of National Statistics, can be used to revalue historical money amounts. Applying the index, and using the current pound/euro conversion rate, values in pounds sterling prevailing in most of the years 1820 to 1880 should be increased by a factor of from 90 to 120 to convert to approximate current amounts in Euro.

QUOTATIONS

Quotations are generally in their original form; spelling, punctuation and the use of capitals have not been modernised.

THE LEAHYS
IN IRELAND
1796-1846

CHAPTER I

PATRICK LEAHY – COUNTRY SURVEYOR

"Memorialist is now an active, steady, experienced man, … has been closely engaged in public and private surveys for twenty-seven years … and fully possessing the requisite qualifications for constituting an experienced Engineer".

Patrick Leahy, 1823[1]

A "steady and deserving" surveyor

Although the family name Leahy (in Irish Ó Laochdha) was common in counties Cork, Limerick and Tipperary in the post-Gaelic period, it cannot be said with any certainty where in Munster the name originated; there appears to be no mention of anyone of the name in either the Gaelic genealogical compilations of the early period, in the Annals, or in the *Leabhar Muimhneach*, a genealogical corpus in Irish which was updated in the eighteenth century.[2] While the more remote ancestry of Patrick Leahy is therefore obscure, it is clear that he was born around 1780[3] into a family which probably lived on the Vere Hunt estate in County Limerick. The first Vere Hunt, an officer in the Cromwellian Army, arrived in Ireland in 1657 and later settled at Curragh Chase, about eleven miles west of Limerick city on the road to Askeaton.[4] The head of the family in the early years of the nineteenth century was Sir Vere Hunt (1761–1818), grandfather of the poet Aubrey de Vere. An active and generous landlord who enlarged and improved the estate at Curragh Chase and strove for many years to develop another family property at Glengoole in the barony of Slieveardagh in County Tipperary,[5] Sir Vere has been described as "a Protestant landlord of the better type, who evidently got on well with his tenants and with Catholics generally".[6] His diary[7] provides a fascinating picture of life in both rural and urban Ireland in the early years of the nineteenth century, as well as a detailed account of his own estate management and other activities. There are numerous references to members of the Leahy (sometimes Lahy or Lahey) family from which the following picture may be assembled:

- John Leahy died at Curragh Chase in April 1811, having lived there for seventy or eighty years, the last twenty as a pensioner, and was "an old, tried and valuable domestic".

- Tom Lahy was reared and brought up on the estate but left there in 1808 "with his father's instruments of surveying on his back".
- Mick Lahy [also Lahey], father of Tom, was employed occasionally by Vere Hunt on small surveying assignments in County Tipperary and is referred to as "the unfortunate Mick Lahey", because of a drink problem.
- Pat Leahy was also employed as a surveyor by Vere Hunt on projects in County Tipperary but was "steady and deserving", in contrast to Mick Lahey, "his unfortunate relation".

It is possible therefore to suggest, although not with great confidence, that both Mick and Pat Leahy were sons of the John Leahy who died at Curragh in 1811, and that their emergence as land surveyors in County Tipperary in the first decades of the nineteenth century was linked to Vere Hunt's ambitious plans to develop his estate in that county.

New Birmingham
According to Samuel Lewis, Sir Vere Hunt was "struck with the favourable situation" of New Birmingham, formerly Glengoole (Gleann an Ghuail), "contiguous to the coalmines of the Killenaule district" and "used every effort to raise it into manufacturing importance".[8] Having drawn up plans to settle his estate in the area with tenants from Curragh Chase and tradesmen and others from Limerick, Dublin and London, he advised the Lord Lieutenant of his plans in May 1802 and sought support from the Government for what he considered to be "a Public Benefit":

> I have commenced the Building of a Town on my Estate in the County of Tipperary according to the Plan which will be herewith delivered ...
> It is intended to be principally inhabited by English Manufacturers, who it is presumed, considering the particular local advantages of the Place, may be easily encouraged to settle there, provided the Establishment is sanctioned by Government.
> The Situation is unquestionably the most favourable in the kingdom, it is in the Centre of the Richest and most plentiful Part of it, midway between the Capitals of Dublin and Cork, and abounding with every Convenience to induce Settlers of all Descriptions to give it a Preference.
> The Works which have been Commenced with Spirit are at present in very great forwardness, and my Efforts shall be redoubled on the Slightest Assurance of meeting that progressive Support from Government which his Excellency may be pleased from time to time to Consider the Undertaking entitled to, as one of a Public Benefit.

Unless it has the Appearance of being so sanctioned and Countenanced I fear it cannot be carried on to that Extent as to make it a National Object, neither would Sufficient Stability be attached thereto to create that Confidence, and those Ideas of Security which at present it may be necessary to impress on the English Mind.[9]

Having supported the Union in his capacity as a member of the Irish Parliament for Askeaton, and because he believed that his plans would help to convince the people of the benefits of the Union, Vere Hunt appeared to be confident of Government support for the development of his model town. His contacts with major figures in the Dublin administration led him to believe that they would respond favourably to his requests to locate a military barracks, a post office and other public buildings in the town[10] and, in addition, his ambitions were inspired by misplaced hopes for a canal link to the area. But New Birmingham did not prosper as Vere Hunt had planned - it was "comparatively deserted" in 1837, with only about 50 houses and a population of 298.[11]

The nature and extent of Patrick Leahy's involvement in the early development of New Birmingham is not recorded. By his own account, he began his career in land surveying as a young man in 1796-97[12] but he has left no record of his education, training or early experience, nor any clues as to which of the more senior members of the land surveying profession he may have been apprenticed, or with whom he may have worked as an assistant. In 1805, when he married, he had settled at Fennor, just south of Urlingford and only four miles from Vere Hunt's new town. However, specific references to him in Vere Hunt's diary do not occur until August 1813 when he was employed to prepare plans for a bridewell which was to be built at New Birmingham at a cost of not more than £200.[13] Leahy's contribution to the project is, however, far from clear. Vere Hunt had agreed a site for the bridewell with John Neville,[14] an architect, in July 1813 and had sent a plan and estimate, presumably Neville's, to the county treasurer seeking a presentment for the work. But there was opposition among the grand jury to the cost and scale of the proposed building and it was then that Leahy was engaged to come up with a plan for a less costly one. However, a week later, Neville was directed by Vere Hunt to produce an elevation and estimate for the building and did so on 29 August. The bridewell was duly completed and put into service; it was a modest structure, with legal accommodation, a day room, two cells, a female room, two yards and apartments for the keeper.[15] Like most of the other buildings at New Birmingham, including the catholic chapel[16] and the small police station, it should probably be credited to Neville rather than to Leahy.

Leahy's early work
Development projects at New Birmingham would not have required the full-time services of a land surveyor, engineer or architect in the early years of the nineteenth century and

Chapter I — Patrick Leahy Country Surveyor

while it must be assumed, therefore, that Leahy would have been available for other professional engagements, no record of his other work in those years seems to have survived. The *Dictionary of Land Surveyors and Local Map-makers of Great Britain and Ireland, 1530-1850* does not record any activity by Leahy before 1810 nor does *Plantation Acres,* the definitive work by J H Andrews on the Irish land surveyor and his maps.[17] The *Alphabetical List of Irish Architects, Craftsmen and Engineers from the 15th to the 20th century* describes Leahy as "unimportant" and gives no details of any engagements; he is not mentioned in *A Biographical Dictionary of Civil Engineers in Great Britain and Ireland, 1550-1830*; and his name does not appear in the *Directory of British Architects, 1830-1914* which lists more than 11,000 architects who practiced in Great Britain at any time during those years.[18]

The absence of documentation on Leahy's early work must, however, be seen in context. Land surveyors were very numerous and widely dispersed in Ireland at the beginning of the nineteenth century before official maps became available, but most of their output was privately commissioned and often involved no more than two customers, landlord and tenant.[19] The surveyors' everyday work consisted of preparing manuscript maps, plans and surveys which were needed in connection with the management, subdivision and development of large estates, litigation between property owners and tenants, land valuation and the sale and letting of property; documents produced for such purposes would rarely have been engraved or published but would have been retained with

A nineteenth-century print showing a land surveyor and his assistants holding a short chain, possibly the two-pole Irish chain (Brocas, PD 2177 TX, courtesy of the National Library of Ireland).

other estate records or in solicitors' offices. While a remarkable number of maps and surveys of this kind have survived and are now held in the National Library of Ireland and other repositories, no example of Leahy's earliest estate survey work is to be found in these collections. Surveyors of the period might also expect to be engaged from time to time to prepare surveys and maps to support new road proposals for submission to the grand jury but, again, surveys and plans of this kind would generally have been retained by the individual concerned with the presentment rather than by the secretary of the grand jury or by any other public body. The fact that there are relatively few surviving examples of local work of this kind by early nineteenth century surveyors is therefore understandable.

At national level, the first significant assignments which arose for surveyors at the beginning of the nineteenth century derived from an Act of 1805 under which the Post Office authorities were authorised to "to procure and employ ... proper and efficient persons" to survey and make maps of the roads on which the mails were carried and to suggest improvements to achieve the "most level and most convenient" routes.[20] This work was assigned to four surveyors, one of whom, William Larkin, supported by a team of assistants, surveyed 1,740 of the total of 2,761 miles of road involved.[21] Larkin's surveys and manuscript maps included a number relating to the Dublin–Cork route, one of them (dated 1810) covering the section from Durrow via Urlingford to Cashel, and running through the area in which Leahy had been living and working.[22] Larkin was also responsible three years earlier for the survey and mapping of a number of other road sections in County Tipperary. It is tempting to speculate that he may have engaged Leahy to assist in some of this work but there is no indication on the completed maps, or any other documentary evidence, to support this; besides, the fact that Leahy never subsequently advertised any involvement of his with the mail coach road surveys, as he did in the case of other projects, would lead one to dismiss this possibility.

"several proposed extensions of Inland Navigation"

Leahy claimed to have worked on "several proposed extensions of Inland Navigation in Ireland"[23] and that an extension of the Grand Canal in 1811 for the Directors General of Inland Navigation was one of the "several public improvements through the County Tipperary" which he had projected.[24] It seems clear, however, that he was never directly employed by the Directors General but worked on canal schemes between 1809 and 1812 only as an assistant to Thomas Townshend (c.1771–1846); the latter was an experienced engineer who had worked with John Rennie on the design and construction of canals in England before coming to Ireland in 1802 to work, initially, as an assistant engineer on the planning and construction of the Royal Canal.[25]

Ireland was gripped by a form of canal mania in the decade following the establishment in 1800 of the Directors General of Inland Navigation, with a fund of £500,000 granted by the Irish Parliament for the promotion of inland navigation.[26] Among the proposals which emerged was "the Colliery Canal", a scheme promoted by the Queen's County Canal Company in 1803 for extending the Grand Canal from Athy to Castlecomer and Kilkenny, and a scheme designed by David Aher in 1802 for a 33-mile-long canal from Tipperary town to Carrick-on-Suir, with a branch canal from the summit level at Ballydoyle, near Cashel, to New Birmingham, roughly 15 miles away to the east.[27] There was a revival of interest in these two sets of proposals later in the decade when efforts were made to establish Clonmel (which was already linked to Waterford by the 25-mile-long Suir Navigation) as the hub of a major canal network.[28] In 1807, Lord Cahir (later to become the first Earl of Glengall) submitted plans prepared by George Joyce, engineer, for a canal from Cahir to Clonmel.[29] A year later, the inhabitants of the towns of Clonmel, Cahir, Cashel, Tipperary and Templemore, and of the barony of Slieveardagh, petitioned the Government to extend the Suir navigation northwards through County Tipperary to link up with the Grand Canal[30] and in early 1809, memorials were submitted by Kilkenny city and county interests seeking to have the Grand Canal extended to serve the Castlecomer collieries and the city of Kilkenny.[31]

Having considered the various proposals, the Directors General took the view that it would be more desirable "as a National Object" if the Tipperary and Kilkenny schemes could be linked by constructing a canal "from a proper level of the Suir across the Vale to Kilkenny, forming a great communication by a Navigable Canal through that part of the country which appears to be favourable for the Undertaking and is now totally without such an accommodation"; such a canal would serve both the Castlecomer and Slieveardagh collieries and link them to the Suir navigation at Clonmel.[32] In May 1809, they endeavoured to persuade Lord Cahir and the Kilkenny interests to develop a scheme on these lines[33] but decided, a few months later, to take the initiative themselves by engaging Thomas Townshend to carry out surveys of possible canal routes. Townshend agreed to start work on 1 August[34] but, in early October, when his engagement with the Bogs Commissioners was about to commence, he reported that the work had already been delayed by bad weather and because he had been "unavoidably absent from the surveys for a few weeks".[35] In these circumstances, he was allowed to employ a surveyor at a fee of one guinea a day to assist him[36] and it was then that Leahy, presumably because of the local knowledge which he had gained while practicing in the area, was taken on; there is no evidence of any previous professional association between the two men nor any evidence that Leahy had been consulted by those who had been promoting the different canal schemes through County Tipperary in the previous ten years.

Townshend reported in November 1809 that he had taken the levels and completed the surveys for a canal from Athy via Castlecomer to Kilkenny, but he was urged by the Directors General to press on with the work of tracing a line "through the Vale which lies between the Castlecomer Hills and the Suir"[37] because that was "the measure which the Board most particularly desire to have under their consideration as soon as may be".[38] In March 1810, when Townshend submitted "the results of his searches" for such a line, together with five sections of the different lines he and Leahy had surveyed, he was told that the Directors General would speedily give his findings their full consideration and was asked to send on his cost estimates for the lines.[39] There seems to have been no further communication between Townshend and the Directors General until the following November when he told them that he had competed his surveys of all of the different lines he had laid out on their instructions and sought directions as to the scale of the maps he was to provide; he had already made rough maps on a scale of four inches to the mile and was told to produce separate surveys of each line at this scale, to save time.[40]

The early work carried out by Townshend and Leahy may have provided the basis for a report of February 1810 from the Directors General to the Lord Lieutenant which stated that a canal which would serve Roscrea, Templemore, Urlingford, Thurles, Holy Cross, Cashel, Cahir and Clonmel would be practicable, "with cuts of small dimensions and expense suitable to the particular objects of the adjacent towns or of the proprietors of lands, mines, manufactures, etc".[41] However, given the number and variety of canal proposals they had received, the Directors General recommended to the Lord Lieutenant in May 1810 that John Killaly, who had been the Grand Canal Company's principal engineer since 1798, should be engaged to conduct further more general surveys as a framework for future canal investment.[42] When Killaly began work in September, his instructions included a specific direction that he should "complete the surveys which have been in part made between the River Suir and the Grand Canal" and in the following December, after he had worked his way southwards from Monasterevan into Tipperary, he was urged to search for a practicable line "through the midst of the Vale which lies between the Castlecomer and Roscrea Hills" and taking in the many towns along the river Suir.[43]

The fact that the instructions given to Killaly duplicated those given to Townshend more than one year earlier suggests that the Directors General cannot have been very happy with the progress made by the Townshend-Leahy team or with the reports and surveys which they submitted. The assignment on which they were engaged certainly took an inordinately long time - the accounts for fees and expenses which they submitted suggest that their work on "the several lines between the Barrow and the Suir" in Queen's County and in Tipperary and Kilkenny continued intermittently for over two years up to November 1811, during which time both of them were also working on surveys for the Bogs Commissioners.[44] Over and above the delay, serious difficulties also arose in dealing with Townshend's accounts, including the payments claimed by Leahy. Townshend was told, for example, in July 1812 "that the Account of Mr Patrick Leahy, Surveyor, commenced 9th July 1810, nearly one month previous to the date of Mr Townshend's Account" and that "there were several days spent by the Surveyor in "laying on the features of the County, Towns etc on Map"; moreover, he was asked to explain the fact that "the Surveyor was not employed in the Field assisting Mr Townshend in the actual surveys which was the object of the Board's permission of 9th October 1809".[45]

Chapter I — Patrick Leahy Country Surveyor

The records which have survived do not allow for an assessment of what exactly Townshend and Leahy may have produced in terms of reports and plans. Drawings, dated 1810 and bearing Townshend's name, for Grand Canal extensions to Castlecomer and Goresbridge on the Barrow, and from Maryborough (Portlaoise) to Mountrath and Roscrea, are extant,[46] but his plans for the canal south of Roscrea or for a canal from Castlecomer to North Tipperary are not. However, Leahy's 1818 map of the barony of Slieveardagh[47] shows the "Grand Canal Extension as measured by Townshend and me in 1810" and six years later, his map of the Killenaule coal district[48] depicted what he described as the extension of the Grand Canal "as Laid out and Levelled by Townshend and me in 1811 for ye Directors of Inland Navigation".

Unlike Townshend, Killaly completed his assignment for the Directors General with commendable speed. In October 1811, he submitted a report and plans for a canal running from Monasterevan to Mountrath and Roscrea, and from there to Clonmel, passing close to Thurles, Cashel and Cahir; with improvements in the existing navigation to Carrick-on-Suir, this would have involved a total of 94½ miles of new or improved waterway.[49] Killaly thought that this scheme would be "highly advantageous to the country and perfectly practicable" even though it was estimated to cost £630,000. From an engineering point of view, it was to be a complex scheme, with 13 locks between Roscrea and Cashel alone, and another 30 between Cashel and Clonmel, 15 of them in one flight about two miles west of Clonmel.

Killaly's line[50] bore little relationship to that proposed by Leahy and Townshend which would have run very close to the main street of Killenaule and within a few hundred yards of New Birmingham, from which it was to continue north-eastwards towards Freshford,[51] thus meeting the wishes of the Directors General for a link between the Suir valley and the Castlecomer area. However, Killaly did not overlook the case for serving the "Slieve Ardagh Colliery" where, he was informed, "there is an inexhaustible vein of coal and culm, the trade in which, particularly the latter, would be immense as the passion for liming the ground is so prevalent". For this reason, he suggested that a short branch might be constructed to serve the colliery and this "could be made at small expense". Killaly's survey was thought by the Directors General to be "an exposition of what promises to be advantageous and practicable" and the Lord Lieutenant approved of a proposal to print the report and plans and to distribute them to persons who might be interested in the development of the country.[52] But, while Leahy continued into the 1820s to make the case for a canal serving the Killenaule area, no practical steps were taken to advance the project after 1812, although it was considered in 1813 by the Committee on Inland Navigation in Ireland[53] and was given some further consideration in 1825[54] by which stage the canal age was already drawing to a close.

Working with the Bogs Commissioners

Different views were held about the competence of the surveyors who practised in Ireland in the first decades of the nineteenth century. As late as 1824, the Lord Lieutenant claimed that "neither science, nor skill, nor diligence, nor discipline, nor integrity" could be found among the land surveyors or engineers of Ireland sufficient to enable them to be recruited to the staff of the new Ordnance Survey organisation.[55] However, this sweeping dismissal of the Irish surveyors was hardly justified; Alexander Nimmo was probably nearer the mark when he told a parliamentary committee in the same year that the term "common surveyor is a very wide expression, because we have them of all degrees of skill".[56] Where Leahy may have fitted into the spectrum can only be a matter of conjecture at this stage, but it seems clear that he had reached a reasonable level of proficiency and a certain standing in the profession by January 1810 when he joined the large team of surveyors engaged to work for the Bogs Commissioners to whom, in September 1809, the Government had assigned the monumental task of mapping roughly one million acres of bogland in Ireland and suggesting possible drainage, reclamation and development works.[57] The Commissioners initially employed seven district engineers (later increased to ten), including most of the major figures in Irish engineering at the time, and left it to these to recruit suitably qualified surveyors to assist them.

David Aher (1780-1841), manager of the Castlecomer colliery, was the engineer assigned to oversee the survey of some 36,000 acres of bogland in Counties Laois, Kilkenny and Tipperary, including bogs in the Littleton, Urlingford, New Birmingham and Johnstown areas with which Leahy must have been very familiar.[58] However, instead of becoming part of Aher's team, Leahy was one of three surveyors taken on by Thomas Townshend,[59] with whom he had already worked on canal surveys, and who was initially given responsibility for District No 6, made up of 34,500 acres in Counties Offaly and Westmeath. Based on the work done by Leahy and the other two surveyors in his team, Townshend was able to submit his report and maps of the district in February 1811.[60] One month later, on the recommendation of John Rennie, Townshend was given a second assignment which involved mapping and reporting on 65,000 acres of bogland in five counties surrounding Lough Neagh.[61]

In April 1812, after the Commissioners had presented a paper to Parliament setting out the estimated cost of completing work in each district,[62] Townshend assured them that he would submit his report by the end of October.[63] However, in September, the Commissioners were complaining that "several months had elapsed since they had any communication from Mr Townshend", and their concern was heightened when they learned that he had been appointed Resident Engineer and Surveyor of the Witham Navigation Company, with responsibility for improving the 36-mile waterway from Lincoln to the sea.[64] Townshend was censured at this stage for having absented himself from his district without permission, and for the fact that progress had fallen very far short of what had been expected, and he was directed to complete the necessary maps and reports without delay.[65]

Two of the three surveyors on Townshend's team (John Thomas and James Browne) had taken up other employment by March 1811[66] and, when Townshend's own absence in England and his work for the Directors General of Inland Navigation are taken into account, it is clear that the burden of carrying out the survey and mapping work in the counties adjoining Lough Neagh must have fallen almost entirely on Leahy. However, the seven maps and three cross-sections of the River Bann which were eventually submitted by Townshend in March 1813 bore his name only and did not acknowledge the contribution Leahy had made to the work.[67] But Townshend's final report did recognise Leahy's special efforts: "should the Commissioners think proper to distribute any rewards above the usual allowance, I take the liberty of recommending Mr Leahy to their notice, for his exertions and abilities in the surveys".[68] The Commissioners failed to act on this recommendation, partly, perhaps, because they regarded Townshend's maps and report "in so many respects unsatisfactory and incomplete" after such an expense of time and money; the documents were returned to him in August 1813 so that he could make the necessary additions and amendments, and with a warning that failure to complete this task by November 1813, at his own expense, would leave the Commissioners with no option but to "express their opinion very fully to Parliament upon the whole matter".[69]

While the surviving documents do not allow a judgment to be made on the extent to which the criticisms of the Lough Neagh report and maps, as originally submitted in March 1813, may have been attributable to defects in Leahy's survey and mapping work, it seems more likely that Townshend personally was at fault given his virtual abandonment of the district in the second half of 1812, and the subsequent rush to write up his report and complete the maps without further field work. Leahy himself had no doubt about the quality of his survey work which, he stated, "was executed Trigonometrically, under like principles as the Grand Trigonometrical Survey of England by General Roy, Col Mudge and Col Williams" and with great accuracy and expedition.[70] On a personal basis, he looked on the three years he had spent working on the bog surveys as the best days of his career. He regarded Townshend as his "first encourager into life" and, nearly 20 years after their association had ended, he told him that he had "never felt happy in his professional life but with you". He still had fond memories of Townshend's young family, including Susan, "a plump rosy cheeked favourite" of his, and took pleasure in visiting the family in Birmingham in the 1830s.[71] He followed the career of John Thomas, one of Townshend's original surveyors, with interest but seems to have lost touch with the other surveyor, James Brown, although he described him in 1830 as "my old friend and pot companion".[72]

Leahy's engagement with the Commissioners lasted for a total of 696 days between January 1810 and March 1813, a longer period of service than most of the 40 surveyors involved. He was paid a fee of one guinea a day in salary and allowances which, with some minor amounts for expenses, brought him a total of £791 14s.[73] Claiming that "in hopes of meeting some reward beyond his salary" he had "exerted himself in an extraordinary degree in the execution of these laborious admeasurements", he applied to John Leslie Foster, Chairman of the Commissioners,[74] for payment of the reward recommended by Townshend, but was told that the available funds were exhausted,[75] the Commissioners having actually overspent their budget by £453.[76] Ten years later, "in days of happy administration to Ireland", he restated his case in memorials addressed to the Lord Lieutenant, Marquis Wellesley, and Henry Goulburn, the Chief Secretary, "as the only patrons on which he now rests his hope". Sadly, however, his confidence "that the justice of his claim will be recognised" at a time "when merit has been found to meet encouragement" was misplaced.[77]

Of the three surveyors employed by Townshend on the bog surveys, Leahy was the only Irishman and the only one whose career and reputation in civil engineering did not advance rapidly afterwards: John Thomas went on to superintend the construction of a new Royal Dockyard at Sheerness in the Thames estuary, designed by John Rennie in 1813, a project on which he was engaged until 1830; and James Brown became resident engineer under Rennie on the construction of the Admiralty Pier at Holyhead Harbour (completed in 1824), before working under Thomas Telford for the Holyhead Road Commissioners and under Sir John Rennie on the Hartlepool Docks until 1840.[78] One can only speculate as to whether Leahy may have tried and failed to secure similar engagements in Britain when his bog surveys were completed, or whether he remained in Ireland by his own choice to pursue a career in land surveying, as did most of the other surveyors who had been engaged on those surveys. In advancing his career, however, he did attempt - but with no real success - to profit from his engagement with the Bogs Commissioners by publicising the fact that he had distinguished himself in their service. His advertisements in the *Clonmel Advertiser* in 1818 highlighted his work with the Commissioners and his commendation by their engineer[79] and, in 1824, he was describing himself as "Surveyor to the Commissioners for Draining and Improving the Several Bogs, Loughs and Waste Lands throughout Ireland".[80] In 1828, some 15 years after the assignment had been completed, he still thought it worthwhile to draw attention to his involvement in it, although by then he had assumed the title of "Assisting Engineer and Surveyor" to the Commissioners.[81] And later still, in 1834, more than 20 years after the event, he relied again on Townshend's commendation when seeking to be engaged by the Dublin Castle authorities on a mapping assignment.[82]

"a most exquisite Instrument"

So that his surveys for the Bogs Commissioners would be strictly accurate, Leahy "procured a most exquisite Instrument from that celebrated artist, Troughton of London[83] but actually contrived by ... himself, by which, with the assistance of a small set of accompanying Tables, contrived also by [himself], he is enabled to ascertain, by inspection only, the distance of any object within a distance of Thirty miles with such degree of precision as to surprise the most celebrated Engineers now in Great Britain".[84] In 1821, believing that his invention was then "prepared for practical application", he submitted it to the Chief Secretary, Charles Grant, "under the hope that should my slender exertions afford me a place among the benefactors of mankind, my production or invention will not remain unrewarded",[85] and he followed up by presenting a statement of the *Description and Use of the New Tables of Inaccessible Distances*:

> The New Tables of Inaccessible Distances will exhibit, from a bare inspection only, the Distances of all objects within view at Land or Sea within a distance of twenty Irish miles at least, with great accuracy. This performance is now independent of any of the known methods practised by Gentlemen in Military or Naval departments. The contrivance of these New Tables is such, and the mode of using them so simple and easy, that their operation is within the reach of almost every individual who can only read and write, as nothing more is wanting than the measure of an angle, from which alone, the distance can be had by an inspection into the New Tables.
>
> The great advantage derived from this invention by the Military Engineer (or any soldier who can read or write) is obvious in itself, as these tables alone will supply all the elaborate and tedious operations in calculation and projection, which must otherwise be resorted to, in crossing great Rivers, viewing the movements of an enemy, and ascertaining the distances of Cities, Towns, Forts and Garrisons in possession of an enemy.
>
> The Mariner and Civil Engineer in various situations and conditions will feel equally indebted to this invention which will supply the operations resulting from Canons of Sines, Tangents and Secants as well as projections.[86]

When Chief Secretary Grant consulted Edward Wilson of Thurles, Chief Magistrate of the Police, about the invention, he was assured that Leahy "was a most useful surveyor and of very good character" and that Wilson himself had viewed a practical demonstration of the instrument and tables with which Leahy "had proved to a nicety the exact distance to an object in a field that I pointed out".[87] Grant, however, appears to have taken no further action before leaving office in December 1821, forcing Leahy to raise the matter with the new administration at Dublin Castle in January 1822.[88] He set out his case once again in a formal Memorial to the new Chief Secretary, Henry Goulburn, in July 1823 but received neither the reward nor the encouragement to which he felt entitled.[89] Some time before that, a submission of his invention to the Board of Longitude[90] did, at least, provide some encouragement – the Board, according to his own account, "acknowledged its merits but extended no reward".[91]

One could hypothesize that Leahy's "exquisite instrument" may have resembled a tacheometer, a form of theodolite which could be used in conjunction with a graduated staff to determine quickly the distance and elevation of a distant object, without using a chain or separate levelling instrument. But tacheometry was a continental speciality and the first practical instrument based on the tacheometric principle which could be applied directly and successfully to geodetic work is generally believed to have been invented in 1824 by an Italian engineer, Ignazio Porro (1801–1875).[92] Taking account therefore of Leahy's emphasis on the need to measure only a single angle, the absence of any reference by him to a staff, and his insistence that distances of 20 to 30 miles could be measured accurately, it seems most likely that his instrument was a form of rangefinder rather than a tacheometer. The instrument can hardly have been as effective as Leahy made it out to be; there is no mention of it in writings on the history of surveying and, if the claims made for it were substantiated, one would expect him to have patented it and to have publicised it more widely. Moreover, Edward Troughton, who made the instrument for Leahy and who for many years was the leading English producer of scientific instruments (including tacheometers for export), would surely have realised its potential and developed it commercially.[93]

Return to Tipperary

When Leahy returned to County Tipperary after completion of his service with the Bogs Commissioners in 1813, he took up residence in the Thurles area and appears to have resumed his links with Sir Vere Hunt who engaged him and John Neville in April 1815 to lay out one of the streets of New Birmingham.[94] He may also have been engaged in connection with some of the presentments for road improvement works which Vere Hunt regularly obtained from the grand jury. However, although his colliery was reported to be in a promising state at the beginning of 1814,[95] Hunt's diary records that he was struggling, two years later, without money, to make progress in the building of the town; times were bad, he noted, and tenants were unable to pay their rents due to falling prices in the years after Waterloo.[96] Leahy cannot, therefore, have expected to receive a continuing flow of commissions from Vere Hunt who died in 1818 with his ambitions for the town unfulfilled.

In 1813, Leahy applied for the job of making a new grand jury map of County Tipperary. Vere Hunt, who was in Clonmel for the Assizes, noted in his diary on 7 August that "the unfortunate Mick Lahy came to me as usual in an happy state of intoxication ... looking for the survey of the county and for money to drink"; Hunt wrote that "I would not injure my character by recommending him" for the assignment, but went on to record that "Pat Lahey, another surveyor, solicits me on the same subject and as I consider him steady and deserving, I promise him my vote and support, he promising to give his unfortunate relation employment in the work if he gets it".

Chapter I — Patrick Leahy Country Surveyor

There was obviously a great need for a new survey of the county at the time because, according to a local magistrate, the Down Survey - made by Sir William Petty in the 1650s - was still being used as the basis of local taxation in Tipperary although "it was a very unfair survey".[97] In addition, the county treasurer had to admit as late as 1824 that, for roads purposes, the grand jury had nothing better than a 1799 map which was "considered so imperfect and inaccurate as never to be referred to and does not specify the number of inches to the mile".[98] However, the grand jury did not go ahead with the new survey, although proposals for a county map were again invited in 1814;[99] it may be that they decided to await the outcome of the debate at national level about the need for a complete re-survey and valuation of the country for local taxation and administrative purposes. In any event, although Thomas Townshend had apparently recommended his employment to the grand jury,[100] it seems unlikely that even if the project had gone ahead a country surveyor like Leahy would have been engaged for an assignment worth, perhaps, £1,500, in preference to one of the better-known engineers and surveyors who were being commissioned to prepare other county maps in the first decades of the century.

Leahy continued to work mainly in North Tipperary throughout the 1820s. For most of this period, he is likely to have been engaged by private clients on the routine work of a country surveyor with, perhaps, occasional projects of an engineering nature, including small building projects and road maintenance and improvement works financed by the grand jury. In 1818, he was advertising regularly in the local press, offering his services "as

> **AS ENGINEER & LAND-SURVEYOR,**
> PATRICK LEAHY respectfully begs leave to acquaint the Nobility, Gentry, and Public, that he now resides in the Town of Thurles, where future Commands in his Professional Line will meet with due Attendance. Those who wish for reference will please enquire at the Library of the Dublin Society House for the Third Report of the Commissioners on the Drainage of the Bogs of Ireland, printed by order of the House of Commons, page 162, in which Work is also contained his Trigonometrical Survey of a great portion of Ulster, conducted under Mr. Townsend, an English Engineer.
> ☞ Such as will please to favour him with their Calls, will meet unusual Satisfaction in the accuracy and neatness of his Works in general—and where Drainage, Irrigation, Planting, Subdividing, or Valuation comes under consideration, his Opinion will be found highly instructive.
> ⁂ Terms—10d. per Acre for Townlands surveyed into ordinary-sized Farms, and but 6½d. per Acre for general Surveys only.   Thurles, Oct. 1, 1818.
> *An* APPRENTICE *wanted.*

Patrick Leahy's advertisement in Clonmel Advertiser, 10 October 1818.

In search of fame and fortune: the Leahy family of engineers, 1780-1888

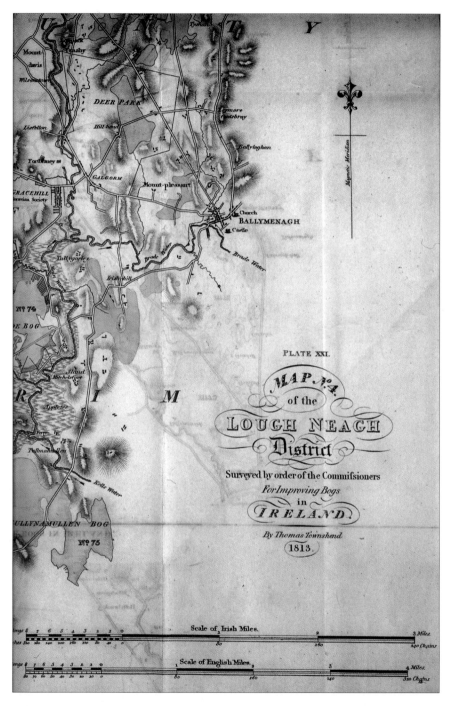

Section of one of the maps of the Lough Neagh District submitted by Thomas Townshend to the Bogs Commissioners in 1813, based on survey work by Patrick Leahy (engraved for publication on a reduced scale in the Third Report of the Bogs Commissioners, HC 1813-14 Vol VI (130), Plate XXI, Map No 4).

Chapter I — Patrick Leahy Country Surveyor

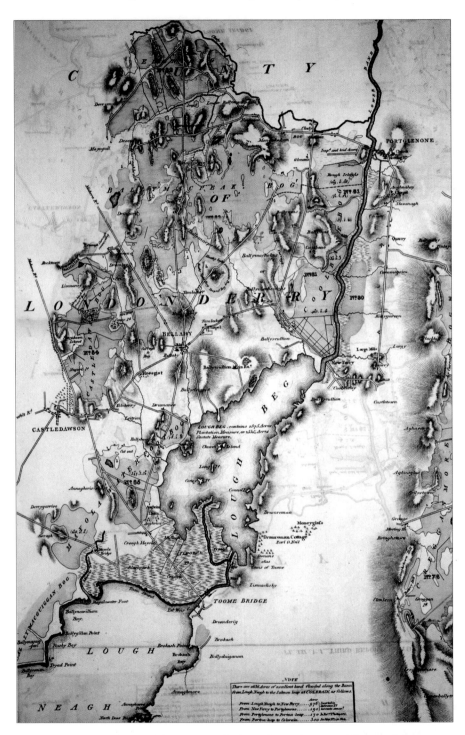

Map of the bogs in counties Derry and Antrim, north of Lough Neagh, surveyed by Patrick Leahy in 1811-13 for the Bogs Commissioners (engraved for publication on a reduced scale in the Third Report of the Bogs Commissioners, HC 1813-14 Vol VI (130), Plate XXI, Map No 4).

In search of fame and fortune: the Leahy family of engineers, 1780-1888

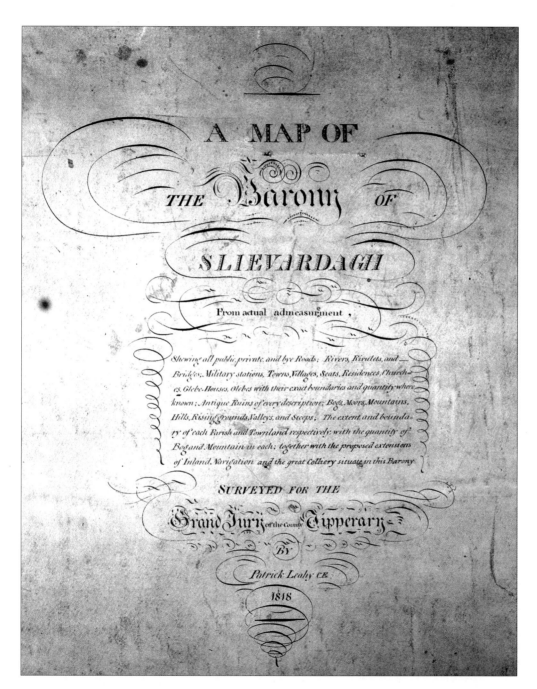

Title of "A Map of the Barony of Slievardagh ... surveyed for the Grand Jury of the County Tipperary by Patrick Leahy CE, 1818" (Archives of the Geological Survey of Ireland, BY 7-2-24-001).

Table of Mineralogical Reference and Geological Description included in "A Map of the Barony of Slievardagh … surveyed for the Grand Jury of the County Tipperary by Patrick Leahy CE, 1818" (Archives of the Geological Survey of Ireland, BY 7-2-24-001).

In search of fame and fortune: the Leahy family of engineers, 1780-1888

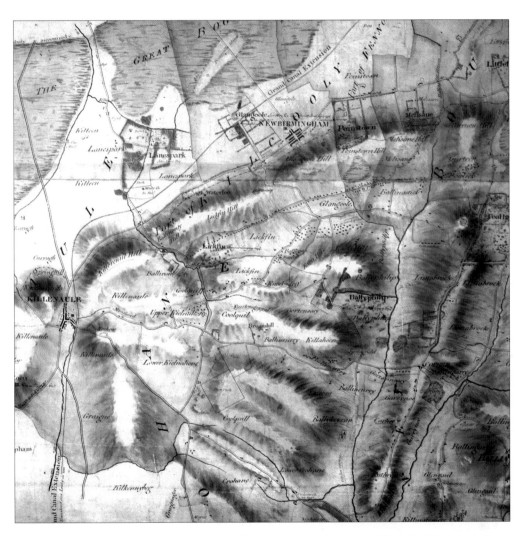

Section of Patrick Leahy's map of the Barony of Slievardagh, 1818, showing Killenaule, New Birmingham, the coalpits and the proposed extension of the Grand Canal (Archives of the Geological Survey of Ireland, BY 7-2-24-001).

Chapter I — Patrick Leahy Country Surveyor

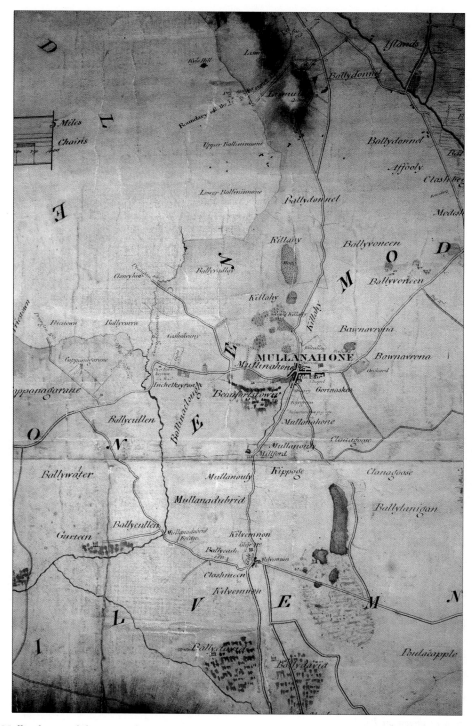

Mullinahone and district, as shown on Patrick Leahy's map of the Barony of Slievardagh, 1818 (Archives of the Geological Survey of Ireland, BY 7-2-24-001).

In search of fame and fortune: the Leahy family of engineers, 1780-1888

Kilcooly, as shown on Patrick Leahy's map of the Barony of Slievardagh, 1818 (Archives of the Geological Survey of Ireland, BY 7-2-24-001).

Chapter I — Patrick Leahy Country Surveyor

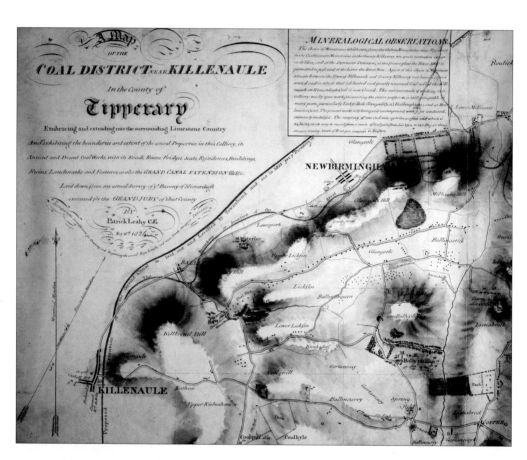

Section of "A Map of the Coal District near Killenaule in the County of Tipperary" by Patrick Leahy CE, 1824, showing the line of coalpits running north-eastwards from Killenaule and the "Proposed Extension of the Grand Canal as laid out and levelled by Townshend and me in 1811" (NLI 16 I 17[1-2]), courtesy of the National Library of Ireland).

In search of fame and fortune: the Leahy family of engineers, 1780-1888

Cover of the book of 51 watercolour maps of the city and part of the corporation lands of Waterford by P Leahy & Sons, 1831-32 (Waterford City Archives, M/PV/018).

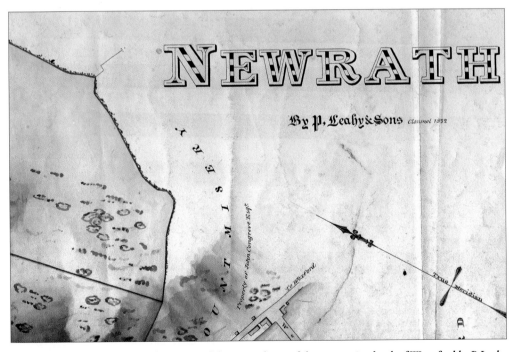

Section of one of the 51 watercolour maps of the city and part of the corporation lands of Waterford by P Leahy & Sons, 1831-32 (Waterford City Archives, M/PV/018).

Chapter I — Patrick Leahy Country Surveyor

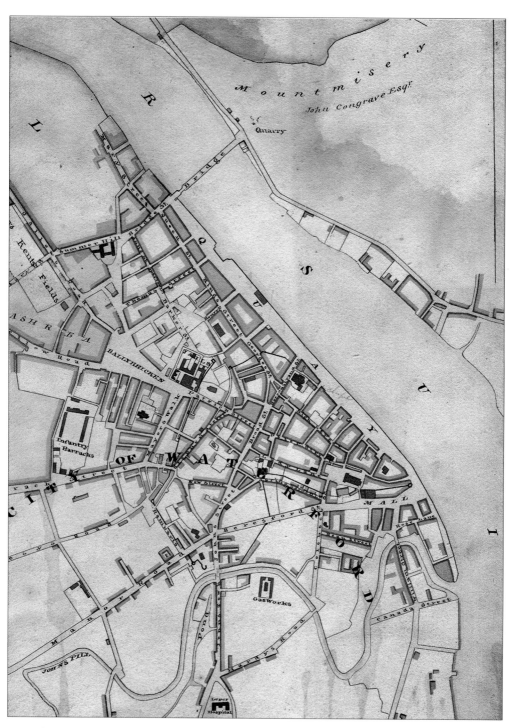

Part of the manuscript general map of the city centre of Waterford prepared by P Leahy & Sons, 1831-32 (Waterford City Archives, M/PV/018).

In search of fame and fortune: the Leahy family of engineers, 1780-1888

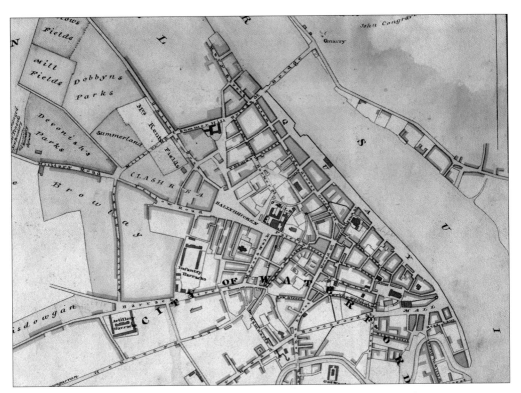

Part of the manuscript general map of the city centre of Waterford prepared by P Leahy & Sons, 1831-32, showing the Artillery Barracks, the Infantry Barracks and the Jails (Waterford City Archives, M/PV/018).

Chapter I — Patrick Leahy Country Surveyor

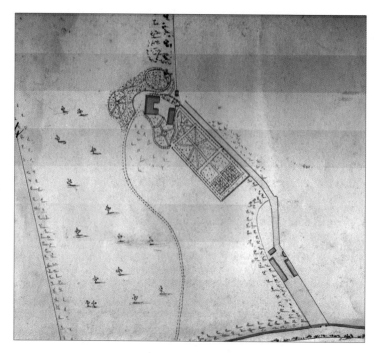

Detail of house and grounds at Woodstown from the book of watercolour maps of the city and part of the corporation lands of Waterford by P Leahy & Sons, 1831-32 (Waterford City Archives, Ref M/PV/018).

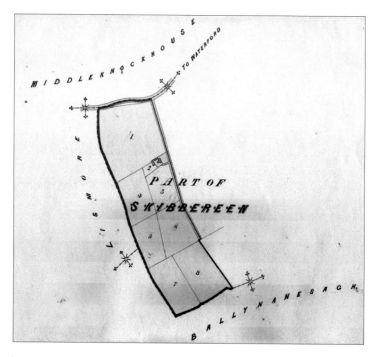

Map of Waterford Corporation land in the townland of Skibbereen, south-west of the city, showing eight numbered fields for which areas (in statute measure and plantation measure) were shown in a separate table of reference (Waterford City Archives, Ref M/PV/018).

In search of fame and fortune: the Leahy family of engineers, 1780-1888

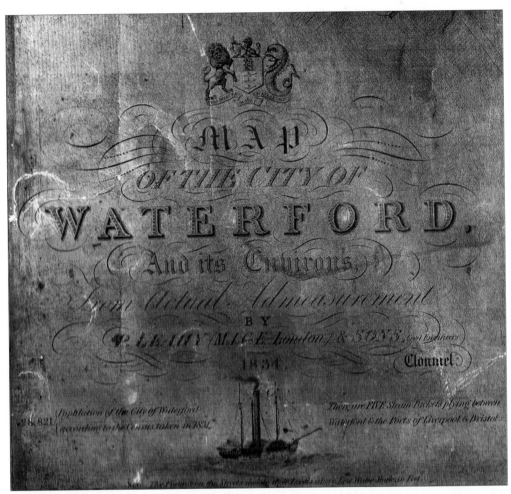

Title of the engraved Map of the City of Waterford and its Environs by P Leahy & Sons, 1834 (Waterford City Archives, Ref M/PV/035).

Chapter I — Patrick Leahy Country Surveyor

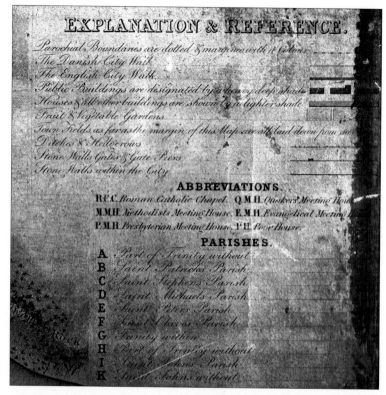

Table of Explanation and Reference from the engraved Map of the City of Waterford and its Environs by P Leahy & Sons, 1834 (Waterford City Archives, Ref M/PV/035).

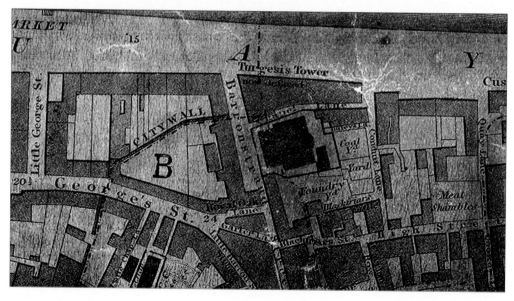

Part of Waterford city centre as shown on the engraved Map of the City of Waterford and its Environs by P Leahy & Sons, 1834 (Waterford City Archives, Ref M/PV/035).

In search of fame and fortune: the Leahy family of engineers, 1780-1888

SECTIONS

of the

STREETS SEWERS &c

of the

CITY OF WATERFORD

For the Mayor Sheriffs and Citizens

OF SAID CITY

By P Leahy & Sons

Civil Engineers

1833

Title page of the book of "Sections of the streets, sewers etc of the City of Waterford" prepared for Waterford Corporation by P Leahy & Sons, 1833 (Waterford City Archives, Ref M/PV/036).

Comments and recommendations by Patrick Leahy on the gradients of the Waterford sewers and the need for cleansing them (included in the book of sections of the sewers prepared for Waterford Corporation in 1833 (Waterford City Archives, Ref M/PV/036).

Chapter I — Patrick Leahy Country Surveyor

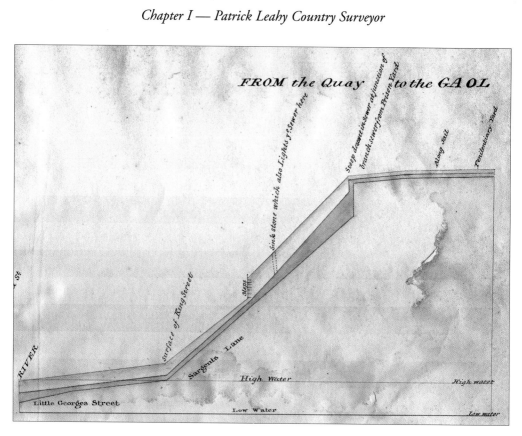

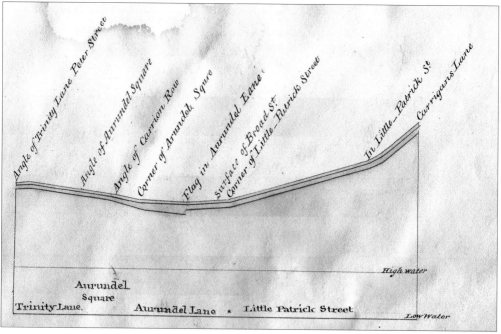

Sections of two of the sewers in Waterford prepared for Waterford Corporation by P Leahy & Sons in 1833 (Waterford City Archives, Ref M/PV/036).

engineer and land surveyor", highlighting his work with the Bogs Commissioners and his commendation by their engineer, and promising that "Such as will please to favour him with their Calls will meet unusual Satisfaction in the accuracy and neatness of his Works in general – and where Drainage, Irrigation, Planting, Subdividing, or Valuation comes under consideration, his opinion will be found highly instructive".[101] His charge for survey work was 10 pence per acre "and but 6½ pence per acre for general surveys only". However, it appears that in spite of his willingness to function as a jack-of-all-trades, he failed to make a living from his professional activities alone partly, perhaps, because he faced competition from a number of other surveyors, including the prominent Dublin surveyor, John Longfield, who was actively engaged in the preparation of manuscript maps of estates in Tipperary in the first decades of the century.[102] Thus, Leahy was forced to supplement his income by keeping a shop in Thurles in which he sold groceries and whiskey.[103]

Estate surveys and maps

Using, presumably, the new instrument and survey methods which he had developed during his engagement with the Bogs Commissioners, Leahy prepared a number of estate surveys and maps of parts of County Tipperary, some of which have survived. In 1818, he completed his most significant cartographic project: *A Map of the Barony of Slieveardagh . Surveyed for the Grand Jury of the County Tipperary by Patrick Leahy CE, 1818.*[104] This large one-sheet coloured map, measuring approximately 212cm by 124cm, was drawn to a scale of four inches to one Irish mile (1:20,160), the scale which the Bogs Commissioners' engineers had been required to use in the preparation of their base maps. Manuscript barony maps of this kind dating from before 1830 are quite rare, the best known, perhaps, being the set of maps of the baronies of County Kilkenny prepared for the grand jury between 1812 and 1824 by David Aher and his partner, Hill Clements.[105] It is tempting to see the Slieveardagh map as some sort of imitation of this Kilkenny mapping programme and even to speculate that Leahy may have had an involvement in the survey work in Kilkenny;[106] however, there is no evidence to support this and Leahy, in the several surviving letters in which he describes his career, never mentioned an association with Aher.

Leahy's Slieveardagh map is a particularly detailed one and covers an area extending approximately 26½ statute miles from north to south and 15½ miles from east to west, taking in the villages of Urlingford, Killenaule, New Birmingham, and Mullinahone; only a very limited part of this area had been covered by the map prepared by David Aher in 1811 for the Bogs Commissioners.[107] It professed to show all public and private roads, bridges, military stations, rivers, bogs, moors and mountains, and all of the coal pits forming the "Great Colliery of Slieveardagh." In addition, towns, villages, gentlemen's seats and residences, churches, glebe houses, "antique ruins of every description" and "the dwellings of the poor" were shown, together with the boundaries and areas of each parish and townland. Larger houses were shown "in elevation" (in profile rather than in plan), a practice which, according to J H Andrews, was then one of the "survivals from an earlier epoch," while the map also employed a variety of different scripts, each with its own significance; this, to quote Andrews again, was another "long-lived but otherwise pointless and aesthetically dubious habit among late eighteenth-century surveyors".[108] Nevertheless, the Slieveardagh map was an attractive and technically competent one, which probably ranks among the better examples of the art of cartography as practised at the time by Irish land surveyors. It includes an exceptionally early "Geological Description"[109] and is likely also to have been of significant practical value for it included a large table of "Mineralogical Reference" listing the various townlands which lay within the coal district "with such remarks annexed to each as may readily serve to point out the respective lands now explored and working, their several quantities, and the names of the proprietors vested with their several royalties".

Cartouche on a map of the town and lands of Thurles, in distinct denominations, farm holdings and enclosures, the estate of the Earl of Llandaff, surveyed by Pat Leahy in 1819 and copied by John Longfield, Harcourt St, Dublin, in 1827 (Tipperary Libraries, Thurles).

In search of fame and fortune: the Leahy family of engineers, 1780-1888

In 1819, Leahy completed a manuscript map of the town of Thurles and some of the surrounding lands which formed part of the estate of the Earl of Llandaff. Almost 400 properties were delineated and numbered on the map which was drawn to a scale of 20 perches to one inch. Separate documentation, linked to the map, provided the names of the individual tenants or leaseholders, together with a summary description of each property.[110] The original map does not appear to have survived but an apparently accurate, although rather inelegant copy, made in 1827 by John Longfield (or, more likely, by one of his assistants), is extant.[111]

Leahy used his Slieveardagh map in 1824 as the basis of *A Map of the Coal District near Killenaule in the County of Tipperary*,[112] drawn to a scale of four inches to one Irish mile, and with co-ordinates shown in numerals along the top and in letters along the sides, so as to allow ready identification of different locations. The map covers some 60 square miles and includes a panel of "Mineralogical Observations" describing the ancient mode of working the colliery by "excavating the entire surface as is still perceptible in many parts" and the more modern method of "boring and underground work, which is yet conducted extremely unskilful (sic)". It was accompanied by more detailed notes providing an assessment of the potential of the coalfields and suggestions for improving the area's transport facilities, including specific proposals for the extension of the Suir Navigation from Two-Mile-Bridge, near Clonmel, to Killenaule, at an estimated cost of £30,000. The notes were signed *Pat Leahy CE, Dublin* but nothing has emerged to explain why Leahy was in Dublin when this work was completed or why he spent some time there in 1826 when he prepared the tables of weights and measures referred to below.

In 1828, Leahy completed a further map, survey and detailed rental reference book of the Killenaule Estate, then the property of Lawrence Waldron and comprising an area of 3,339 acres.[113] For each of the 693 separate parcels of land, the tables of reference set out the names of the lessees and undertenants, the area in English measure and in Irish measure, and a description of the holdings sufficient "to enable a person to arrive at a knowledge of their present state, as well as to form an estimate of the individual and comparative values of each holding". The original map, which has not survived, was based on a survey which, according to Leahy, had been "executed from Magnetical, verified by Trigonometrical operation". Different colours, shading and handwriting were used to distinguish individual subdivisions and the map conformed "to every possible purpose of utility and utterly precludes the necessity of any future admeasurement". An accompanying report set out a detailed development plan for the estate, with observations on the minerals and soils to be found there, notes on the natural advantages of the district, and suggestions for improvement works of various kinds - new roads, drainage and irrigation schemes (a subject on which Leahy claimed to have great expertise, given his service with the Bogs Commissioners), the division and fencing of commonage and the establishment of mills, a brewery and other industries.

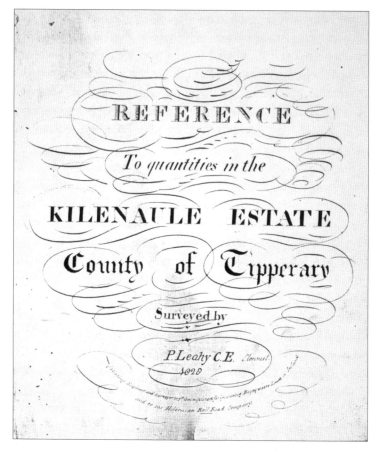

Title page of Rental Reference Book for the Killenaule Estate, prepared by Patrick Leahy in 1828.

Application for appointment as county surveyor

For Leahy and his contemporaries, real hopes of building a secure future for themselves must have arisen when legislation was enacted in 1817 to provide for the appointment by the Lord Lieutenant, "when and as soon as persons can be found properly qualified", of a county surveyor for each county, with salaries of up to £600 a year.[114] Richard Griffith thought that the new office would be "a very arduous one and requiring great personal exertion"; it was clear also that candidates would need to have a rare combination of surveying, architectural and engineering skills if they were to perform all of the duties involved which, as Griffith saw them, would include the detailed mapping of baronies, the design of public buildings and bridges, and the maintenance of roads.[115] In a radical departure from the patronage system, the Act provided that no one was to be appointed to these important new posts until he had been examined by a board of at least three civil engineers set up by the Lord Lieutenant, and had obtained a certificate from the board of his fitness to hold the office.

In search of fame and fortune: the Leahy family of engineers, 1780-1888

In December 1817, persons who wished to offer themselves for appointment were invited by public advertisement to present themselves for interview in Dublin before a panel of three distinguished practitioners in civil engineering and architecture - Thomas Telford, the Scottish engineer renowned for his extensive work on canals, roads, bridges and harbours; Major Alexander Taylor, another Scotsman, who built the military road through the Wicklow mountains between 1800 and 1809; and Francis Johnston, architect to the Board of Works and designer of many public buildings, including the GPO in Dublin. Unfortunately, the list of those who presented themselves for interview is no longer available[116] but Patrick Leahy was certainly among them and was recommended by "the Nobility and Gentry of the County Tipperary".[117] In all, according to the Chief Secretary, "no less than 95 persons applied"[118] but many of these, in the view of a contemporary observer, were men who, in the expectation of so many appointments taking place, had been induced "to assume the title of Engineer, who never dreamed of it before, as if a few months acquirements could fit them for such a situation".[119] Having interviewed all comers, the interview board was forced to report in January 1819 that "upon the whole of this painful and tedious investigation, finding so very few of the great number who presented themselves possessed of the necessary experience and practical knowledge in road-making and bridge-building", it could not certify the necessary number of candidates to allow the Act to be brought into operation.[120]

Faced with this embarrassing verdict on the calibre of the emerging civil engineering profession in Ireland, Parliament was forced to suspend the 1817 Act[121] and to replace it by an Act which required that all proposals made to the grand juries should in future be accompanied by maps, plans and estimates "signed by some known Surveyor, Engineer or Architect, or by some other competent person".[122] But even this modest measure of reform, which would have enhanced the employment prospects of men like Leahy, was repealed in July 1819, leaving it to the grand juries to muddle along, more or less as they had done before, for another 15 years.

Emergence as a civil engineer
Civil engineering began to be recognised as a separate profession around the end of the eighteenth century when a growing number of those who were engaged on major programmes of public works described themselves as civil, as distinct from military, engineers. In the early decades of the nineteenth century, there was increasing use of the term *engineer* instead of *surveyor* as the emerging profession developed new prestige and status in Britain, leading to the foundation of the Institution of Civil Engineers in 1818; the trend was similar in Ireland where canal building and the extensive public works programmes of the 1820s generated additional demand for men with engineering skills. The decision of the Government in 1824 to establish the Ordnance Survey of Ireland, initially with a military staffing organisation, to prepare large-scale maps of both urban and rural areas, had obvious implications for the employment prospects of the existing land

surveyors who felt that "their bread and butter was being taken away from them";[123] this must have influenced many of them to opt, where they could do so, for a future in engineering rather than surveying. But it took time for a clear distinction to emerge between engineering and surveying: there were no recognised tests of entitlement to describe oneself as a civil engineer and no restriction on the use of the initials CE - in fact "there was nothing except conscience, or fear of ridicule, to prevent the initials being flaunted by anyone who took a fancy to them".[124]

Leahy's newspaper advertisements sought engagements of an engineering nature, in addition to surveying work, as early as 1818[125] and he used the initials CE in the same year on his map of the barony of Slieveardagh. However, it is not apparent that, apart from road survey work, he had any real experience in engineering when he decided, ahead of many of his contemporaries, to make the transition, at least on paper, to civil engineering. Nevertheless, in his own mind, he had all of the necessary qualifications when he submitted this interesting self-assessment of his situation in support of an application for Government employment in 1823:

> Memorialist is now an active, steady, experienced man, having arrived to his Forty-fourth year; has been closely engaged in public and private surveys for twenty-seven years; has had a peculiar turn for executing maps in the most approved stile of the day; is well conversant in Abstract and Mixed Mathematics, and fully possessing the requisite qualifications for constituting an experienced Engineer ...[126]

His objective at that stage was to join the small team of engineers (Killaly, Bald, Nimmo and Griffith) which had been appointed by the Lord Lieutenant in 1822 to manage Government-funded programmes of road construction and other public works, mainly in the more remote areas where distress was being experienced arising from famine, rural unrest and minor uprisings. In making his application, he relied once more on his commendation by Townshend for his work on the bogs surveys and expressed "the strongest hope that Irish merit, producing the recommendation of an English Engineer, will no longer be left unrewarded".[127] He argued also that with his "discoveries and improvements", his new methods of surveying sea coasts, harbours, loughs etc would bring great savings to the Government, as his surveys would be executed "in half the time, and for a fourth part of the expense as would attend such surveys if executed by others". In addition, having just taken up residence in Clonmel, he suggested that he was "conveniently situated" to act as District Engineer for counties Kilkenny, Waterford, Tipperary and Cork, and that "much time and trouble could be spared Mr Killaly in the exercise of his duties, and public money judiciously applied" if he were to be so appointed. But these arguments made little sense: the programme of relief works did not apply to Waterford and Kilkenny; Richard Griffith already had responsibility for County Cork; and

Killaly's district consisted mainly of County Clare, with only a relatively small adjoining area of Tipperary. And so, although he offered references from several members of the nobility of Tipperary and Waterford, Leahy was told that the authorities at Dublin Castle were "not aware that there was any opening for employment by the Government".[128]

"Indispensably necessary" tables of weights and measures

Evidence of Leahy's willingness to enter new fields and to explore a wide range of options to boost his earnings is provided by a letter which he sent to the Chief Secretary in July 1826 seeking support for the publication of "a small practical essay on the late Act for assimilating the currency, weights and measures, with suitable tables for practical use".[129] The tables, he told the Chief Secretary, were prepared while he had "some leisure hours in Dublin last March" and were already on sale at Bentham & Hardy's of 24 Eustace Street, Dublin. They had been well received by "some of the most enlightened, as well as distinguished personages in the Metropolis", and had "received the approbation of the public". The need for such tables had arisen from the fact that the currencies of Great Britain and Ireland were assimilated with effect from the previous January and, to make the transition even more difficult for consumers and business generally in Ireland, new Imperial standards of weights and measures had been introduced at much the same time.

In seeking some "encouragement towards the expenses incurred in the publication", Leahy seems to have overlooked the fact that a number of similar works were already available. For example, *Coldwell's Tables for Reducing Irish Money into British Currency*, first published in 1825, appeared in a second edition in 1826, with a declaration from the publisher that "the rapid sale of the first edition proves the utility of these tables, and their superiority over any others that have yet appeared".[130] Leahy, however, seems to have had no doubt about the merits and relevance of his own tables because "his practical pursuits for nearly thirty years in some of the higher departments of his profession as land-surveyor has given him such acquaintance with the commercial and agricultural transactions of his country as enables him to form a correct opinion of the necessity and essential use of the Tables". On this basis and because he had "rendered an important public benefit", he optimistically urged the Chief Secretary to order one hundred pounds' worth of copies (1,000 copies), presumably for use by the public service in Ireland. However, a curt note on his letter indicated the official view that this "does not appear necessary".

Few copies of Leahy's 68-page publication appear to have survived but, fortunately, a copy is held in the Haliday Collection of pamphlets at the library of the Royal Irish Academy.[131] In addition to 11 tables of conversions, the publication set out the text of relevant sections of the Weights and Measures Act of 1825[132] with observations on each section and general commentary on the transition to the new standards. Some of the tables relate to straightforward operations such as converting Irish acres into English acres, but more complex conversions were also provided for, eg converting "the rent of One Irish

> NEW AND GENERAL
> TABLES
> OF
> **WEIGHTS & MEASURES;**
> WITH
> AMPLE CALCULATIONS,
> AND
> SUITABLE EXAMPLES UNDER EACH HEAD;
> All reduced agreeable to the
> ACT OF THE 6TH GEO. IV. CHAP. 12.
>
> INDISPENSABLY NECESSARY FOR
> *Landlords, Tenants, Agents, Mayors, Magistrates,*
> *Merchants, Traders, Mechanics, Artists, &c.*
>
> BY PATRICK LEAHY,
> Civil Engineer, &c.
>
> DUBLIN:
> PRINTED BY BENTHAM & HARDY, 24, EUSTACE-ST.
> Sold by all the Booksellers.
> 1826.

Title page of Patrick Leahy's New and General tables of Weights & Measures, published in 1826 (courtesy of the Library of the Royal Irish Academy).

Acre in the old currency to One English acre in British currency", or "reducing rates of carriage by the Irish mile to a proportional rate for one English mile". Each table was followed by a number of worked examples and, in some cases, by convenient shortcuts.

In search of fame and fortune: the Leahy family of engineers, 1780-1888

In the context of a study of his career as a land surveyor, the most interesting section of Leahy's booklet is his commentary on land surveying practice in Ireland, a subject on which very few of the many hundreds of Irish land surveyors recorded their views. Although his tables provided conversions, correct to three decimal places, from Irish into English acres, he urged that the introduction of the statute acre to Ireland should be followed by "an admeasurement with a four-perch English chain of 22 yards long [ie Gunther's chain],[133] undertaken or performed by a correct land-surveyor". He argued that the English chain was a more accurate lineal measure than the chain usually used in Ireland (two Irish perches, or 14 yards, long); the English chain had 100 links while the Irish chain had 50 links and, according to Leahy, the surveyor was more liable to make mistakes in measurement and in computation when using the latter. He went to explain that the Irish surveyor had not adopted a four-pole chain, because-

> the four-pole Irish chain was too long [at 28 yards], the tension was too great, and its sagg, or centrepetal force on the middle so great, as to form the chain when held up between two persons – drawn tight, into a parabolic curve, rather than a straight line. The four-pole chain, therefore, was too long for land-surveying, except along roads, where lineal measure only was the object in view. The two-pole chain, on the other hand, was too short and straight lines could not be laid out, nor kept so direct, this chain being only fourteen yards from handle to handle; but the English four-perch chain is that which men of skill and knowledge will confess to be a measure best adapted for nice operations.

He concluded by recommending that all land surveyors should "furnish themselves with the English four-pole chain, abandon the old measure, and perfectly submit to that which is now by law established" instead of continuing to make their surveys in Irish measure and then converting them by his own or similar tables into English measure.

The Limerick and Waterford Railway

Patrick Leahy was already in his mid-fifties when Ireland's first railway became operational in 1834 and, like the rest of his generation of engineers, he had no formal training in railway engineering or in the technology of steam. Since 1810, he had been advocating a canal, rather than a railway, to serve the Killenaule coalfield and as far as steam locomotives were concerned, it would not have been surprising if he had aligned himself with the views of the Duke of Wellington, who was quoted as saying "I see no reason to suppose that these machines will ever force themselves into general use". It seems clear, however, that Leahy was associated with the proposed Limerick & Waterford Railway in 1825-26, one of the first railway projects to be promoted in Ireland. His introduction to railway engineering was almost certainly fortuitous and his role in that particular project was probably a limited one. Nevertheless, his involvement in the new branch of engineering, and in railway promotion itself, was to last for over 20 years.

On 31 May 1826, when the first Irish Railway Act received the Royal Assent,[134] the Hibernian Railway Company was authorised to build a line from Limerick to Waterford "with several branches therefrom in the County of Tipperary". One of these was to run northwards from Cappagh, between Cahir and Tipperary town, through Cashel to Thurles. A subsidiary branch was planned from a point a few miles north of Cashel, running "nearly in the line at one time surveyed for an extension of the Grand Canal and connected by a short branch with inclined planes to the collieries of Killenaule".[135] The Limerick-Waterford route had been surveyed in 1825 by Alexander Nimmo and the original plans deposited at Westminster were signed by him and by Benjamin Meredith, a surveyor.[136] The plans do not acknowledge any involvement by Patrick Leahy in the survey work although ten years later, when the scheme was revived, he claimed that it was he who had "projected" it.[137] While that claim may not entirely fit the facts, it seems clear that without some involvement in the project, Leahy could hardly have described himself publicly, as early as 1828, as "Assisting Engineer and Surveyor to the Hibernian Rail Road Company".[138]

Alexander Nimmo, who had primary responsibility for the railway survey, had been one of the Bogs Commissioners' district engineers from 1811 to 1813 and, even if he had not come into direct contact with Leahy during that period, he would probably have known of his satisfactory service with the Commissioners. Nimmo's subsequent work in Ireland related to piers and harbours, and to roads in the west of the country, and he is unlikely to have had any direct local knowledge of the Tipperary area. In such circumstances, it would have been entirely logical for him to engage Leahy on a freelance basis to assist in the railway survey, especially because Leahy had already mapped the barony of Slieveardagh and had surveyed possible canal routes in the area. It seems likely, therefore, that the plans for the branch lines in County Tipperary were prepared by Leahy and this would be consistent with his claim in 1828 to have laid out the "the line of Rail Road" to the Killenaule colliery.[139] This conclusion is supported by the fact that the branch line plans differ in style from the plans for the Limerick-Waterford main line and are remarkably similar to the Slieveardagh map.

"no public works or surveys of account have offered"

By 1830, Patrick Leahy had been in practice as a land surveyor and civil engineer for over 30 years and had reached his fiftieth year. He was obviously a competent land surveyor and cartographer but was only one of many practicing in this field at the time. He had not succeeded in coming to notice or gaining commissions at national level and, with the exception of the Slieveardagh map, had no major local projects to his credit. He was holding himself out also as a civil engineer and seeking engagements in the engineering field but, while he had been involved in survey and preliminary design work for canals and a railway scheme, his name had not come to be associated with any completed infrastructure or construction project of significance. Apart from his engagement between 1810 and 1813 with the Bogs Commissioners, he had failed to break into the small group of engineers who regularly obtained well-paid Government commissions between then and 1830: as he put it himself, "no public works or surveys of account have offered [in Ireland] since that time".[140] He had hoped that legislation to divide up the joint ownership of bogland areas and to provide for their drainage would have followed the Commissioners' reports, thus opening up "a new field … in that neglected country for scientific labours" – but that had not happened. He had been told that there was no opening for employment on Government road schemes and Colonel Colby, Director of the Ordnance Survey, had rejected his application for employment on the triangulation and survey programme because "no vacancy existed".[141] An application to Richard Griffith for employment on the Boundary Survey had also been unsuccessful, prompting Leahy to note that "Mr Griffith has several old favourites since the time of the Bog Surveys whom he still retains and he requires much formality by way of recommendation, or letters from the Nobility and Gentry to which alone he pays attention on such occasions".[142]

In failing to gain worthwhile Government employment in the 1820s, Leahy was in much the same position as the majority of Irish surveyors and engineers who had to accept the fact that, with the exception of those schemes on which Richard Griffith was engaged, the design and direction of virtually all of the major engineering projects commissioned in Ireland in the first decades of the century were assigned to engineers based in England or Scotland, or to the small number of English and Scottish engineers who based themselves in Ireland – mainly Alexander Nimmo, William Bald and John Killaly. Thus, in 1830, Leahy was forced to seek employment in England for himself and two of his sons who by then had joined him in his surveying and engineering practice.

REFERENCES

1. NAI, CSORP 1823/6319, Memorial of 31 July 1823 from Patrick Leahy
2. I am grateful to Fergus Gillespie, Chief Herald of Ireland, for his advice on this point.
3. NAI, CSORP 1823/6319, Memorial of 31 July 1823 from Patrick Leahy, states that he had then "arrived to his forty-fourth year"; Leahy's death notice in the *Cape Town Mail*, 12 October 1850, states that he was then aged 70; the earliest baptismal register for Adare Roman Catholic parish (where Leahy was probably born) dates from 1832.
4. Curragh Chase House was destroyed by fire in the 1940s; the estate was acquired by the State in 1957 and is now a Forest Park of about 550 acres but only the outer shell of the house remains; see Joan Wynne Jones, *The Abiding Enchantment of Curragh Chase – A Big House Remembered*, 1983.
5. William Nolan, "A Public Benefit: Sir Vere Hunt, Bart. and the Town of New Birmingham, Co. Tipperary, 1800-18", in *Surveying Ireland's Past: Multidisciplinary Essays in Honour of Anngret Simms*, Geography Publications, Dublin, 2004, page 415
6. Edward MacLysaght, "Survey of Documents in Private Keeping, First Series", *Analecta Hibernica*, No 15, 1944, page 389
7. Microfilm copy, National Library of Ireland, P.5527, 5528

Chapter I — Patrick Leahy Country Surveyor

8. Samuel Lewis, *A Topographical Dictionary of Ireland*, Volume I, London, 1837, page 208
9. NAI, OP143/19, letter of 8 May 1802 from Vere Hunt
10. William Nolan, "Literary Sources", in *Irish Towns: A Guide to Sources*, Geography Publications, Dublin, 1998, page 164
11. Samuel Lewis, *A Topographical Dictionary of Ireland*, Volume I, London, 1837, page 208
12. NAI, CSORP 1822/2313, letter of 11 January 1822 from Patrick Leahy, states that he had by then devoted twenty-five years of his life to surveying for public authorities and for many private gentlemen.
13. VHD, 4 and 18 July; 9, 11, 22 and 28 August 1813
14. A son of Neville's, also John, born in County Limerick in 1813, was to become one of Patrick Leahy's county surveyor colleagues in 1840.
15. *Seventh Report of the Inspector General on the General State of the Prisons of Ireland*, HC 1829 Vol XIII (10); by 1839, the New Birmingham bridewell was "a disgrace to the county" and the county surveyor was directed by Dublin Castle to make a presentment to have it improved (NAI CSORP 1839/27/10069).
16. K Whelan, "The Catholic Church in County Tipperary, 1700-1900", in *Tipperary: History and Society*, ed. William Nolan, Geography Publications, Dublin, 1985, pages 215-255
17. *The Dictionary of Land Surveyors and Local Map-makers of Great Britain and Ireland, 1530-1850* (Second Edition), The British Library, 1997; J H Andrews, *Plantation Acres*, Ulster Historical Foundation, 1985
18. Rolf Loeber, *An Alphabetical List of Irish Architects, Craftsmen and Engineers from the 15th to the 20th century*, Irish Georgian Society, 1973; *A Biographical Dictionary of Civil Engineers in Great Britain and Ireland, Volume 1: 1550-1830*, Thomas Telford Publishing, 2002; *Directory of British Architects 1834-1914* (2 vols), Continuum, London, 2001
19. For a detailed assessment of the role of the early land surveyors, see J H Andrews, *Plantation Acres*, Ulster Historical Foundation, 1985 and *A Paper Landscape*, (second edition), Four Courts Press, Dublin, 2002; see also Jacinta Prunty, *Maps and Map-Making in Local History*, Four Courts Press, Dublin, 2004.
20. An Act to amend the Laws for improving and keeping in Repair the Post Roads in Ireland …, 45 Geo. III, c. 43
21. Peter O' Keeffe, *Ireland's Principal Roads AD 1608 – 1898*, National Roads Authority, Dublin, 2003, pages 133 - 140
22. NLI, Irish Road Maps, 15 A 6, Map No 53
23. NAI, CSORP 1823/6319, Memorial of 31 July 1823 from Patrick Leahy
24. NAI OP 1834/184, letter of 10 May 1834; see also NAI CSORP 1846 W 5222, letter of 17 March 1846, in which Leahy claims to have served under the Directors General of Inland Navigation.
25. *A Biographical Dictionary of Civil Engineers in Great Britain and Ireland, Volume 1: 1500-1830*, Thomas Telford Publishing, London, 2002, page 712
26. An Act for Granting to His Majesty the Sum of Five Hundred Thousand Pounds for promoting Inland Navigation in Ireland … 40 Geo III c. 51 (Ir)
27. NAI, OPW5HC/6/525, 525A, 526, engineering drawings of canal from Tipperary to Carrick on Suir
28. Rev William P Burke, *History of Clonmel*, 1907, pages 195 - 197
29. NAI, OPW 1/5/4/2, DGIN Letter Book No 2, Southern District, letter of 29 May 1807 to Lord Cahir
30. NAI, OP 255/2, report of 23 February 1808 from the Directors General of Inland Navigation; *Third Report from the Committee on Inland Navigation in Ireland*, HC 1812-1813 (284) VI
31. NAI, OPW 1/5/1/9, DGIN Minute Book No 9, page 217
32. NAI, OPW 1/5/1/10, DGIN Minute Book No 10, pages 18 and 22; OP 279/5, letter of 28 April 1809 from the Directors General to the Lord Lieutenant
33. NAI, OPW 1/5/4/3, DGIN Letter Book No 3, Southern District, letters of 3, 5 and 30 May 1809
34. NAI, OPW 1/5/1/10, DGIN Minute Book No 10, page 141
35. NAI, OPW 1/5/1/10, DGIN Minute Book No 10, page 261
36. ibid
37. NAI, OPW 1/5/1/10, DGIN Minute Book No 10, page 285
38. NAI, OPW 1/5/4/3, DGIN Letter Book No 3, Southern District, letter of 6 November 1809
39. NAI, OPW 1/5/4/11, DGIN Minute Book No 11, page 24
40. NAI, OPW 1/5/4/11, DGIN Minute Book No 11, page 331
41. NAI, OP 297/3, report dated 17 February 1810; see also OP 279/5 and OP 297/3
42. NAI, OPW 1/5/1/11, DGIN Minute Book No 11, page 87
43. NAI, OPW 1/5/1/11, DGIN Minute Book No 11, pages 236 and 344
44. The accounts of the Bogs Commissioners show that Leahy was not paid by them for any survey work in the months of October, November and December in 1810 or 1811 and it may be that he concentrated on canal work in these periods.

In search of fame and fortune: the Leahy family of engineers, 1780-1888

45. NAI, OPW 1/5/1/10, DGIN Minute Book No 13, page 197
46. NAI, OPW Architectural and Engineering Drawings, OPW5HC/6/480 and 481
47. Archives of the Geological Survey of Ireland, Dublin, BY 7-2-24-001
48. A Map of the Coal District near Killenaule in the County of Tipperary, by Patrick Leahy CE, 11 May 1824, NLI 16 1 17 [1-2]
49. *Inland Navigation: Papers relating thereto, July 1812*, Part VII, report of 31 August 1811, HC 1812 Vol V (366)
50. Map included in *Inland Navigation: Papers relating thereto, July 1812*, HC 1812 Vol V (366)
51. This line is shown on Leahy's 1818 map of the barony of Slieveardagh held in the Archives of the Geological Survey of Ireland, Dublin, BY 7-2-24-001.
52. *Inland Navigation: Papers relating thereto, July 1812*, HC 1812 Vol V (366)
53. *Third Report from the Committee on Inland Navigation in Ireland*, HC 1812-13 Vol VI (284)I
54. NAI, CSORP 1825/11817, reports by John Killaly and other papers relating to the extension of the Grand Canal southwards from Monasterevan
55. Letter from the Lord Lieutenant, Marquis Wellesley, to the Duke of Wellington, 17 February,1824, quoted in J H Andrews, *A Paper Landscape* (second edition), Four Courts Press, Dublin, 2002, page 21
56. *Report from the Select Committee on the Survey and Valuation of Ireland*, Appendix A, Minutes of Evidence, HC 1824 (445) VIII
57. *Reports of the Commissioners Appointed to Enquire into the Nature and Extent of the Several Bogs in Ireland and the Practicability of Draining and Cultivating them*, HC 1810 Vol X (365) (First Report); 1810-1811 Vol VI (96) (Second Report); 1813-1814 Vol VI (130) (Third Report); 1813-1814 Vol VI (131) (Fourth Report); see also Arnold Horner, "Napoleon's Irish Legacy: the Bogs Commissioners, 1809-14", in *History Ireland*, Vol 13, No 5, September/October 2005
58. NAI, 1137/77, Minute Book of the Commissioners for the Improvement of the Bogs in Ireland, 1809-1813 (hereafter BC Minutes), 3 January 1810; *Third Report*, Appendix No 2, report dated 3 July 1811
59. Townshend transferred from the employment of the Royal Canal Company to take up a district engineer post with the Bogs Commissioners in October 1809 (BC Minutes, 19 September and 9 October 1809).
60. BC Minutes, 26 February 1811; *Second Report*, Appendix No 7
61. BC Minutes, 12 February and 8 March 1811
62. *Estimate of the Sum that will be necessary to complete the Surveys of the Districts*, HC 1812 Vol V (94)
63. BC Minutes, 29 September 1812
64. *A Biographical Dictionary of Civil Engineers in Great Britain and Ireland, Volume 1: 1500-1830*, Thomas Telford Publishing, London, 2002, page 712; in June 1811, when Townshend was offered a permanent position by the Birmingham and Worcester Canal Company, the Directors General had insisted that he should complete his contract with them before taking up another post (BC Minutes, 18 June 1811).
65. BC Minutes, 1 and 8 December 1812.
66. BC Minutes, 28 February 1812 incorporating the accounts for 1810 and 1811
67. *Third Report*, Appendix No 10 and Plates XVIII to XXVII; a set of the original maps on the scale of four inches to one mile is held at NLI, 16 E 1-6 and 16 D 8-22.
68. *Third Report*, Appendix No 10
69. BC Minutes, 10 August 1813
70. NAI, CSORP 1823/6319, Memorial of 31 July 1823 from Patrick Leahy; General William Roy (1726-1790) was regarded as the father of the Ordnance Survey in England while General William Mudge was Superintendent of the Survey from 1798, when he succeeded Colonel Edward Williams, until his death in 1820.
71. Birmingham City Archives, MS 3192/Acc1941-031/713, letters of 31 October 1830 and 19 December 1832 from Patrick Leahy to Thomas Townshend
72. ibid
73. *Fourth Report*, Appendix No 1
74. John Leslie Foster (1780-1842), a Dublin barrister and former MP, was appointed Chairman of the Bogs Commissioners following the death of General Charles Vallancey in August 1812.
75. NAI, CSORP 1823/6319, Memorial of 29 October 1823 from Patrick Leahy
76. *Fourth Report*
77. NAI, CSORP 1823/6319, Memorials of 31 July and 29 October 1823 from Patrick Leahy
78. *A Biographical Dictionary of Civil Engineers in Great Britain and Ireland, Volume I, 1500-1830*, Thomas Telford Publishing, London, 2002 contains entries on Brown (page 81) and Thomas (page 699); the latter was a relative – possibly a brother-in-law – of Townshend's.

79. *Clonmel Advertiser*, 10,17, 24 October 1818
80. A Map of the Coal District near Killenaule in the County of Tipperary, by Patrick Leahy CE, 11 May 1824, NLI 16 1 17 [1-2]
81. Killenaule Estate Rental Reference Book, 1828, NLI microfilm, Pos 8873
82. NAI, CSORP 1834/903, letter of 21 February 1834 from Patrick Leahy
83. In the early decades of the nineteenth century, Edward Troughton (1753-1835) was the leading English maker of scientific instruments, including those used by navigators and surveyors; from 1826 onwards, he and his partner, William Simms, traded as Troughton & Simms.
84. NAI, CSORP 1823/6319, Memorial of 31 July 1823 from Patrick Leahy
85. NAI, CSORP 1820 L 110, letter of 19 January 1821 from Patrick Leahy
86. NAI, CSORP 1820 L 110, memorandum dated 29 January 1821
87. NAI, CSORP 1820 L 110, letter of 28 January 1821 from Edward Wilson
88. NAI, CSORP 1822/2313, letter of 11 January 1822 from Patrick Leahy
89. NAI, CSORP 1823/6319, Memorials of 31 July and 29 October 1823 from Patrick Leahy
90. Established in 1714 in response to pressure to encourage the finding of more reliable means of navigation, the Board of Longitude, until its abolition in 1828, occasionally made small awards for new or improved navigational instruments or procedures.
91. NAI, CSORP 1823/6319, Memorial of 31 July 1823 from Patrick Leahy
92. William Ford Stanley, *Surveying and Levelling Instruments*, London, 1890, page 330
93. I am grateful to J H Andrews and Dr R C Cox for their assistance in assessing the possible nature of Leahy's invention.
94. VHD, 5 April 1815
95. VHD, 1 January 1814
96. VHD, 31 December 1815
97. *Report from the Select Committee on the Survey and Valuation of Ireland*, Appendix B, evidence of Thomas Lannigan, HC 1824 Vol VIII (445)
98. This map was completed in 1799 by Neville Bath, Engineer, at a cost of £1500 - see *Report from the Select Committee on the Survey and Valuation of Ireland*, HC 1824 Vol VIII (445), Appendix H, No 26.
99. J H Andrews, *Plantation Acres*, Ulster Historical Foundation, 1985, page 355 and 363, note 75
100. NAI, CSORP 1823/6319, Memorial of 31 July 1823 from Patrick Leahy
101. *Clonmel Advertiser*, 10,17, 24 October 1818
102. The Longfield Map Collection (NLI, 21 F 46 and 47) includes several maps by John Longfield and his associates of estates in County Tipperary; see J H Andrews, "The Longfield Maps in the National Library of Ireland: An Agenda for Research", in *Irish Geography*, Vol 24 (1), 1991, pages 24-34.
103. Evidence given by Leahy at an inquiry conducted by a committee of the Privy Council on 16 June 1837, *Clonmel Advertiser*, 21 June 1837; NAI, Council Office Papers, Box 44
104. Archives of the Geological Survey of Ireland, Dublin, BY 7-2-24-001
105. J H Andrews, "David Aher and Hill Clements's Map of County Kilkenny 1812-24", in W Nolan and K Whelan (eds), *Kilkenny: History and Society*, Geography Publications, Dublin, 1990, pages 437 - 463
106. Aher told the *Select Committee on the Survey and Valuation of Ireland* in 1824 (HC 1824 (445) VIII) that he had three or four assistants on the survey work in Kilkenny - Clements, his partner, two of his sons, and an unnamed fourth assistant.
107. *Third Report*, Plate V, Map of the Southern Part of District No 6
108. J H Andrews, *Plantation Acres*, Ulster Historical Foundation, 1985, pages 176 and 231
109. For the development of geological cartography generally in Ireland, see Gordon L Herries Davies, *Sheets of Many Colours: The Mapping of Ireland's Rocks 1750-1890*, Royal Dublin Society, 1983.
110. See James Condon, "Mid-Nineteenth Century Thurles - The Visual Dimension", in *Thurles: The Cathedral Town, Essays in honour of Archbishop Thomas Morris*, Geography Publications, Dublin, 1989, page 82.
111. Tipperary Libraries, Thurles, *A Map of the town and lands of Thurles, in distinct denominations, farm holdings and enclosures, the estate of the Earl of Llandaff, surveyed by Pat Leahy in 1819 and copied by John Longfield, Harcourt Street, Dublin in 1827*; the National Library of Ireland holds a substantial collection of manuscript maps by John and William Longfield and their associates, mainly dating from the early decades of the nineteenth century, and including copies of maps made by others not connected with the firm, but the Thurles map is not among them.
112. A Map of the Coal District near Killenaule in the County of Tipperary, by Patrick Leahy CE, 11 May 1824, NLI 16 1 17 [1-2]

113. Killenaule Estate Rental Reference Book, 1828, NLI microfilm, Pos 8873
114. An Act to provide for the more deliberate Investigation of Presentments to be made by Grand Juries for Roads and Public Works in Ireland, and for accounting for Money raised by such Presentments, 57 Geo III, c.107
115. Notes respecting the duties of County Surveyors written in the month of January, 1819, included in Appendix F to the *Report from the Select Committee on the Survey and Valuation of Ireland*, HC 1824 Vol VIII (445)
116. NAI, OP 494/20, 1817, "Applications for position of County Surveyor", contains only a letter from a Naas lady to William Gregory, Under-Secretary at Dublin Castle, seeking an appointment for her son; OP 490, 1817, "Situations and Relief", does not contain any letters of application for appointment as county surveyor.
117. NAI, OP 1834/184, letter of 10 May 1834 from Patrick Leahy
118. Parliamentary Debates (First Series), Vol XXXVIII, 6 May 1818
119. William Grieg, *Strictures on Road Police* (sic), Dublin, 1818
120. *Report of the Board of Civil Engineers which sat at No 21 Mary Street, Dublin, from 23rd December 1817 to the 19th January, 1818*, HC 1818 Vol XVI (2)
121. An Act to suspend, until the End of the present Session of Parliament, the Operation of an Act made in the last Session of Parliament, to provide for the more deliberate Investigation of Presentments to be made by Grand Juries for Roads and Public Works in Ireland, and for accounting for Money raised by such Presentments, 58 Geo III, c.2
122. An Act to provide for the more deliberate Investigation of Presentments to be made by Grand Juries for Roads and Public Works in Ireland, and for accounting for Money raised by such Presentments, 58 Geo III, c.67
123. Col Sir Charles Close, *The Early Years of the Ordnance* Survey, David & Charles Reprints, 1969, page 121; J H Andrews, *A Paper Landscape: The Ordnance Survey in Nineteenth –Century Ireland* (second edition), Four Courts Press, 2002, pages 11 - 21
124. J H Andrews, *Plantation Acres*, Ulster Historical Foundation, 1985, page 253
125. Advertisements in *Clonmel Advertiser*, 10, 17, 24 October 1818
126. NAI, CSORP 1823/6319, Memorial of 31 July 1823 from Patrick Leahy
127. NAI, CSORP 1823/6319, Memorials of 31 July and 29 October 1823 from Patrick Leahy
128. NAI, CSORP 1823/6319, note on memorial of 29 October 1823 from Patrick Leahy
129. NAI, CSORP 1826/14473, letter of 19 July 1826 from Pat Leahy
130. *Coldwell's Tables for Reducing Irish Money into British Currency*, second edition, printed by T Coldwell, 21 Batchelor's Walk, Dublin, 1826; see also *Coldwell's Tables of the Weights and Measures*, Dublin, 1826; *Tables for reducing Irish Money into British Currency*, Simms and McIntyre, Belfast, 1826; and *Tables of Currency, Coins, Weights and Measures etc*, J Charles, Dublin, 1826
131. Patrick Leahy, Civil Engineer etc, *New and General Tables of Weights & Measures with ample calculations and suitable examples under each head*, Bentham and Hardy, Dublin, 1826
132. 6 Geo. IV, c.12
133. Edward Gunther (1581-1626), an English astronomer and mathematician, invented his surveyor's chain in 1620; it was 66 feet long (equivalent to 22 yards or four perches or poles), and was divided into 100 links.
134. 7 Geo. IV, c. xxxix
135. Report of Alexander Nimmo CE to the Secretary of the Hibernian Railway Company, *Dublin Philosophical Journal*, No III, February 1826
136. House of Lords Record Office London, Deposited Plan, 1826, HC/CL/PB/6/plan 1826/61
137. NAI, CSORP 1834/184, letter of 10 May 1834
138. Killenaule Estate Rental Reference Book, 1828, NLI microfilm, Pos 8873
139. ibid
140. Birmingham City Archives, MS 3192/Acc1941-031/713, letter of 31 October 1830 from Patrick Leahy to Thomas Townshend
141. ibid
142. ibid

CHAPTER II

A FAMILY BUSINESS DEVELOPS, 1830-34

"I can now, thank God, turn out a strong party into the field for Levelling or Surveying".

Patrick Leahy, 1830[1]

Children of Patrick and Margaret Leahy
Patrick Leahy married Margaret Cormack, a native of Gortnahoe, County Tipperary, with Father Michael Meighan officiating, at Gortnahoe, probably in 1805.[2] The couple were living at Fennor, near Urlingford, when their eldest son, Patrick, was born on 31 May 1806 and baptised at Gortnahoe a few days later.[3] Although the family probably continued to live in the area until the beginning of 1810, the register at Gortnahoe does not record the baptism of any later Leahy children.[4] A daughter called Helena (also known as Ellen) was born and baptised at Portglenone, County Antrim, c.1813, while her father was serving with the Bogs Commissioners,[5] and two other children, Denis and Margaret, were probably born before that in one of the midland or northern counties. Younger members of the family (Edmund, Matthew, Susan, Anne and Alice) are likely to have been born in Thurles after the family returned there in 1813 but dates of baptism have been traced for only two of them, Susan (Susanna), born in 1819[6] and Alice, born in 1823.[7]

Of the five girls born to Patrick Leahy and his wife, Alice is presumed to have died in childhood because of the complete absence of references to her in any subsequent family papers or other documents. Margaret was the only sister who married.[8] With few career opportunities available to middle-class women until the later decades of the nineteenth century, the three unmarried sisters, Helena, Anne and Susan, lived unobtrusively together for much of their lives, and were dependent on financial support from their brothers to meet living expenses in London and elsewhere in England from the 1850s until their deaths. None of them appears to have achieved independent status at any stage: when Ellen (Helena) died in 1863, the occupation column of her death certificate described her as "Daughter of Patrick Leahy, Civil Engineer (deceased)"[9] and when Susan died 20 years later, the death notice in *The Times* described her as "sister of the late Archbishop of

Cashel".[10] As late as 1905, nearly 55 years after her father had died, Anne's death certificate described her as a woman of independent means but still added that she was "Daughter of Patrick Leahy, a Civil Engineer (deceased)".[11]

The Pride of the Family
Patrick Leahy, the eldest member of the family, was described in one of his obituaries as "the pride and ornament - the decus, spes et gloria – of his family".[12] In contrast to the lives of his sisters, his career is well documented arising from distinguished service in the Catholic Church which culminated in his appointment as Archbishop of Cashel and Emly in 1857.[13] In August 1826, when he was 20 years old, and had read the entrance course for Trinity College, Dublin and a general course of classics,[14] Patrick began his studies for the priesthood at St Patrick's College, Maynooth where he pursued courses in theology and philosophy. He was one of a small number of students called to give evidence at an inquiry conducted in Dublin Castle by the Commissioners of Irish Education in 1826 into the affairs of the College, including the attitudes and backgrounds of the students.[15] Among other things, he told the Commissioners[16] that he had received his education principally at a classical day school in Thurles, where the fee was a guinea per quarter, and subsequently under a master at the Free School in Clonmel - an endowed school set up originally in 1685 for the education of the sons of Protestant freemen but which, in the nineteenth century, "held out temptations to the better-class Catholics" according to the town's historian.[17] Leahy also told the inquiry that his father was "engaged in the surveying and engineering department" and, if an 1875 obituary can be relied on, he seems to have participated in that work himself during holiday periods when, for example, he was said to have assisted in a mapping assignment in the early 1830s.[18]

After his ordination in June 1833, Leahy served for two years as a curate in Knocklong, Co Limerick before moving in 1835 to Thurles where he was to live for most of the following 40 years, serving first as a curate, then from 1837 to 1847 as professor at the newly opened St Patrick's College, later as President of the College, and finally as Archbishop. During this period, he can have had little direct contact with the other members of the family, most of whom had left Tipperary by 1834 but, as will be seen in Chapter XI, he and some of the family were regularly in correspondence in the 1850s and 1860s about family and other matters.

Brother engineers
The other three Leahy brothers, Denis, Edmund and Matthew, who are assumed to have been born c.1811,[19] c.1814[20] and c.1818,[21] respectively, all chose to follow their father's profession of land surveyor/civil engineer. While there is no record of the early education they received, it is likely to have been broadly similar to that of their older brother but with more emphasis on mathematics; their father recorded in 1830 that he had "paid every

attention to their education, both Classical and Mathematical"[22] to equip them for their chosen careers.

There was no university course in engineering in Ireland before November 1841 when Trinity College, Dublin, inaugurated a two-year diploma course, with John MacNeill (soon to become Sir John) as Professor of Civil Engineering.[23] Thus, "in the absence of an established school, and a dearth of professional books, a mathematical education, with a knowledge of surveying, were principally relied on" as the basis of the formation of an engineer, according to the historical sketch of engineering in Ireland presented by the President of the Institution of Civil Engineers in Ireland in 1860.[24] However, while aspiring engineers could acquire the necessary knowledge of mathematics through attendance at general degree courses at the university, relatively few of the Leahys' contemporaries appear to have done so.[25] In addition to a groundwork in theoretical knowledge, the aspirant engineer needed to build up his practical knowledge by contact with senior members of the profession, visits to construction sites, the study of treatises on the various branches of civil engineering and the perusal of reports on particular projects and schemes. Moreover, a career in engineering usually involved a period of apprenticeship in the office of an established member of the profession, leading on to a position as assistant, and to independent private practice at a relatively young age. Very often, however, it was a case of "natural aptitude supplying the place of regular professional training", as was said of Barry D Gibbons, a well-known contemporary of the Leahys.[26]

None of the Leahy brothers received a university education or other formal instruction to equip them for engineering careers; instead, according to their father, they were instructed by him "in his own line of business".[27] In effect, they learned on the job, as did many others, working initially as apprentices or junior assistants to their father. Edmund, however, regularly boasted[28] that he had been a pupil of Alexander Nimmo, the eminent Scottish engineer, who had been a professor of mathematics, navigation and surveying, and rector of Inverness Academy, before working extensively on public projects in Ireland between 1811 and 1832. Throughout his years in Ireland, Nimmo continued his educational activity by engaging a large team of young men as junior or trainee engineers, surveyors and valuers[29] and it would have been a natural development for Edmund and Denis Leahy, having probably come in contact with Nimmo in 1825-26 when their father was involved in the Limerick & Waterford Railway survey, to seek to join that team when they grew older. While there is no evidence of an association in Ireland between the younger Leahys and Nimmo,[30] Denis was later to claim that he had been involved with him in laying out the Limerick & Waterford line; however, as he was no more than 15 years old when the original surveys for that line were made, his contribution to the work cannot have been a very significant one.[31]

Chapter II — A family business develops, 1830-34

A family firm in Clonmel

By July 1823,[32] the Leahy family had moved to Clonmel, perhaps because it was then - before the division of Tipperary into two ridings in 1838 – the county capital, where the grand jury met at the Spring and Summer Assizes to deal with the fiscal business of the county which was of central importance to the employment prospects of local engineers and surveyors. Apart from this, Clonmel was the largest town in the county, with a population of 15,590 in 1821, more than twice the number living in Thurles.[33] It was a thriving industrial and commercial centre and, since 1815, the headquarters of Charles Bianconi's rapidly expanding coaching business. Here, Patrick Leahy continued to pursue his surveying and engineering business and, after the introduction of the £10 householder franchise in 1831,[34] he was registered under the designation "civil engineer" in the list of about 500 voters in the Borough.[35] The Leahy residence was at 18 Anne Street, a pleasant cul-de-sac of 24 three-storey houses built in the early 1820s and described at the time as "a street of very eligible houses".[36]

Anne Street, Clonmel, where the Leahy family resided from 1823 to 1834.

When Denis joined his father's engineering practice around 1828 it became known as "Messrs Leahy & Son". When Edmund joined the firm some years later, the title was changed to "Messrs P Leahy & Sons" and, later again, when Denis left to practise on his own account, the title "Messrs P & E Leahy, Civil Engineers" was used. Little evidence of the work of the firm between 1828 and 1830 has survived but it is probable that it continued to be involved with relatively minor projects, mainly in County Tipperary.

Denis Leahy, however, seems to have had some involvement in the building of St. Patrick's College, Thurles. The foundation stone of the college, which has been described as "one of the finest examples extant in the country of institutional buildings of the pre-famine period",[37] was laid in July 1829[38] but the building was only partially complete when it opened in 1837.[39] While Denis Leahy has been described as the engineer for the project,[40] the plans were prepared by the architect, Charles Frederick Anderson, and as there was no general contractor, the works were supervised by a local priest, Rev Thomas O'Connor, who was to become first President of the College;[41] the precise nature of Leahy's involvement is therefore difficult to establish.

Procession of the Bishops & Clergy to St Patrick's College, Thurles, at the Synod held on 25th August 1850 (NLI PD 1696 TC, courtesy of the National Library of Ireland).

Railway work in England
The railway era began in England with the opening of the Liverpool and Manchester railway in September 1830 – the first railway designed to carry passengers as well as freight and to rely on steam locomotives (instead of horses) for motive power. With the immediate success of the line, new railway companies were quickly formed, many of them in the Liverpool and Manchester areas, to construct additional lines.[42] Alexander Nimmo was one of the engineers called on to assist the promoters of several of the new companies in the selection and planning of routes: he was appointed engineer to the Liverpool and Leeds company in September 1830, and held appointments also as consulting engineer with the Manchester, Bolton & Bury, the Preston & Wigan, the St Helens & Runcorn Gap, and the Birkenhead & Chester railway companies.[43] But Nimmo was still heavily involved in

Chapter II — A family business develops, 1830-34

the roads and other public works schemes which he had been directing in the Western District of Ireland since 1822 (and was, indeed, the subject of criticism at the time on foot of his management of these works), with the result that he was obliged in 1831 to devote no less than 325 days to the affairs of the District.[44] It was fortunate, therefore for Patrick Leahy, who had almost despaired of finding suitable employment opportunities in Ireland at the time, that Nimmo was obliged to engage a number of assistants and to delegate much of his English railway work to them.

Writing from Hayward's Hotel, Manchester, in October 1830 to his former mentor, Thomas Townshend, Leahy told him that he had been called on by Nimmo to survey the proposed line of railway from Liverpool to Leeds, that the survey had reached an advanced stage and that they hoped "to have all ready for this Session of Parliament". He wrote also that Nimmo planned to replace the Bolton and Bury Canal by a branch line of the Manchester & Liverpool Railway and that his own railway work was likely therefore to take more time than had initially been foreseen. He informed Townshend that he had with him "two sons who are as capable and able to do business as ever I was when you lavished so much praise on my humble but diligent performance"; the two young men were "equally clever in the office" and he thanked God that he could now "turn out a strong party into the field for Levelling or Surveying".[45]

By March 1831, Leahy had completed the surveys for the Liverpool & Leeds line[46] but the extent of the further railway survey work he and his sons may have carried out in 1830-32 cannot readily be established; the indications are, however, that they were not continuously engaged in England during those years. Similarly, the extent to which their railway work may have involved close contact between the younger Leahys and Nimmo can now only be a matter of conjecture. Nonetheless, in later years, both of them regularly cited their association with Nimmo in attempting to advance their careers. Denis, for example, when in need of a reference in 1861, was able to call on Sir John Hawkshaw (1811-91), who was then one of the most prominent consulting engineers in England,[47] to testify that he and Denis were in Nimmo's office in early life; and a testimonial provided at the same time by John Scott Russell, another eminent consulting engineer, also referred to the early association between Denis and Nimmo.[48] Edmund, in attempting to boost his status as a railway engineer in the 1840s, boasted that he had earned about £500 a year on his English railway surveys[49] and declared that he had first become involved in railways when "I had the good fortune of being under the instructions of my late lamented friend, Mr. Nimmo; and I enjoyed the advantage not only of being his friend but, to use the common phrase, of being his pet".[50]

Maps and surveys of Waterford city
Following Nimmo's death in January 1832, Patrick Leahy seems to have had difficulty in obtaining further employment for himself and his sons in England. Thomas Townshend

had given him a letter of introduction to John Urpeth Rastrick, one of the most important civil and mechanical engineers of the early railway era, but Rastrick told Leahy in December 1832 that "his business in the line of surveying etc was now over", presumably because he was temporarily concentrating on his locomotive engineering work.[51] Leahy then decided to return to Ireland, telling Townshend that there was "no business now appearing here sufficient to warrant encouragement" and suggesting to him that if, in "your wide field of acquaintance anything should occur in which either myself or my family-assistants could be called into action" he might drop him a line at Clonmel.[52] However, Townshend was heavily engaged as contractor on major canal engineering projects in the Birmingham area at the time and could find no place for the Leahys in his team.

The most substantial body of work undertaken by the Leahys in Ireland in the years 1831 to 1834 was commissioned by Waterford Corporation which, in November 1830, had decided to treat with a competent person to make accurate surveys and maps of their extensive properties in and around the city.[53] Writing from Liverpool where he was still engaged on railway work, Patrick Leahy submitted proposals in the following March for the preparation of these surveys and maps, distinguishing the boundaries and quantities in every leasehold, giving a full description of each property and including tables of reference in alphabetical and numerical order. Streets, lanes and alleys were to be laid down to their exact dimensions and "church spires and all conspicuous objects [were to be] ascertained and delineated from trigonometrical observation as a verification of the admeasurement". The admeasurement itself was to be calculated with strictest accuracy in English measure and then converted to Irish measure, and the maps were to be "executed in the neatest style of modern fancy and taste".[54] The Corporation was to provide mearsmen, with information on lessees and tenants, who would attend and point out the boundaries to the surveyors, while Leahy was to provide chainmen at his own expense.

Leahy's proposals for undertaking this work, and for preparing "a skeleton map of the city for the purpose of its internal improvement or further extension", were accepted in April 1831 and the work seems to have been completed a year later when he had received fees totalling £550.[55] The coloured skeleton map produced by "P Leahy & Sons" was drawn on a single large sheet on a scale of 40 inches to one mile (1:1584); it depicted the built-up area in fine detail, highlighted key buildings such as the Artillery and Infantry Barracks, the Workhouse and Asylum, the City and County Gaols and the Town Hall, and included the results of extensive depth-soundings in the river Suir.[56] The associated book of 51 watercolour maps of individual townlands included tables of reference providing information on relevant leases, an index of lessees and acreages held, and detailed comments on each property.[57] These maps formed the basis for the administration of the Corporation's corporate property holdings for many years; tracings were made under the supervision of the Borough Surveyor in 1884[58] and, together with the originals, were in use for decades after that.

Chapter II — A family business develops, 1830-34

Following an outbreak of cholera in Waterford, the Corporation decided in September 1832 to procure a map of the streets and lanes, showing all of the public sewers, water pipes and gas pipes[59] and, once again, the commission went to Messrs Leahy. For a fee of £500, they had prepared by March 1833 a book of eight maps at various scales, with comments on the history of the city sewers and suggestions for their possible development, including the provision of an interceptor sewer to replace the many different discharges to the river.[60] There was also a general map dated 1834 showing the city sewers and "drawn for the purpose of making improvements on their directions, extensions and construction".[61]

Finally, in 1834, the Leahys prepared a one-sheet *Map of the City of Waterford and its Environs,* on a scale of approximately 20 inches to a mile (1:3168) which was engraved by James Neele & Company of London.[62] The map covered an area of more than two square miles comprising the entire built-up area – then largely on the southern side of the River Suir – and the surrounding undeveloped lands. It was prepared "from actual admeasurement" by Patrick Leahy, with the assistance of Edmund and, according to an 1875 obituary, Patrick Leahy junior, while a clerical student at Maynooth, also participated in

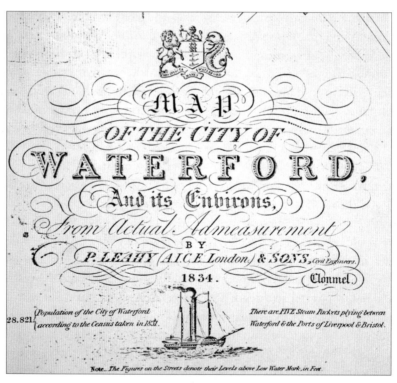

Cartouche on the engraved Map of the City of Waterford and its Environs by P Leahy & Sons, 1834, showing the arms of the city and the paddle-steamer Nora Creina.

the work.[63] An attractive cartouche in the bottom left corner of the map includes a note of the population of the city, as enumerated at the 1831 Census, and a drawing of the paddle-steamer *Nora Creina* which had been carrying passengers, goods and cattle three times a week from Waterford to Bristol since its launch at Birkenhead in 1826.[64] While it appears that this map was prepared by the Leahys on their own initiative rather than as a commissioned project, the Corporation decided to purchase four copies in November 1834.[65]

In the years preceding the Ordnance Survey, few Irish towns or cities were mapped to the standard or on the scale of the Leahys' maps. For Waterford itself, the only previous printed town plans were those prepared by Charles Smith in 1746 and by Bernard Scalé in 1764.[66] The six-inch Ordnance Survey map of Waterford was not published until January 1843, the first large-scale town plan of Waterford (11 sheets at a scale of 1/1056) was not published until 1872, and the first 25-inch maps of Waterford city and county did not emerge until 1903-05.[67] Thus, there were no official maps to supplant the Leahys' large scale maps until 1872 and their engraved map of 1834 was the best available single-sheet map of the entire city on a reasonably large scale until the 25-inch map was published some 70 years later.

When he failed in 1833 in a bid to be engaged for yet another Waterford mapping project, this time involving the preparation for the local Harbour Commissioners of a chart of the port and harbour, Leahy complained bitterly to the Commissioners of Public Works about what he considered to be interference by them in the selection process. The Commissioners, however, denied that they had done anything to prejudice his bid: they had merely supplied, on request, a reference for another of the applicants, Samson Carter, whom they considered to be "able and useful for many operations connected with engineering and surveying".[68] In the event, Carter and Noblett St Leger, both of whom were to become county surveyors in 1834, were given the assignment and had completed it by 1835.[69]

Mapping Clonmel and the Baronies of Tipperary
In parallel with their work in Waterford, and also apparently as a private speculation, the Leahys prepared a detailed map of the borough of Clonmel and, although the map was "in the hands of a London engraver of first talent" by March 1831, copies of the original manuscript map were apparently made and sold to Charles Bianconi and other local gentlemen who, according to Leahy, preferred these copies to engravings.[70] This is borne out by the fact that a copy of the map, dated 1832, held at South Tipperary County Museum was made for John Bagwell, one of the largest property owners in the town,[71] while another copy, also dated 1832, held at Waterford City Archives, was made for Thomas Taylor of Birdhill.[72] The Clonmel map, which measures approximately 60cm by 42 cm, was on a scale of one inch to 20 perches (1:3960) and shows individual house plots

Chapter II — A family business develops, 1830-34

as well as all of the public buildings, churches and meeting houses, and nurseries, orchards, and vegetable and fruit grounds. The old town wall was delineated as was the "New boundary of Borough agreeable to the Reform Bill". The only private residence whose occupier was identified was Melview House, the home of the prominent industrialist, David Malcomson, which suggests that he too may have purchased a copy of the map.

When Commissioners appointed to inquire into the affairs of the municipal corporations in Ireland sat in Clonmel in October 1833, Patrick Leahy, who was one of a very small number of townspeople, other than the Mayor, Town Clerk and other officials who attended, produced yet another map for the information of the Commissioners.[73] This map does not seem to have survived but, according to Leahy, it had been traced by his son, under his direction, from an engraved copy of Sir William Petty's survey of County Tipperary and delineated the different ecclesiastical and administrative boundaries, as ascertained by the Leahys, as well as the boundaries of the Commons of Clonmel. Patrick gave evidence to the effect that the Commons – a mountain area in County Waterford owned by the Corporation and on which the freemen of the town were entitled to graze cattle and cut turf – amounted to upwards of 8,000 acres according to his computation, and was critical of Petty's admeasurements which, in his experience as a surveyor, he had not always found to be accurate. As no map of the Commons was otherwise available to the Commissioners, they commended Leahy for his evidence which they found extremely useful and interesting; one of them commented that "there was a satisfaction and pleasure in examining such a witness" while his colleague remarked that Leahy was obviously "a useful member of society".[74] When the report on Clonmel was presented to Parliament in 1835, Leahy's evidence, attributed to "an eminent surveyor", was incorporated in full.[75]

> **Patrick Leahy & Sons,**
> CIVIL ENGINEERS.
> **Surveyors, Valuers of Lands,**
> &c. &c.
>
> BEG leave to state, that owing to the circumstance of there having been engaged in England for the past two or three years, by which they have been obliged to decline much of their business here, have now made such arrangements as will enable them, in future, to devote close and prompt attention to their professional business in this country.
>
> Having for several years past, and more particularly during their Surveys through various parts of England, directed their attention to Agricultural, Mineral, and Geological enquiries, they are persuaded that various highly useful and beneficial branches of manufacture could be fully established here by the united efforts of the proprietors and occupiers of the soil alone, without the aid or introduction of British capital.
>
> To stimulate the inactive spirit of the country to its internal resources, the advertisers propose tendering their feeble efforts, by adding to all their future Surveys and Valuations, detailed reports on the local and natural advantages of each property; embracing the nature and quality of the soil and subsoil, with the manufacture best suited for each situation.
>
> As to the accuracy of their Surveys and beauty of their Drawings, in the estimation of the most celebrated English Engineers, "they may be equalled, but certainly not excelled."
>
> Their terms will be moderate, their attention unremitting, and their drawing and reports highly useful and satisfactory.
>
> Commands addressed to their office at CLONMEL, will meet the most prompt attention.
>
> 18, Anne-street, 6th Sept., 1833.

An advertisement in Tipperary Free Press by Patrick Leahy & Sons, 6 September 1833.

In 1833, the Leahys were seeking engagements in civil engineering, surveying and valuing, with "commands" to be addressed to the firm's offices at 18 Anne Street, Clonmel. An advertisement in the *Tipperary Free Press* in September of that year claimed that "As to the accuracy of their Surveys and beauty of their Drawings, in the estimation of the most celebrated English engineers, they may be equalled but certainly not excelled".[76] It indicated that "their terms will be moderate, their attention unremitting, and their drawings and reports highly useful and satisfactory". In addition, potential clients were assured that the Leahys proposed "adding to all their future Surveys and Valuations detailed reports on the local and natural advantages of each property; embracing the nature and quality of the soil and subsoil, with the manufacture best suited for each situation". This was, in effect, what Patrick Leahy had already done in the case of his 1824 and 1828 surveys of the Killenaule area.

Along with his newspaper advertising, Patrick Leahy was continuing to apply directly for Government assignments in the 1830s. Emboldened, perhaps, by his success in winning mapping contracts in Waterford, and noting that the Ordnance Survey had only just begun to produce six-inch maps of the northern counties where their work had begun in 1824,[77] he proposed to the Lord Lieutenant in 1834 that he should be engaged to produce a set of maps of each police district in County Tipperary.[78] He argued that such maps would "enable Chief Constables and other public officers to proceed in the exercise of their respective duties with greater certainty, facility and dispatch than has been hitherto practicable" and that if copies were deposited at the Chief Secretary's Office, the officials there would be "perfectly enabled to point out the situation of each Police Barrack and station, as well as knowing the relative situation of every Parish within the County". He suggested also that separate maps of each of the baronies in the county might be prepared showing physical features and roads, and all of the police barracks and military stations marked in "some one striking colour, say a red colour, to render them perfectly conspicuous and make the entire a military map".

To support his proposals, Leahy explained that in the absence of any published map of the county he had regularly been called on by the police to assist them, and he claimed that "where conspiracy and combination calls for the assistance of the strong arm of the law", the administration of the law had often been defeated by the absence of maps and local sketches. In addition, he believed that the work arising from the commutation and composition of tithes under an Act which had been passed in 1832 was being frustrated and delayed because of the absence of maps showing the parish and other boundaries.[79] Leahy guaranteed that he would be able to produce maps of the 11 baronies of Tipperary quite quickly and "at a cheap rate", because he had in his possession, based principally on his own work, "a very minute map" of the county, indicating the various civil and ecclesiastical divisions. However, his proposals were turned down because the Lord Lieutenant did not have power to incur the expense involved and, in any case, the fact that changes were proposed in the police districts meant that "this would not appear to be the time for the undertaking".

Chapter II — A family business develops, 1830-34

Edmund's activities

At the end of 1826, concern about the slow pace of the work of the Ordnance Survey, which had been established in 1824 with a military staffing organisation, led to the recruitment of "Civil Assistants" as draftsmen and surveyors – generally young men of talent but with little experience.[80] Edmund Leahy, who was then 19 years old, took up one of these positions on 19 August 1833 and was assigned to work in District No 5 for which Captain M A Waters of the Royal Engineers was responsible.[81] Leahy joined a mixed team of up to 18 Royal Sappers and Miners/Civil Assistants working in the parish of Upper Moville, County Donegal,[82] under Lt W E D Broughton and was engaged in "plotting and drawing" rather than "surveying" which occupied most members of the team. He was paid four shillings a day, which was the maximum rate of pay for all but 23 of the 155 Civil Assistants employed at the time. At the end of December 1833, Captain Waters reported that "every branch of the fieldwork has been delayed throughout the whole of the present month by incessant Storm and uninterrupted Rain, and the Drawing also has been in consequence thrown back owing to the impossibility of examining the work plotted on the Ground". Whether because of the harsh working conditions caused by the inclement weather, or for other reasons, Leahy resigned his position in mid-January 1834, after only five months' service.[83]

Edmund Leahy seems to have been preparing and submitting drawings for engineering projects in his own name even before his brief engagement with the Ordnance Survey – indeed, a surviving drawing, dated 1833, relates to a project which, if it had been completed, would have guaranteed him a place in the annals of Irish engineering. Entitled *Elevation of an Iron Suspension Bridge, Designed for passing the Valley of the Honor at Grange between Clonmel and Fethard*,[84] this shows on a scale of 1:120, a proposal for a suspension

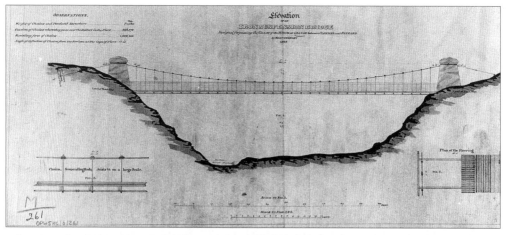

Elevation of an Iron Suspension Bridge, prepared by Edmund Leahy in 1833, to carry the Clonmel-Fethard road over the River Moyle (OPW5HC/6/261, courtesy of the National Archives of Ireland).

bridge with a span of 180 feet between the mid-points of 20-feet-high towers, intended to carry a public road over a river where it flowed through a narrow gorge. A study of the first edition of the six-inch map of Tipperary (published in 1843) suggests that the bridge was designed to carry what is now Regional Road R689 over the Moyle river, a tributary of the Anner, in the townland of Baptist Grange, about three miles south of Fethard where *Albert Bridge* appears on the map. As designed by Leahy, the bridge deck was to be suspended 35 feet above the river by chains and rods weighing 72,000 pounds, and the entire structure was to be anchored in the rock at each side by "cast iron ballast plates". In 1833, when this proposal was put forward, the design and construction of suspension bridges was still in its infancy. An American, James Finlay, had been the first to design and build such bridges early in the nineteenth century. In England, Captain (later, Sir Samuel) Brown led the way by completing the Union Bridge across the Tweed in 1820 and this was followed by Thomas Telford's spectacular bridge carrying the London-Holyhead mail coach road over the Menai Strait in 1826. Although a light suspension bridge, for pedestrians only, had been constructed in the grounds of Birr Castle in 1826, another 15 years were to elapse before Ireland's first road bridge of this kind, designed to carry the Glengariffe–Kenmare road over the Kenmare River, was completed.

It is most unlikely that Edmund Leahy had any real understanding in 1833 of the complex issues that arise in the design of suspension bridges or that he had conducted tests of various arrangements of chains, linkages and anchorages before completing his design, as both Brown and Telford had done. His proposed bridge was probably based on earlier plans by another engineer and, although a copy of the drawings found its way to the Office of Public Works, it is not possible to establish whether the scheme was ever put forward to that office or to the grand jury as a serious proposal, or was merely the product of a young man's imagination. Unfortunately, the local newspapers of the period are of no value in attempting to establish the facts in cases such as this because, in considering presentments, it was the practice of the grand juries until 1834 to retire to the grand jury room, from which the public and press reporters were excluded. Moreover, the records of South Tipperary County Council do not contain any information on the bridge which was built before 1843 at the location for which Leahy's suspension bridge was designed, or on the date of its replacement.[85]

The collection of OPW Architectural and Engineering drawings held at the National Archives of Ireland also includes Edmund Leahy's *Map and Sections of a new line of Road to pass Market Hill on the Road between Clonmel and Fethard, Designed, Surveyed, Leveled (sic) and Drawn by Edmund Leahy And now executing under the directions of P Leahy, Associate of the Society (sic) of Civil Engineers London, & Sons*.[86] The plans, which are dated 1834, relate to a new section of road, approximately 1,000 yards long, about one mile south of Fethard, running east of the existing road but lower down the hillside, and with a cutting 13 feet deep at the summit level so as to reduce the maximum gradient from 1

Chapter II — A family business develops, 1830-34

in 10 to 1 in 17. The completed new road (now part of Regional Road R689) was shown clearly on the six-inch Ordnance Survey map, sheet 70, for which the survey work was carried out in 1840-41.

Professional recognition

The Institution of Civil Engineers, founded in London in 1818 with Thomas Telford as its President, was the first organisation of its kind in the world but membership grew slowly in the early years, influenced by Telford's view that "talents and respectability are preferable to numbers" and that "too easy and promiscuous admission" would inevitably lead to "incurable inconveniences".[87] When Patrick Leahy applied for membership of the Institution in 1834, he was recommended, from their "personal knowledge," as a proper person by William Cubitt, Richard Townshend, a nephew of Thomas Townshend, and John Thomas who had worked under Townshend on the bog surveys. His application was approved at a meeting of the Council of the Institution, presided over by Telford, and he was elected as an Associate Member on 25 March 1834.[88] Although this category of membership was designed to cater for persons "whose pursuits constitute branches of engineering but who are not Civil Engineers by profession" rather than for persons "who are or have been engaged in the practice of a Civil Engineer",[89] Leahy's admission as an associate of this prestigious body was still a significant achievement for an engineer based

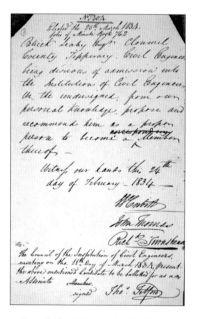

Record made by Thomas Telford in March 1834 of Patrick Leahy's admission as an associate of the Institution of Civil Engineers, London, and Leahy's signed undertaking to observe the Regulations of the Institution (Archives of the Institution, London).

in Ireland, and a major enhancement of his professional status. Three years after his admission, there were still only seven members or associates of the Institution with Irish addresses among the total of 294.[90]

The great majority of engineers working in Ireland in the early 1830s did not belong to any professional organisation and engineering standards and education in the country were seen to be in need of considerable improvement.[91] Colonel Sir John Fox Burgoyne, Chairman of the Commissioners of Public Works, complained in 1835 that the engineering field had been invaded by "persons without education or skill who had frequently been employed in operations of importance, resulting in bad or injudicious works, wasteful or fruitless expenditure and a certain degree of discredit to the country".[92] To give "respectability to the profession" and to demonstrate "the improved position assumed by the profession", a group of Irish engineers, most of whom were employed by public authorities, decided therefore in 1835 to form the Civil Engineer's Society of Ireland, with Burgoyne as its first President.[93] Edmund and Patrick Leahy were listed as foundation members of the new society but were unable to attend its first meeting in the Custom House, Dublin, on 6 August 1835.[94] Edmund was among the attendance at the second meeting in April 1836 when he presented a number of items to the Society's library.[95] These included a map of Waterford – probably a copy of that made by his father and himself in 1834 - and a copy of a treatise on steam engines by Rev Dionysius Lardner,[96] an indication, perhaps, that Leahy had already developed a special interest in the power of steam. There were two further meetings of the society in 1836, at one of which Denis Leahy was admitted to membership, and no further meetings until August 1844 when the society transformed itself into the Institution of Civil Engineers of Ireland. There is no record of any involvement by the Leahys in these subsequent meetings, or in any other activities of the Society.[97]

REFERENCES

1. Birmingham City Archives, MS 3192/Acc1941-031/713, letter of 31 October 1830 from Patrick Leahy
2. CDA 1847/6, note in Latin signed + P Leahy; the civil parish of Fennor, with the parishes of Buolick and Kilcooly, formed the Catholic parish of Gortnahoe (now Gortnahoe and Glengoole) from at least 1704 - see booklet prepared for the Solemn Dedication and Blessing of the Church of Saints Patrick and Oliver, Glengoole, 17 March 1976.
3. Gortnahoe Baptismal Register, entry for 3 June 1806
4. Personal communication from Very Rev John J Lambe PP, May 1987
5. CDA 1863/79, Ordo of Archbishop Patrick Leahy for 1863, entry for 28 November, records that he acted as sponsor for his sister Helena at her baptism in Portglenone for which the earliest surviving baptismal register dates from 1864; her death certificate gives her age as 49 when she died in November 1863.
6. Entry in Thurles Parish Baptismal Register for 5 November 1819
7. Entry in Thurles Parish Baptismal Register for 27 February 1823
7. For details, see Chapter XVII
9. Death Certificate issued by the General Register Office, RD Thanet, Vol 2a, page 450
10. *The Times*, 26 April 1883
11. Death Certificate issued by the General Register Office, RD Thanet, Vol 2a, page 685

Chapter II — A family business develops, 1830-34

12. *The Limerick Reporter* & *Tipperary Vindicator*, 29 January 1875
13. Christopher O'Dwyer, *The Life of Dr Leahy, 1806-1875*, MA thesis, St Patrick's College, Maynooth, 1970; CE 28 January 1875; Dom Mark Tierney, OSB, *Calendar of the papers of Dr Leahy, Archbishop of Cashel, 1857-1875*, NLI Special List 171
14. *Eighth Report of the Commissioners of Irish Education Inquiry*, HC 1826-1827 Vol XIII (509), Appendix 55
15. Jeremiah Newman, *Maynooth and Georgian Ireland*, Kenny's Bookshops and Art Galleries, Galway, 1979, pages 212-213
16. *Eighth Report of the Commissioners of Irish Education Inquiry*, HC 1826-1827 Vol XIII (509), Appendix 55; there were 19 catholic fee-paying schools in Thurles in the 1820s, as well as schools established by the Christian Brothers, the Ursuline Sisters and the Presentation Nuns - see An Bráthar Seán Ó Dúgáin, "The Christian Brothers in Thurles", in *Thurles: The Cathedral Town, Essays in honour of Archbishop Thomas Morris*, Geography Publications, Dublin, 1989, page 223.
17. Rev William P Burke, *History of Clonmel*, Clonmel, 1907, page 149; see also, Michael Quane, "The Free School of Clonmel", *JCHAS* Vol LXIX, No 29, January - June 1964, page 1
18. *The Limerick Reporter* & *Tipperary Vindicator*, 29 January 1875
19. In completing the 1861 Census of England form (RD Islington, RG09/135/15/29) Denis gave his age as 49.
20. CO 48/392, letter of 21 August 1858 from Edmund Leahy gives his age as 20 in 1834.
21. Mathew was 23 years of age in 1841 according to the evidence of his father in *Report of the Commissioners appointed to revise the Several Laws under or by virtue of which Moneys are now raised by Grand Jury Presentment in Ireland*, Appendix B, HC 1842 Vol XXIV (386).
22. Birmingham City Archives, MS 3192/Acc1941-031/713, letter of 31 October 1830 from Patrick Leahy
23. R C Cox, *The School of Engineering, Trinity College Dublin: A Record of Past and Present Students 1841-1966*, Fourth Edition, 1966, page 1; R B McDowell and D A Webb, *Trinity College Dublin 1592-1952: An Academic History*, Cambridge, 1982, page 180
24. Address of M B Mullins, President, *TICEI*, Vol VI, 1863, page 1
25. Robert Mallet (1810-1881) was one of the exceptions, studying at Trinity College, Dublin, between 1826 and 1830 (R C Cox, "Robert Mallet, Engineer and Scientist", in *Mallet Centenary Seminar Papers*, Institution of Engineers of Ireland, Dublin, 1982, pages 1 - 3).
26. M B Mullins, "Memoir of Barry Duncan Gibbons", *TICEI*, Vol VII, 1864, page 168
27. Birmingham City Archives, MS 3192/Acc1941-031/713, letter of 31 October 1830 from Patrick Leahy
28. *Second Report of the Tidal Harbours Commissioners*, Appendix B, evidence taken in Cork on 10 September 1845, HC 1846 Vol XVIII (692); CE 22 October 1845; IRG Vol. I, No 52, 27 October 1845
29. *Report from the Select Committee on the Survey and Valuation of Ireland*, Appendix A, Minutes of Evidence, HC 1824 Vol VIII (445)
30. Maria Edgeworth's *Tour in Connemara and the Martins of Ballinahinch* (edited by Harold Edgeworth Butler, London, 1950), a series of letters written in the 1830s, refers to several of Nimmo's young associates but not to any Leahys.
31. NAI, CSORP 1836/1579, Memorial of Denys C Leahy
32. NAI, CSORP 1823/6319, Memorial of 31 July 1823 from Patrick Leahy
33. W E Vaughan and A J Fitzpatrick, *Irish Historical Statistics: Population, 1821-1921*, Dublin, Royal Irish Academy, 1978, Table 10
34. Parliamentary Elections (Ireland) Act, 1829, 10 Geo IV c.8
35. *Third Report from the Select Committee on Fictitious Votes, Ireland*, HC 1837 Vol XI (480), Part II, Appendix No 6, Voters Registered in Boroughs since the passing of the Irish Reform Act
36. Elizabeth Shee and S J Watson, *Clonmel: An Architectural Guide*, An Taisce, Dublin, 1975, page 43
37. Kevin Whelan, "The Catholic Church in County Tipperary, 1700-1900" in *Tipperary: History and Society*, ed. William Nolan, Dublin 1985, pages 215 - 255; the College was the venue for the National Synod of Bishops in 1850 and was considered as a possible location for the Catholic University when it was being established in 1854.
38. J M Kennedy, *A Chronology of Thurles, 580 - 1978*, (revised and updated), Cavan, 1979
39. ibid
40. Rev Philip Fogarty, "St Patrick's College, Thurles", MS in St Patrick's College Thurles
41. Father Francis Ryan, "St Patrick's College, Thurles", in *Capuchin Annual*, 1960, page 319; Rev Christy O'Dwyer, "The Beleaguered Fortress: St Patrick's College, Thurles, 1837-1988", in *Thurles: The Cathedral Town, Essays in honour of Archbishop Thomas Morris*, Geography Publications, Dublin, 1989, page 237
42. Ernest F Carter, *An Historical Geography of the Railways of the British Isles*, Cassell, London, 1959, pages 37 - 41
43. Ted Ruddock, biographical note on Alexander Nimmo, in *A Biographical Dictionary of Civil Engineers in Great Britain and Ireland, 1500- 1830*, Thomas Telford Publishing, London, 2002, page 483

44. ibid
45. Birmingham City Archives, MS 3192/Acc1941-031/713, letter of 31 October 1830 from Patrick Leahy
46. Waterford City Archives, letter of 23 March 1831 from Pat Leahy, included in minutes of meeting of 19 April 1831, Waterford Corporation Minute Book
47. Memoir of Sir John Hawkshaw in *PRICE*, Vol 106
48. CO 295/215, testimonials submitted with letters of 12 and 15 February 1861 from Denis Leahy
49. *Cork Southern Reporter*, 30 July 1844
50. CE, 22 October 1845; IRG Vol I, No. 52, 27 October 1845
51. Birmingham City Archives, MS 3192/Acc1941-031/713, letter of 19 December 1832 from Patrick Leahy
52. ibid
53. Waterford City Archives, Waterford Corporation Minute Book, minutes of meeting of 23 November 1830
54. Waterford City Archives, Waterford Corporation Minute Book, letter of 23 March 1831 from Pat Leahy, included in minutes of meeting of 19 April 1831
55. Waterford City Archives, Waterford Corporation Minute Book, minutes of meetings of 19 April and 13 September 1831 and 23 April 1832
56. Waterford City Archives, M/PV/029
57. Waterford City Archives, M/PV/018
58. Waterford City Archives, MPV/019
59. Waterford City Archives, Waterford Corporation Minute Book, minutes of meeting of 5 September 1832
60. Waterford City Archives, M/PV/036
61. M/PV/016, listed in 1985 as being in the Waterford City Library but not, apparently, transferred to the City Archives
62. Waterford City Archives, M/PV/035; it appears that two separate engravings were made, on one of which Patrick Leahy used the designation "MICE" to which he was not entitled, while he used "AICE" on the other.
63. *The Limerick Reporter* & *Tipperary Vindicator*, 29 January 1875
64. Bill Irish, *Shipbuilding in Waterford: A historical, technical and pictorial study*, Wordwell Ltd, Bray, 2001, page 24; the *Nora Creina*, moored at Waterford, had already appeared in an 1830 Bartlett print.
65. Waterford City Archives, Waterford Corporation Minute Book, minutes of meeting of 5 November 1834
66. J H Andrews, *Plantation Acres*, Ulster Historical Society, 1985, page 345
67. J H Andrews, *A Paper Landscape* (2nd edition), Four Courts Press, Dublin, 2002, Appendix G
68. NAI, OPW 1/1/4/4, Public Works Letters No 4, page 22
69. Samson Carter and Noblett St Leger, Civil Engineers, *Waterford Harbour, Surveyed for the Commissioners for Improving the Port*, engraved by Keeble and printed by S H Hawkins, London, 1835
70. Waterford City Archives, Waterford Corporation Minute Book, letter of 23 March 1831 from Pat Leahy, included in minutes of meeting of 19 April 1831
71. South Tipperary County Museum, Clonmel, 1989.469, *A Map of Clonmel from Actual Admeasurement by P Leahy and Sons, 1832, for John Bagwell Esq*; the map at Clonmel is a copy redrawn in the County Council's Drawing Office from an original held at the Bagwell Estate Office but which cannot now be located.
72. Waterford City Archives, *A Map of Clonmel from Actual Admeasurement by P Leahy and Sons, Clonmel, 1832 for Thomas Taylor*, Bird Hill (photocopy)
73. *Clonmel Advertiser*, 23 October 1833
74. ibid
75. Appendix to the *First Report of the Commissioners of Municipal Corporations in Ireland, Part I, Report on the Borough of Clonmel by William Hannah and Maurice King*, HC 1835 Vol XXVIII
76. *Tipperary Free Press*, 14 September 1833
77. The six-inch maps of County Tipperary were not produced until October 1843, based on survey work carried out in 1840-41.
78. NAI, CSORP 1834/903, letters of 21 and 27 February 1834 from Patrick Leahy and response of 26 February from Sir William Gosset
79. Composition for Tithes Act, Ireland, 2&3 Will. IV, c. 119; the implications of this legislation for mapping and survey work are discussed in J H Andrews, *Plantation Acres*, Ulster Historical Foundation, 1985, pages 373 - 378.
80. J H Andrews, *A Paper Landscape: the Ordnance Survey in Nineteenth-Century Ireland*, (second edition) Four Courts Press, 2002, pages 61 - 65; Colonel Sir Charles Close, *The Early Years of the Ordnance Survey*, David & Charles Reprints, 1969, pages 105-106, 113

Chapter II — A family business develops, 1830-34

81. NAI, OS 2/14, Ordnance Survey Letter Register Vol II; as in the case of many of the documents described in the Register, the three items relating to Leahy's employment (OS 3/4812, 5161 and 5173) are no longer extant and the register is the only known record of them.
82. By coincidence, the action in Brian Friel's play *Translations* (Faber and Faber, London, 1981) takes place near the end of August 1833 in the townland of Ballybeg in County Donegal where a group of Ordnance Survey military staff had set up camp; on this, see J H Andrews, "Notes for a Future Edition of Brian Friel's Translations", in *The Irish Review*, No 13, Winter 1992-93.
83. NAI, OS 1/9, Monthly Returns of the Ordnance Survey of Ireland for the Months of January to December 1833
84. NAI, OPW5HC/6/261
85. Personal communication from Michael O'Malley, Senior Engineer, South Tipperary County Council, who states that the present bridge, which is still known locally as *Albert Bridge,* is a composite steel/concrete structure, which may have been built in the 1940s or 1950s.
86. NAI, OPW5HC/6/167
87. J G Watson, *The Institution of Civil Engineers: A Short History,* Thomas Telford Ltd., London, 1988, page 8
88. Election papers in the archives of the Institution of Civil Engineers, London
89. J G Watson, *The Institution of Civil Engineers: A Short History,* Thomas Telford Ltd., London, 1988, page 13
90. Membership list, Transactions of the Institution of Civil Engineers, Vol II, 1838; Leahy was still in membership in 1841 (Membership list, Vol III, 1839-1842); the next membership list available at the Institution is dated 1851, a year after Leahy's death.
91. *Reports from the Select Committee on Public Works, Ireland,* HC 1835 Vol XX (329) (573)
92. Presidential Address to the Society of Engineers, August 1835, Minute Book, ICEI Archives, Dublin
93. ibid
94. Minutes of the Proceedings of the Society, ICEI Archives, Dublin
95. ibid
96. Rev Dionysius Lardner, *The Steam Engine familiarly explained and illustrated,* Fifth Edition, London 1836
97. Edmund, Denis and Matthew never applied for membership of the Institution of Civil Engineers, London (personal correspondence with Mrs M K Murphy, Archivist, December 1988).

CHAPTER III

FIRST CORK COUNTY SURVEYORS, 1834

"the earliest instance of a competitive examination for public office (which) would always be creditable to Ireland"

Sir T A Larcom[1]

The evils of the grand jury system

By the beginning of the 19th century, the grand juries - the forerunners of today's county councils - were responsible under a series of Acts passed during the previous two centuries for a wide range of judicial and fiscal (or civil) business in their counties. The latter included the construction and maintenance of public roads and bridges, the provision and maintenance of courthouses, bridewells and gaols, the construction of minor marine works and, in some counties, the operation of lunatic asylums, infirmaries and fever hospitals. Expenditure on these functions and services amounted to about one million pounds per year in the early 1830s.[2]

The grand jury for each county was made up of 23 members of landed society, all selected for the purpose by the High Sheriff at his discretion, subject only to having one member from each barony. It assembled twice each year at the county courthouse for the Spring and Summer Assizes and transacted all of the fiscal business relating to the county in the few days before the Assizes Judges were due to arrive to begin the criminal business. Presentment sessions, composed of the magistrates for each barony were held in the weeks before the assizes to screen applications for presentments for new works or for road maintenance contracts. Where presentments were approved by vote of the grand jury, and fiated (in effect, approved) by the judge, the works could be carried out either by the person who obtained the presentment or his contractor. When the works were completed, an affidavit was sworn and submitted to the grand jury which then levied the amount of the presentment on the county cesspayers and recouped the person concerned.

In the early decades of the nineteenth century, the presentment system involved gross malpractice, inefficiency and waste. Public money was frequently misapplied, according to one commentator,[3] much of it was either pocketed or wasted, according to another,[4] and the system as a whole had been "productive of fraud, perjury and demoralisation", in the words of a third.[5] Corruption and jobbery were widespread, with grand jurors themselves seeking presentments for works of personal advantage and works which would benefit their friends and associates. The granting of presentments to named individuals involved obvious favouritism and led to a situation in which, according to Alexander Nimmo, road repairs were often carried out "by a class of persons who make a trade of it, as a market for the labour of their poor peasantry; the latter were not, properly speaking, paid for what work they do, but have the amount allowed by their landlords as a set-off against the rent of their holdings".[6] The grand jury itself was transient in nature, being dissolved at the end of each assizes, and it had no administrative or engineering organisation for vetting the need for particular projects and the plans and cost estimates, which were often inflated. Similarly, the arrangements for the supervision of work in progress, and for the inspection of the finished work, were unsatisfactory, with local landowners and others, who might well have an interest in the works or the contracts, acting as supervisors, overseers or conservators on behalf of the grand jury.[7] As a result, Daniel O'Connell could speak of "families who had made fortunes, and purchased estates, out of their jobs in road-making, though the roads they made were now not to be distinguished from the surrounding bogs".[8]

Pressure for reform
After the failure of the efforts made in 1817-18 to reform the grand jury system and to provide for the appointment of a county surveyor for each county, successive Governments were reluctant in the 1820s to take up the matter again although pressure for reform of the system continued. In 1825, a committee appointed by the County Cork grand jury reported that the appointment of a "properly qualified district engineer" for each of the two ridings into which the county had been divided in 1823 was essential if the execution of public works was to be improved.[9] This view was endorsed by a parliamentary committee in 1827 which noted that "a considerable school of engineering" was said to be growing up in Ireland as a result of the various programmes of public works.[10] Again, in 1830, the Select Committee on the State of the Poor in Ireland drew attention to the "admitted necessity for a more effective reform than has been as yet applied to correct the acknowledged imperfections and abuses of the present system of grand jury presentments"; the need to put an end to "the system of favouritism in the disposal of presentments by adopting an open system of contract"; and "the want of local engineers to direct and advise" on presentments.[11]

There were also some useful contributions to the debate in pamphlets published by public-spirited individuals and landowners. In 1831, for instance, Lord Carbery proposed the appointment by the Government of a county engineer for each county (or more than one for larger counties) who would operate under the control of the grand jury and would have "general superintendence" of all roads and bridges; in addition, this corps of "really active, efficient officers" would carry out periodic inspections, without notice, of all works being done by contractors under presentments, and certify the satisfactory completion of the work and its conformity with the presentment.[12] A much more limited role for a new category of professional officers was suggested in a pamphlet published two years earlier which described in great detail how the work of making and repairing roads might be improved; the organisational arrangements would not be altered dramatically but "properly qualified and instructed" men would be "constantly going through the county" so as to "point out to the Overseers of Roads the right system of making and repairing roads".[13] John Killaly, one of the most experienced canal and road engineers in the country, told the Chief Secretary in January 1830 that there was not, in his judgment, "any one measure, as regards the interests of the country, more deserving of the attention of the legislature" than the reform of the grand jury system and, in particular, the elimination of "the ills resulting from road jobbing".[14]

The failure of successive Governments to respond to the many recommendations for the reform of the grand jury system was due, in part, to a reluctance to incur the wrath of the landed gentry by reducing their effective domination of the grand jury and its affairs. The situation changed, however, with the formation by Earl Grey of a new Whig cabinet in November 1830, with Edward Stanley as Chief Secretary for Ireland. The Whigs, in contrast to the Tories, favoured reform and Stanley (the future 14th Earl of Derby, and three times Prime Minister) was himself responsible, during his three years in the Irish Office, for a considerable number of reforming measures. By 1833, he was ready to tackle the vexed question of grand jury reform believing, as he put it, that all classes were by then united as to the evils with which the system was subjecting the country.

The Grand Jury (Ireland) Act, 1833

Edward Stanley came before Parliament in February 1833 with a statement of his proposals for reforming the grand jury system and gained general support for them.[15] Nevertheless, when the Bill which was to become the Grand Jury (Ireland) Act, 1833 was brought before the House of Commons some

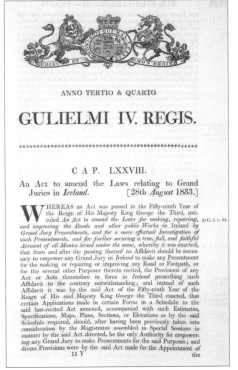

The Grand Jury (Ireland) Act. 1833, which created the office of county surveyor.

months later, most of the Irish members sought to have it postponed to the next session so as to give more time for consideration of the new arrangements it proposed.[16] Daniel O'Connell was among those who supported this approach, notwithstanding his dislike of the existing system under which "favourites of the grand jury were fattened".[17] But the Government were determined to push the Bill through and succeeded in doing so in August 1833, against the wishes - and the votes - of many of the Irish members.

The case for the new Act[18] was set out clearly in the preamble which stated that the existing provisions for the consideration of presentments were not adequate "to secure the needful investigation of the disbursement of the public moneys and the due and economical performance of the works carried out by the grand juries" who had not "sufficient time to deliberate and examine into the several presentments laid before them". To eliminate these defects, the Act provided that the presentment sessions in each barony would in future be composed of a number of the major cesspayers (not more than twelve and not less than five) in addition to the local magistrates, and that these sessions should receive and consider all applications for presentments for public works in the area, with county at large presentment sessions dealing only with the relatively small number of matters of concern to the county as a whole. All presentments would still have to be approved by the grand jury itself before any work could go ahead, but the jury was to be deprived of its earlier power to initiate projects and to alter presentments. In addition, the jury would have to sit in public when dealing with presentments, instead of retiring to the grand jury room, from which the public was excluded, as was the practice until then.

To enable this new and somewhat more representative county government system to operate effectively, the 1833 Act reintroduced provision for the appointment of a county surveyor for each county, except Dublin. The appointment procedure was similar to that of the 1817 Act; in particular, a panel of persons whose qualifications had been examined into and certified by a board of three civil or military engineers was to be set up, with specific appointments from the panel being made by the Lord Lieutenant. This novel feature of the Bill seems to have attracted very little debate during its passage through Parliament.

Selecting the first surveyors
On 3 April 1834, the task of conducting the examination of candidates for appointment as county surveyor was assigned to the Chairman of the Commissioners of Public Works, Colonel John Fox Burgoyne, his colleague, Commissioner John Radcliffe, like Burgoyne a former army engineer, and the Commissioners' Engineer and Architect, Jacob Owen.[19] This arrangement subsequently attracted criticism for a variety of reasons in professional and other circles: according to a printed statement circulated in the mid-1850s,[20] one of the examiners - Owen - was "merely an architect" and a significant number of the successful candidates were said to have been assistants, apprentices, or private pupils of his in the Office of Public Works; besides, Radcliffe was said to be "unknown to the public or

to the profession as a civil or military engineer".[21] But Thomas Larcom, who was himself a Commissioner of Public Works from 1846 to 1853 and Under-Secretary at Dublin Castle from 1853 to 1868, defended the arrangements made in 1834, arguing that engineering talent in the 1830s - especially in Ireland - was not abundant and "it would not have been possible at the time to have found three other men combining the necessary qualifications who could have been called on to do the work unpaid" as the 1833 Act provided.[22]

When applications for the new posts were invited in 1834, Patrick Leahy, then aged 54, and his son Edmund, who was only 20, were quick to enter the lists but Denis Leahy appears not to have done so, even though his experience of engineering at that point would have been superior to that of his younger brother.[23] Burgoyne's recollection, more than 20 years later, was that about 50 candidates came forward for examination but one of the candidates thought that the number was much higher - 150 or upwards.[24] An even higher figure of 239 was quoted, but without identifying a source, in the *Tipperary Free Press* in 1834.[25] In any case, the process got underway on 1 April 1834 when a circular was issued by the Chief Secretary's Office to "the numerous candidates" notifying them of the dates on which they should attend for interview, "bringing specimens of your work and testimonials as to character and ability from professional and other gentlemen".[26] Candidates were advised that they would have to undergo a very strict examination: they would have to show "a correct knowledge and practical experience in land and road surveying and mapping; bridge-building, masonry and other works; the drawing up of specifications and estimates; as well as the details of the measurement and execution of the several descriptions of work connected with road engineering". In addition, "some knowledge of builders' business" was considered necessary.

Interviews at the Custom House

The two Leahys and the other candidates duly attended for examination and interview at the offices of the Commissioners of Public Works at the Custom House, Dublin, on 23 April 1834 and, according to Patrick Leahy, underwent "during a period of 14 successive days … a most rigorous course of examination in the various departments of Theoretical and Practical Science requisite to constitute a County Engineer".[27] Thomas Jackson Woodhouse, the first Antrim surveyor, provided a somewhat similar account of what transpired[28] as did several of the other surveyors when they gave evidence many years later to a House of Commons Select Committee: Charles Lanyon, who became Kildare county surveyor in 1834, confirmed that the examination went on every day for a fortnight; Henry Stokes, the first Kerry surveyor, asserted that the examination was "an elaborate one"; and Henry Brett, who was appointed in December 1834 to King's County (Offaly), recalled examinations which had gone on for some 12 days, with written questions and answers on education and previous experience, practical engineering matters, surveying, architecture, bridge-building, road maintenance and improvement, algebra, geometry, trigonometry, mechanics and hydraulics.[29] All of this, according to one local newspaper, "should teach parents the propriety of giving their children a good mathematical education" if they were to qualify for permanent well-paid appointments like that of county surveyor.[30]

The Leahys were given to understand, according to Patrick, that the result of their examination was creditable but to advance his cause he decided to submit testimonials and recommendations direct to Sir William Gosset, the Under-Secretary for Civil Affairs at Dublin Castle, for inspection by him and by the Lord Lieutenant, together with another document "signed by the Lieutenants and Magistrates of the County Tipperary most numerously"; this latter, he suggested, might not be "the best test of talent" but "in addition to a good examination, such a document, so respectably signed, must ensure further satisfaction in the selection to be made".[31] While the original documents submitted by Leahy are no longer available, it seems clear from his description of them that they were more in the nature of representations on his behalf than testimonials as to his ability and experience. They consisted of:

- a copy of the recommendation made in his favour by the "Nobility and Gentry" of County Tipperary in 1817 when he was previously a candidate for a post as county surveyor;
- a letter from the Earl of Clonmel attesting to his fitness and eligibility to fill the post and a recommendation "in a general way" from the late second Earl of Llandaff whose estates had included the town of Thurles;
- a letter from John Thomas, described by Leahy as "the engineer who built the fort at Sheerness"; Thomas had, in fact, been resident engineer at Sheerness dockyard in the 1820s and had worked with Leahy when they were both members of Thomas Townshend's team on the bog surveys of 1810-1813;
- a letter "declaratory of my being admitted a Member of the Institute [sic] of Civil Engineers in London";
- an "highly worded character" from the Mayor, Sheriffs and Corporation of Waterford.

Like his father, Edmund felt it necessary on completion of his examination to put his case for an appointment directly before the Lord Lieutenant and he submitted a memorial on 10 May 1834 for this purpose.[32] He was not concerned by the fact that he was unable to offer any testimonials or recommendations of his own but claimed to be entitled to share the testimonial submitted by his father and signed by about 40 of the Magistrates and

Deputy Lieutenants of County Tipperary; this claim was grounded on the rather flimsy argument that the works on which the testimonial was based were executed by Messrs P Leahy and Sons. Similarly, Edmund referred the Lord Lieutenant "with pride" to the fact that his own name was mentioned in his father's testimonial from Waterford Corporation.

In advancing his case for appointment, Edmund seems to have been willing to accept any of the posts on offer but Patrick made a special bid for the position of county surveyor in Tipperary; he suggested to the Lord Lieutenant that, "having projected several public improvements" in the county, including canal and railway schemes, his "professional assistance would be infinitely more available in that county than any other candidate".[33] This request, however, took no account of the fact that candidates had been warned in advance that it was not proposed "that gentlemen who may be found eligible should be appointed to counties in which they have been accustomed to reside, or in which they have been chiefly employed".[34]

Fully reliable Information on the outcome of the 1834 competition cannot now be obtained from primary sources as the original documents were either destroyed or mislaid due to the lax filing system in the Chief Secretary's Office, forcing Larcom to report to Parliament in July 1856 that "there is no record of the List returned by the Examiners".[35] However, a memorandum prepared by Larcom stated that after the interview board had examined all-comers, they returned to the Lord Lieutenant, Marquis Wellesley, a list of those they had found qualified, "in the order of their merit, from which the Lord Lieutenant was to select" but, he went on, "it was not the practice of the Lord Lieutenant to *select* but to appoint the first on the list, and so on as vacancies occurred".[36] In the absence of any other system of ranking counties, this might be taken to mean that the person appointed to the Antrim post - Thomas Jackson Woodhouse - was first in the order of merit, and so on down through the alphabetical list of counties. But, as against this, Charles Lanyon who was appointed to Kildare (fourteenth on the alphabetical list) was reported to have achieved second place in the examination. Thus, the full order of merit can only be a matter of conjecture at this stage.

The appointments made in 1834 were the first public appointments in the United Kingdom which were required to be made on the basis of a competitive examination and an objective assessment of merit: the principle of competitive examinations was not adopted for civil service appointments until 1855 and it was not extended to professional posts in the service until the 1870s. Some questions about the appointments procedure appear, nevertheless, to have been raised at an early stage by Feargus O' Connor, the pro-repeal MP for Cork city from 1832 to 1835, but Burgoyne, when contacted by the Chief Secretary about these, assured him that he was "very confident that no part of our conduct throughout that troublesome affair can be reasonably censured by any unprejudiced man".[37] More than 20 years later, in evidence to a House of Commons Select Committee, Burgoyne insisted that while he was Chairman of the Commissioners of Public Works, the

Chapter III — First Cork County Surveyors, 1834

Lord Lieutenant "universally" acted on the order of merit and "no influence was ever used on the part of the Irish Government, nor any influence of any kind on the part of anyone else, to interfere with the examinations or the recommendations".[38]

The first appointments
Having received the examination results from the interview board, the Lord Lieutenant had little time for deliberation as the secretaries of many of the grand juries were pressing him to expedite the appointments, pointing out that any delay would mean that the first series of special presentment sessions under the 1833 Act would have to be postponed. The authorities responded quickly and the appointments were duly made with effect from 17 May 1834; Patrick Leahy was appointed to be the first county surveyor for the East Riding of County Cork and Edmund was appointed at the same time to be surveyor for the West Riding.[39] The appointments were announced in the *Cork Constitution* on 20 May[40] while the *Tipperary Free Press* congratulated the two "talented and respectable engineers" on their success.[41] Patrick was given to understand that he had also been appointed to the separate post of surveyor for the county of the city of Cork which had its own grand jury; the area involved was almost 45,000 acres in extent and it included the city itself and the liberties of the city which encompassed an extensive rural hinterland.[42] Initially, there was some confusion about this second appointment and both Leahy and the secretary of the city grand jury found it necessary to seek directions on the matter from Dublin Castle.[43] However, the matter was soon sorted out and Leahy's tenure of the second post was confirmed. This caused considerable annoyance to many of the city grand jurors who took the view that one man could not properly discharge the duties of the city and county posts and that a separate surveyor should have been appointed for the city.[44]

In all, 32 appointments were made in 1834, one for each county other than Dublin to which the 1833 Act did not apply, and one for each of the ridings in County Cork. These first surveyors were a disparate group. Some of them were young men like Edmund Leahy - probably the youngest to be appointed - while others were much older and more experienced men, including a number who had graduated to civil engineering from the business of land surveying which had gone into decline with the commencement of the work of the Ordnance Survey in 1824. There were a number of Englishmen and a few with railway engineering experience. Some of the new surveyors had been employed by the Board of Works on local projects in the years immediately preceding their appointment[45] and a significant number of others had been trained by Jacob Owen, the Board's Engineer and Architect, who was one of the examiners; these included Charles Lanyon who had married one of Owen's daughters. Others had learned their engineering as part of the large team of engineers, surveyors and valuers maintained by Alexander Nimmo until his death in January 1832. None of them, of course, had any formal third-level engineering qualifications as these did not become available in any of the universities in Britain or

Ireland until ten years later. When asked by a Parliamentary Committee in April 1835 if all of the first group of surveyors were sufficiently qualified, Burgoyne replied that "we found many of them in some degree deficient in particular points"; he went on to say that the interview board had not rejected all who were inefficient but had "recommended the best; we did not think all of them perfectly qualified to the extent we considered desirable".[46] He had no doubt, however, that the examination process would have made the new surveyors "sensible of their deficiencies" and that they themselves would endeavour to correct these.

Taking up duty in Cork

The Leahys lost no time in taking up their new duties in Cork. Patrick was at his post on 22 May, less than a week after his appointment, waited on the secretaries of the city and county grand juries the next day, and immediately made arrangements to attend the series of baronial presentment sessions which were to be held in the following two weeks.[47] He then had to set about inspecting the locations where works were to be carried out and making the necessary measurements, maps and plans, so that the works could be let out on contract – and all of this, as he put it, within the short space of 12 days.[48] After this flying start, he was confident enough to assure the Chief Secretary's Office on 7 June that "the new Grand Jury Act will not be more strictly complied with, nor more practically introduced into operation in any part of Ireland, than in this County".[49] Edmund made an equally impressive start in the West Riding, attending all of the presentment sessions and completing the essential plans and other documentation in good time. The two new surveyors appeared to have settled in well to their posts by the time the assizes for the county opened at the end of July, and they were able to present detailed reports to the grand jury on the condition of the roads, bridges and public buildings and on the need for improvements of various kinds.[50]

Part of Patrick Leahy's first report to the grand jury of County Cork, as published on the front page of The Constitution on 29 July 1834.

Chapter III — First Cork County Surveyors, 1834

Each surveyor was legally responsible to the grand jury for a wide range of public works and services. In practice, his duties related predominantly to the construction, maintenance and improvement of public roads and bridges but his responsibilities also extended to the construction and maintenance of various public buildings, including courthouses and gaols. The title *county surveyor* was, therefore a misnomer, and was regarded as an offensive one by many of the surveyors who, like Patrick Leahy, often used the title *county engineer* in his reports and on plans and specifications.[51] In Cork, where the division into two ridings operated "for the purpose of holding general sessions of the peace and no other",[52] the two county surveyors reported to a single grand jury which dealt with the fiscal and other business relating to the entire county. There were 22 baronies in the county in 1834 - 12 in the East Riding and 10 in the West - and this, coupled with the size of the county and the distances to be travelled, made the duties of the Cork surveyors quite demanding.[53] Attendance at each of the baronial presentment sessions (often referred to as the road sessions), the adjourned sessions, and the Spring and Summer Assizes was itself a time-consuming business: the assizes generally lasted four or five days in mid-March and in late July; the presentment sessions for the baronies took place between mid-January and mid-February, and again between late May and the end of June; and the adjourned sessions were spread over almost another month in each case.[54] On all of these occasions, the surveyor was required to give professional advice and assistance in relation to the numerous applications for presentments, and to make applications for presentments himself, where necessary. The grand jury had no direct labour organisation but relied entirely on the contract system both for carrying out routine maintenance operations and for new works. This meant that the county surveyor had to prepare specifications, plans and tender forms for all approved works before inviting competitive tenders and submitting them for approval to the adjourned presentment sessions and/or to the grand jury. If the works were to cost less than £50, a period of only 30 days was allowed for all of this activity to be completed; in other cases, the surveyor could wait until the next sessions or assizes before submitting the plans and specifications. Once tenders were approved, it was the surveyor's duty to supervise the performance of the works by the contractors, and to certify the payments due to them.

Patrick Leahy's report to the meeting of the grand jury in March 1836[55] gives a good indication of the nature of the demands made on him. On that occasion, a total of 378 applications for works of various kinds which had been approved by the road sessions came before the jury and 14 of these related to what the surveyor regarded as major works. In these cases, he reported that "maps, plans, sections and specifications for these respective works have been prepared at great expense from actual admeasurement and levels". In the case of the remaining 364 applications which related to the repair of roads and other minor works, he had had to prepare specifications "most of which have been elaborately written out by my own hand" and deposit them at the offices of the Secretary in sufficient time to allow for their inspection by those interested in tendering for the works. As only 9 days elapsed between the last of the road sessions for his 12 baronies and the first of the adjourned sessions, Leahy reported that he was "driven to extraordinary exertion" to complete all of this work. And, on top of this, he claimed that it had been necessary for him to travel a total of 3,520 miles in the weeks before the assizes so that he could inspect and report on the state of the roads for which maintenance contracts were in force.

In addition to his biannual reports to the grand jury on the general condition of the roads, bridges and county buildings and on the progress of the county works, each surveyor was required to send regular reports and returns, each in a prescribed form, to the authorities in Dublin Castle who, although they had limited powers to intervene directly, still exercised detailed surveillance over the affairs of the grand juries. The meticulous nature of the reporting requirements is evident from a letter of 9 May 1840 from the Under-Secretary's Office to Edmund Leahy demanding an explanation as to why his report on a contractor's complaint which had been referred to him was not written "on half margin" as required by a circular from the office.[56] Complaints of this kind against Leahy and other surveyors were regularly received in the Chief Secretary's Office but the explanations offered seem to have been accepted without question in most cases.[57]

Another form of supervision was indirectly exercised by the Commissioners of Public Works arising from their functions in relation to the allocation of grants and loans to grand juries for major road and bridge projects and for the construction of courthouses and other public buildings. Where a project of this kind was approved at the assizes, an application to the Commissioners for a loan or grant had to be accompanied by the county surveyor's detailed plans which were generally reviewed by one of the Commissioners' own engineers or, occasionally, by another county surveyor engaged on a fee basis for the purpose. Reviews of this kind often led to suggestions for modifications of the plans which, on occasion, were not welcomed by the surveyors. Patrick Leahy, for example, in 1840, justified his rejection of proposed modifications of his plans for a road scheme in Cork city by telling the Commissioners that, since he travelled that particular road several times a day, his understanding of what was required was superior to that of the Limerick county surveyor who had visited Cork to inspect the site and the plans on behalf of the Commissioners.[58] Similarly, in 1836, Edmund strongly defended his own proposals for a new road through the Beara peninsula, arguing that the alternative line proposed by the Kerry county surveyor would affect a greater number of landowners and was based on inaccurate levels.[59]

In practice, the surveyor had a great deal of independence in going about his duties where grants or loans from central funds were not involved. As the grand jury met only twice each year and had no legal existence between assizes, it was always under pressure to complete its administrative business in the few days before the judge arrived to deal with criminal matters. Its secretary, although usually a very influential and well-connected official, had no supervisory role in relation to the other officers and there was no effective system of independent financial audit of the affairs of the grand juries until 1878. Thus, the arrangements for providing and maintaining roads and other public works depended almost entirely for their effectiveness on the ability, energy and integrity of the individual surveyor.

Salaries

While the abortive 1817 Act had allowed for a salary of up to £600 for the surveyors of the larger counties, the 1833 Act fixed a maximum of only £300 a year, inclusive of all costs and travelling expenses, and without regard to the size of the county.[60] From the beginning, the county grand jury allowed the full £300 to Patrick and Edmund Leahy for their duties in the county but for the city post, the jury decided to allow Patrick only £100 in 1834-35, £150 in 1835-36 and £200-£250 in the following three years,[61] because of their view that one man could not adequately carry out the duties of the two posts. A moiety of the salary was included as a separate item in the estimates presented at the Spring and Summer Assizes each year, and this provided the grand jurors with an opportunity for lengthy discussions of the surveyor's performance and his value to the city. Leahy complained bitterly about this to Thomas Drummond, the Under-Secretary at Dublin Castle,[62] and the issue was referred to the Government's law adviser in 1839 but with no immediate result. However, the salary for the city post was eventually increased to the maximum of £300 a year, giving Leahy a total of £600 for his combined city and county duties.[63] This compared favourably with the salary of the secretary of the county grand jury (£324) and that of the treasurer (£554)[64] and was far greater than the aggregate salary paid to any of the other surveyors who had responsibility of counties of cities in addition to their normal duties.

County surveyors generally considered that their salaries were far too low, bearing in mind the wide range of duties which they had to perform and the lack of any provision for travelling expenses. In his very first report to the county grand jury, Patrick Leahy argued that the considerable expenses he would incur in engaging attendants for surveys and levelling, and in respect of travel and hotel costs, "must inevitably outweigh the salary allowed by law".[65] The same view emerged in the evidence given to the Select Committee on County Cess in 1836: several witnesses supported the case for an increased salary, with the Antrim surveyor, Thomas Jackson Woodhouse, pointing out that travelling expenses alone, including the keep of his two horses, cost him £200 a year or two-thirds of his salary.[66] But the Committee was primarily concerned with the increased amounts which grand juries were obliged to levy by way of cess to meet the additional responsibilities imposed on them by law in the previous 30 years and could not see its way to recommend any increase in salary. The campaign for higher salaries went on for many years, with the surveyors maintaining pressure on the Lord Lieutenant, through letters, memorials and deputations, for salary maxima which would vary with the size of the county, and for allowances to cover travelling and incidental expenses like postage and stationery, but legislation to allow for higher maximum salaries was not enacted until 1861, long after the Leahys had left their county surveyor posts.[67]

Each surveyor was allowed £50 a year to keep an office for the transaction of his business and to employ a clerk to be in regular attendance there.[68] Patrick Leahy complained to the city and county grand juries in March 1836 that he and Edmund had been "miserably situated" for the previous two years and successfully argued that offices should be made available at the new courthouse in Washington Street, Cork, which had just been completed by the Pain brothers.[69] Thereafter, Leahy estimated that two-thirds of his time was spent at the new offices, with the balance devoted to attendance on the grand jury and at the baronial presentment sessions, and the inspection of public works.[70] Edmund also set up his headquarters at the new courthouse, rather than in one of the towns in the West Riding, a decision which was to lead to criticism by grand jurors in subsequent years. He claimed, however, that his duties outside the office absorbed two-thirds of his time.[71]

Assistant surveyors

According to the 1836 report from the Select Committee on County Cess (Ireland),[72] every witness before the committee agreed that it was impracticable for a county surveyor, without assistance, fully to execute his duty, even in the smallest county. The committee therefore recommended - somewhat reluctantly because of their concern about the level of the demands on the county cess - that the surveyors should be allowed, on a trial basis, to appoint assistants who would be paid out of public funds. This was accepted by the Government and provision for the new post of Assistant Surveyor, with a salary of up to £50 a year, was made by law in August 1836.[73] The number of assistants in any county was to be decided by the grand jury, having regard to the size of the county and the extent of the duties to be performed, but the appointments themselves were to be made by the county surveyor from among persons certified to be fit and competent by the same board (effectively, the Commissioners of Public Works) which examined candidates for county surveyorships.

Having considered the appropriate qualifications for the new posts, the Commissioners felt that "it would be very desirable if these situations could be filled by men of a certain degree of education and well conversant with the practical superintendence and mechanical trades adapted to public works ... but such a class of men, thoroughly qualified, are rare in Ireland, and it would be impolitic and unjust to introduce them in any degree from England and Scotland, even if they could be had for the remuneration allotted, which however is not to be expected."[74] In these circumstances, they were prepared to rely largely on the recommendation of the individual county surveyor while prescribing some basic requirements. In general, candidates were to be between 24 and 45

Chapter III — First Cork County Surveyors, 1834

years of age, tolerably robust and of good character, especially as to sobriety; they should at least be able to read and write, have some knowledge of arithmetic and accounts, and some practical knowledge or experience that would be relevant to the duties to be performed. Some years later, in the light of experience, the Commissioners drew attention to the need to appoint practical men instead of inexperienced young men with no knowledge of the business,[75] and the need to avoid as far as possible appointing persons who lived in the barony in which they were to serve; they felt that such persons might be more concerned with increasing their income than with zealous service to the grand jury and that they might have partialities in favour of friends and neighbours.[76] A further direction in 1843 specified that persons connected with families or individuals of influence in the area should not be appointed.[77]

Candidates recommended by county surveyors for appointment as assistants were examined by the Commissioners of Public Works each Friday morning at their offices in the Custom House, Dublin, and certificates in respect of the successful ones were issued to the surveyors who had sent them forward. As surveyor for the East Riding, Patrick Leahy was among the earliest of the surveyors to receive certificates for assistants. His first choice, Denis B Murphy, was certified in January 1837, but for some reason the certificate was cancelled a few months later.[78] Sandford Brunett was certified in February 1837 although the Commissioners were not impressed by him and based their certificate solely on Leahy's opinion.[79] Daniel Donovan was found to be wanting in practical knowledge but, because he was thought to be promising, was certified in the following April.[80] Henry Nelson Bride, who was certified at the same time, seems to have been the most successful of the assistants and held office until 1855.[81] In 1838, the grand jury agreed to allow Leahy an additional assistant[82] but it was not until March 1839 that he received the necessary certificate for James Regan, who was found by the examiners to be intelligent but without much practical knowledge of road works.[83] This left Leahy with four assistants,[84] only one of whom had not attracted some adverse comment from the examiners. Subsequently, he seems to have had difficulty in retaining staff as it was necessary for him to have two new assistants certified in 1841,[85] two more in 1843[86] and a number of others in subsequent years. By mid 1845, there were six assistants in the East Riding: Henry N Bride, Henry Brown, Daniel Donovan, Osborne Edwards, Patrick Dea and James Mitchell.[87]

Edmund Leahy, in the West Riding, fared little better than his father in his choice of assistants. The first candidate he put forward, Jeremiah Sullivan, was initially rejected by the Commissioners who found that "he is but very indifferently acquainted with road-making though he says he has been engaged as a contractor"; besides, Sullivan was "a comfortable farmer, having twenty acres of land" and it would be wrong to appoint him in the district where he lived.[88] The Commissioners relented in May 1837, under pressure from Leahy, and agreed to certify Sullivan, but they warned Leahy that he was taking too much responsibility on himself: "the objection is not to his want of qualification, because it is to be presumed that he might acquire it, but having a farm to attend to, there will naturally be a strong persuasion that he cannot give up the time he ought to the duty of an assistant to the surveyor".⁸⁹ A second assistant, Henry Philip Browne, who was certified for the West Riding in 1837, was still in office ten years later.⁹⁰ In 1839, a third assistant, Charles Branson, who appeared to the examiners to have little knowledge of road works, joined Leahy's team⁹¹ but had left office by 1845.⁹²

Family involvement in county surveyor work
When Patrick and Edmund Leahy took up their positions as county surveyors in Cork in 1834, Patrick's wife, Margaret, their four daughters and their youngest son, Matthew, moved with them from Clonmel. Patrick junior, after his ordination at Maynooth in 1833, was then serving as catholic curate in Knocklong, County Limerick and the fourth brother, Denis, who was in practice as a land surveyor at the time,⁹³ was living at Cabra, near Thurles.⁹⁴ Although Denis had also moved to Cork by 1836 when he was admitted to membership of the Civil Engineers' Society,⁹⁵ he appears to have maintained his connections with his native Tipperary where much of his subsequent work was carried out. Initially, the Leahy family lived "in the demesne of Glenville", a large house set in 56 acres of woodland and farmland near the village of Glanmire.⁹⁶ In 1844-46, they were residents of Bruin Lodge, 109 Lower Glanmire Road, in Cork city.⁹⁷

During their years in Cork, Patrick, Edmund, Denis and Matthew Leahy worked closely together in discharging the county surveyor duties which, strictly speaking, should have been carried out personally by the two officeholders. Denis seems to have worked both with his father and with Edmund, and was regularly the author of maps and drawings of road schemes submitted to the grand jury or to the Board of Works.⁹⁸ Matthew worked occasionally with Edmund on the grand jury business of the West Riding but was engaged primarily in helping his father in the East Riding and in carrying out his functions as city surveyor, especially the latter.⁹⁹ However, neither Denis nor Matthew went forward for certification as assistant surveyors and were not therefore entitled to any salary from the grand jury for the work they carried out. It was obviously more profitable for the family as a whole to arrange matters in this way, bearing in mind the absolute prohibition on the receipt of fees from grand jury work which would have applied to the brothers as assistant surveyors. And, while Patrick's failure to carry out all of his duties in person was regularly criticised by some members of the city grand jury, others saw advantages in allowing the situation to continue; for example, Sir Thomas Deane, the prominent Cork architect, remarked in 1839 that if the Government assigned a separate surveyor to act for the city grand jury, they "would lose the valuable services of Mr Leahy's sons".¹⁰⁰

Denis Leahy was claiming as early as 1836 that he had a great deal of practical experience in working the Grand Jury Acts, having "often inspected for my father over 1,700 miles of roads and laid out over 100 miles of new roads, besides bridges, tunnels, etc. and drawn plans and specifications for all the works which are now executed".[101] He had not applied for appointment as a county surveyor when the positions were first created, but with the experience of grand jury business he had gained in Cork, he sought an appointment in Dublin city or county in May 1836 when it was believed that the provisions of the 1833 Act in regard to county surveyors were about to be applied to these areas.[102] In a memorial to the Lord Lieutenant, he relied, as Edmund had done in 1834, on his father's testimonials from the County Tipperary magistrates "as all professional business was transacted under the title of P Leahy and Sons". The Board of Works advised that Leahy was probably qualified for a position as county surveyor but had not been formally examined, as others had. In addition, they suggested that "as the County of Dublin will be considered a choice station, somewhat difficult to fill, and requiring peculiar qualifications, the Government will be inclined to select for it one of the present surveyors who may be known to be well adapted to it, and that any newly appointed (surveyor) should be placed to inferior counties".[103] In the event, provision for a district surveyor system in County Dublin was not made until August 1844[104] and Denis Leahy was not among those appointed to the three new posts which were then created. He sat for county surveyor examinations held in March and again in December 1838 but on each occasion was held not to be competent to fill the office.[105] It is not possible to establish whether he sat the subsequent examinations in 1846-47 but, if he did so, he was not among those certified by the examiners.[106]

At the Spring Assizes in March 1838, one of the city grand jurors noted that when he called to the surveyor's office, Patrick was not present nine times out of ten, but only his son - and "a more active and intelligent young man I never knew".[107] When another juror observed that, if they were to get a separate city surveyor, they could not get a more competent person than young Leahy, Patrick told the jury that he would be quite happy to retain the city post, and have the county duties assigned to his son, or *vice versa*. Three years later, in 1841, Patrick was able to assure Commissioners who were examining the grand jury laws that his city and county duties were "perfectly compatible" and that he was able to discharge them personally because he had "the assistance of one of my sons with me, aged about 23 years, and who is fully qualified as an engineer" but received neither fees, nor gratuity, nor salary nor emolument from the city or county authorities.[108]

Matthew had been too young and inexperienced to apply for a county surveyor post in 1834 but was a candidate in 1838 when a new series of examinations was announced.[109] In March of that year, when the appointment of a separate surveyor for Cork city was being discussed by the grand jury, Patrick Leahy told them that his son "had been recommended by the Board to the Lord Lieutenant"[110] but this was quite untrue: Patrick had asked only a week earlier that his son's name be added to the list of candidates at the examinations which were to take place the following month[111] and when the first results emerged in mid-May, Matthew was not among those certified to be qualified.[112] With three others, however, his name was included on a second list, published in June, of candidates who might, after some time and with more experience, meet all of the requirements[113] and, with these others, he was invited in December 1838 to attend a further interview "to ascertain if in the interim you have acquired more knowledge of the different branches of engineering, but especially in that of bridge building".[114] But when the results of these examinations were issued in the following March, Matthew had regressed, finding himself on a list of thirteen candidates "whose acquirements", according to the interview board, were "very inferior".[115]

In August 1840, Patrick Leahy again put Mathew forward as a candidate at the next series of examinations[116] but, when Matthew himself expressed interest in an appointment in 1842, he was told that his name had not been included on the list of nine qualified candidates which had been produced by the examiners in December of the previous year.[117] In January 1845, Matthew made a third application for a county surveyor post, "being an engineer for the last 12 years" (ie since the age of 15!) and having been for seven years "employed by my father and brother in carrying on the public works and business of the County of Cork". And, if more than that was needed to demonstrate his eligibility for the situation, he offered to forward references from his "friends and acquaintances", listing Lords Bandon, Mountcashel, Bantry, Shannon, Cork, Bernard, Kingston and Berehaven among these, as well as "most of the influential Magistrates and Gentlemen of the South of Ireland".[118] Leahy was told that, despite his previous failure to qualify, the Commissioners of Public Works would be ready to allow him to enter again at any future examination,[119] and in April 1845, when Edmund wished to appoint him to act as deputy county surveyor for a few months, the Commissioners ruled that he was "sufficiently qualified to fill that office *for a short period*".[120] While it is not possible to determine whether Matthew subsequently re-sat the county surveyor examinations, his name does not appear on the lists of certified candidates forwarded to the Lord Lieutenant following any of these examinations.[121]

"an able and efficient class of county officers"

In 1857, Lieuenant-General Sir John Fox Burgoyne, who was regarded as the head of the engineering profession in Ireland in the 1830s, expressed the personal view that a much better class of gentlemen had been appointed as county surveyors in 1834 than could reasonably have been expected for the remuneration.[122] The Commissioners of Public Works, of which Burgoyne was Chairman from 1831 to 1845, took much the same view in their annual report for 1834, noting that while the duties of the new surveyors were "very laborious and their remuneration small", they had "generally commenced their services with zeal and firmness, accompanied by a degree of skill in their business that will tend gradually to the improvement of the communications and be productive of a considerable saving of expense to the counties".[123] But the Select Committee on County Cess took a somewhat different view in their 1836 report: they noted with concern that the total amount levied in county cess in 1834 had just exceeded one million pounds, they were "deeply convinced that the county rate never can be as economically administered as by those who pay it" and they reported with regret that "the magistrates and cesspayers do not give the same attention to the roads as previous to the passing of the last Act".[124]

Whatever initial misgivings there may have been either in Parliament or among grand jurors about the value of appointing county surveyors, and notwithstanding the doubtful competence of some of the early surveyors and their assistants, the system was to endure, with only minor changes, even after the grand juries themselves were swept away in the reform of county government at the end of the nineteenth century. The establishment of the system had involved, according to Sir Thomas Larcom, "the earliest instance of a competitive examination for public office", a fact which in his view "would always be creditable to Ireland".[125] As the calibre of surveyors generally improved, and as the system itself matured, it made a significant contribution to local development in many parts of Ireland although, as might be expected, performance was uneven, especially in the early years. In 1842, a Royal Commission recommended that the county surveyors should be replaced by district surveyors, operating on the basis of the baronies, with duties similar to those of the assistant surveyors but with higher incremental salary scales; these would be supervised and supported by six circuit engineers to be employed by the Commissioners of Public Works.[126] This recommendation was not accepted by the Government and, when the system was reviewed again by a Select Committee in 1857, the outturn was more favourable to the surveyors: the committee reported that "an able and efficient class of county officers has been formed" and that the new system had been "attended with great public advantage, both as regards the improvement of county roads and bridges, and as regards the economising of county funds".[127]

Whether this general observation was true of Cork city and county while the Leahys worked there is not easy to establish because of the absence of comprehensive and reliable records of the activities of the grand jury between 1834 and 1846. Their contribution to the development of the road network and other infrastructure of the city and county in those years is rarely touched on in any of the published material on the engineering of Cork in the nineteenth century whereas, by contrast, Sir John Benson, who succeeded Patrick as surveyor for the East Riding and for the city in 1846, is regularly credited with building or replacing no less than 31 bridges, including St Patrick's Bridge in Cork city centre, and his churches and other buildings still form a significant part of the city's architecture.[128] Edmund Leahy, in the years after he had left Cork, regularly wrote about his achievements as county surveyor but little reliance can be placed on his extravagant, if not outlandish, claims such as those presented to the Colonial Secretary in 1858:

During the 12 years I held the office, I constructed Harbours, Docks and Piers at Kinsale, Glengariffe, Glandore, Bantry and Berehaven. I designed and made over

In search of fame and fortune: the Leahy family of engineers, 1780-1888

600 miles of new roads, over 200 bridges of stone, iron and wood, one general asylum for insane people, one general county prison and 15 local prisons and session houses, and the Cork & Bandon Railway. Besides, a great quantity of other smaller works, I also directed the repairs and maintenance of all the county works.[129]

The following two chapters draw on the available evidence[130] to assess the Leahys' performance in relation to public roads, bridges and public buildings in Cork. In reviewing their operations, account must be taken of the fact that the county grand jury was reluctant in the 1830s and 1840s to expand its various work programmes because of an overriding concern to avoid increasing the county cess, the rate of which was less than 1% greater in 1843 than in 1824.[131] Making allowance, therefore, for the limited resources available to them, there is *prima facie* evidence that the Leahys performed their duties as county surveyors actively and reasonably satisfactorily for most of the period from 1834 to 1846. In road engineering and new road construction, they made a significant contribution to the infrastructure of the county. In terms of architecture and building, however, their impact appears to have been negligible.

REFERENCES

1. NLI, Larcom Papers, MS 7753
2. *Abstract of Grand Jury Presentments in 1834*, HC 1835 Vol XXXVII (220)
3. Thomas Newenham, *A View of the Natural, Political and Commercial Circumstances of Ireland*, 1809
4. Horatio Townsend, *A General and Statistical Survey of the County of Cork*, Cork, 1815
5. Rev M J Keating of Limerick, *Suggestions for Revision of the Irish Grand Jury Law* (nd)
6. Report of Alexander Nimmo, quoted in *Report of the Select Committee on the State of the Poor in Ireland (Summary Report)*, HC 1830 Vol VII (667); evidence given by Nimmo to a House of Lords Committee on the State of Ireland in 1824, quoted in *Report from the Select Committee on Grand Jury Presentments, Ireland*, HC 1826-1827 Vol III (555)
7. For detailed commentary on grand jury law on road work, see William Grieg, *Strictures on Road Police* (sic), Dublin, 1818.
8. Hansard (Third Series) Vol XIX, 11 July 1833, col 565
9. Reprinted in *Report from the Select Committee on Grand Jury Presentments, Ireland*, HC 1826-27 Vol III (555)
10. *Report from the Select Committee on Grand Jury Presentments, Ireland*, HC 1826-27 Vol III (555)
11. *Report of the Select Committee on the State of the Poor in Ireland (Summary Report)*, HC 1830 Vol VII (667)
12. Lord Carbery, *Observations on the Grand Jury System of Ireland with Suggestions for its Improvement*, London 1831
13. A Country Gentleman, *Hints on the System of Road Making*, Dublin, 1829
14. NAI, OP 972/62, letter of 16 January 1830 from John Killaly
15. Hansard (Third Series), Vol XV, 19 February 1833, col 955
16. Hansard (Third Series), Vol XIX, 11 July 1833, col 561
17. ibid, col 564
18. An Act to amend the Laws relating to Grand Juries in Ireland, 3 & 4 Wm IV, c.78
19. *Returns of the Dates of Commissions issued by the Irish Government for the Examination of Candidates for the Office of County Surveyor in Ireland*, HC 1856 Vol LIII (335)
20. NLI, Larcom Papers MS 7753, *Statement Respecting the Examinations for County Surveyorships Ireland*, undated; it was believed in official circles that the statement emanated from persons connected with the TCD Engineering School.

Chapter III — First Cork County Surveyors, 1834

21. Radcliffe was, in fact, a Peninsular veteran and military engineer who had been employed by the Directors General of Inland Navigation before his appointment as a Commissioner of Public Works.
22. NLI, Larcom Papers MS 7753, letter of 20 May 1857
23. The Chief Secretary's Office list of applicants (CSORP 1833/5635) and other relevant documents are not extant in NAI; it is clear from the available index, however, that applications were received from Patrick Leahy (CSORP 1833/4003) and Edmund (CSORP 1833/4188) although the files themselves are not available in NAI, but the index does not refer to Denis Leahy.
24. Evidence given by Burgoyne and Henry Brett, respectively, to the *Select Committee on County & District Surveyors etc (Ireland)*, HC 1857 Session 2 Vol IX (270)
25. *Tipperary Free Press,* 28 May 1834
26. NAI, CSO Unregistered Papers 1834/720
27. NAI, CSO Unregistered Papers 1834/184, letter of 10 May 1834 from Patrick Leahy
28. *Report from the Select Committee on County Cess (Ireland), Minutes of Evidence,* HC 1836 Vol XII (527)
29. Evidence given to the *Select Committee on County & District Surveyors etc (Ireland),* HC 1857 Session 2 Vol IX (270)
30. *Tipperary Free Press,* 28 May 1834
31. NAI, CSO Unregistered Papers 1834/184, letter of 10 May 1834 from Patrick Leahy
32. NAI, CSORP 1834/2111, letter of 10 May 1834 from Edmund Leahy
33. NAI, CSO Unregistered Papers 1834/184, letter of 10 May 1834 from Patrick Leahy
34. NAI, CSO Unregistered Papers 1834/720, OPW Circular of 1 April 1834
35. *Returns of the Dates of Commissions issued by the Irish Government for the Examination of Candidates for the Office of County Surveyor in Ireland, etc,* HC 1856 Vol LIII (335); Larcom Papers, NLI MS 7753, letter of 20 May 1857
36. NLI, Larcom Papers MS 7753, letter of 20 May 1857
37. NAI, CSO Unregistered Papers, letter of 2 June 1834 from J F Burgoyne
38. Evidence to *Select Committee on County & District Surveyors etc (Ireland),* HC 1857 Session 2 Vol IX (270)
39. NAI, CSORP 1834/2525
40. CC, 20 May 1834
41. *Tipperary Free Press,* 21 May 1834
42. Samuel Lewis, *A Topographical Dictionary of Ireland,* Vol I, London, 1838; in 1840, under 3 & 4 Vict., c.109, the county of the city became coterminous with the Borough of Cork, as defined in 3 & 4 Vict., c.108.
43. NAI, CSORP 1834/2525, letters of 7 and 14 June 1834
44. CC, 24 July 1834
45. *A Return of the Names of Each Engineer Employed under the Commissioners of Public Works in Ireland ... in the years 1832, 1833 and 1834,* HC 1835 Vol XXXVII (536)
46. Evidence given on 8 April 1835 to the *Select Committee on Advances made by the Commissioners of Public Works in Ireland,* HC 1835 Vol XX (573); Burgoyne had expressed similar views in a letter to the Chief Secretary on 2 June 1834, NAI CSO Unregistered papers 1834/494
47. NAI, CSORP 1834/2525, letter of 7 June 1834
48. ibid; Report by Patrick Leahy to the Grand Jury, CC, 29 July 1834
49. NAI, CSORP 1834/2525, letter of 7 June 1834
50. CC, 29 July 1834
51. See, for example, OPW5HC/6/168, Leahy's drawings of the Lower Glanmire Road showing the Intended Improvements, 1840.
52. 4 Geo IV, c.93
53. An additional barony in the East Riding - the barony of Cork - was created from the liberties of the city in 1841.
54. *Return of the Number of Days appointed by the Sheriff for Transacting the Fiscal Business in each County in Ireland,* HC 1844 Vol XLIII (130)
55. CC, 10 March 1836
56. NAI, CSORP 1840/W/6004, letter of 9 May 1840 to Edmund Leahy and his reply of 12 May
57. See, for example, NAI, CSORP 1840/W/6620 and 6528, complaints in April 1840 from contractors in the Ballinadee area alleging partiality and impropriety on the part of Edmund Leahy and one of his assistant surveyors.
58. NAI, OPW 1/1/4/21, Public Works Letters No 21, pages 225 and 255
59. NAI, OPW5HC/6/156
60. 3 & 4 Wm. IV, c.78, section 39; the Grand Jury (Ireland) Act, 1836, 6 & 7 Wm. IV, c.116, which consolidated grand jury law, restated the £300 salary limit in section 41.

In search of fame and fortune: the Leahy family of engineers, 1780-1888

61. CC, 29 July 1834 and 17 March 1838; *Report from the Select Committee on County Cess (Ireland)*, HC 1836 Vol XII (527) Appendix No 11; *Number of County Surveyors and of their Deputies and Clerks, 1834-1839, and of the sums presented for their salaries, office expenses, etc.* HC 1840 Vol XLVII (291)
62. NAI, CSORP 1839/77/95, letter of 19 January 1839 from Patrick Leahy
63. *Report of the Commissioners appointed to Revise the Several Laws under or by virtue of which moneys are now raised by Grand Jury Presentment in Ireland*, Appendix B, HC 1842 Vol XXIV (386)
64. Grand Jury (Ireland) Act, 1836, 6 & 7 Wm. IV, c.116, Schedule S
65. CC, 29 July 1834, report by Patrick Leahy to the grand jury
66. *Report from the Select Committee on County Cess (Ireland), Minutes of Evidence*, HC 1836 Vol XII (527)
67. County Surveyors etc (Ireland) Act, 1861, 24 & 25 Vict., c.63
68. 3 & 4 Wm. IV, c.78, section 41
69. CC, 10 and 12 March 1836
70. *Report of the Commissioners appointed to Revise the several Laws under or by virtue of which Moneys are now raised by Grand Jury Presentment in Ireland*, Appendix B, HC 1842 Vol XXIV (386)
71. ibid
72. *Report from the Select Committee on County Cess (Ireland)*, HC 1836 Vol XII (527)
73. Grand Jury (Ireland) Act 1836, 6 & 7 Wm IV, c.116, section 43; a salary up to £80 was allowed in 1861 by 24 & 25 Vict., c.63, section 7
74. NAI, OPW1/1/9, circular to county surveyors, 26 October 1836
75. NAI, OPW 1/1/9, circular to county surveyors, 3 November 1845
76. NAI, OPW 1/1/9, circular to county surveyors, 20 January 1837
77. NAI, OPW 1/1/9, circular to county surveyors, 16 December 1843
78. NAI, OPW I/1/9, letter from OPW to Patrick Leahy dated 3 January 1837
79. ibid, letter of 9 February to Patrick Leahy
80. ibid, letter of 20 April 1837 to Patrick Leahy
81. *Names of County Surveyors in each County in Ireland and their Assistants, etc.* HC 1863 Vol L (277)
82. NAI, OPW 1/1/9, letter of 20 June 1838 from Patrick Leahy, referred to in letter of 2 July from OPW to Leahy
83. NAI, OPW 1/1/9, letter of 6 March 1839 from OPW to Patrick Leahy
84. *Report of the Commissioners appointed to Revise the Several Laws under or by virtue of which moneys are now raised by Grand Jury Presentment in Ireland*, Appendix B, HC 1842 Vol XXIV (386); Cork Archives Institute, Schedule of Applications for Presentments, Summer 1838
85. NAI, OPW 1/1/9, letters, 15 and 20 July 1841 from OPW to Leahy
86. NAI, OPW 1/1/9, letters of 27 June and 10 August 1843 from OPW to Patrick Leahy
87. Cork Archives Institute, Schedule of Applications for Presentments, Summer 1845
88. NAI, OPW 1/1/9, letter of 20 April 1837 to Edmund Leahy
89. NAI, OPW 1/1/9, letter of 6 May 1837 to Edmund Leahy
90. ibid; *Names of County Surveyors in each County in Ireland and their Assistants*, HC 1863 Vol L (277)
91. NAI, OPW 1/1/9, letter of 6 March 1839 to Edmund Leahy
92. Cork Archives Institute, Schedule of Applications for Presentments, Summer 1845
93. Peter Eden (ed), *Dictionary of Land Surveyors and Local Cartographers of Great Britain and Ireland, 1550-1850*, Kent, 1979
94. Dom Mark Tierney OSB, *Calendar of the Papers of Dr. Leahy, Archbishop of Cashel, 1857-1875*, NLI Special List 171
95. Minutes of the Proceedings of the Society, ICEI Archives, Dublin
96. Evidence of Patrick Leahy in *Report of the Commissioners appointed to revise the several Laws under or by virtue of which Moneys are now raised by Grand Jury Presentment in Ireland*, Appendix B, HC 1842 Vol XXIV (386); by a curious coincidence, the house at Glenville became the residence of Captain Noble W Johnson, the long-serving Secretary of the Cork Grand Jury in 1896 ; the house still exists but is known as Ballinglanna House.
97. *Cork Post Office General Directory, 1844-1845; Slater's National Commercial Directory of Ireland, 1846*
98. See, for example, NAI, OPW5HC/6/385, 157 and 158
99. *Report of the Commissioners appointed to revise the Several Laws under or by virtue of which moneys are now raised by Grand Jury Presentment in Ireland*, Appendix B, HC 1842 Vol XXIV (386)
100. CC, 12 March 1839
101. NAI, CSORP 1836/1579, memorial of Denys C Leahy
102. ibid

Chapter III — First Cork County Surveyors, 1834

103. NAI, CSORP 1836/1579, letter of 25 May 1836 from Henry R Paine, Secretary, Office of Public Works
104. An Act to Consolidate and Amend the Laws for the Regulation of Grand Jury Presentments in the County of Dublin, 7 & 8 Vict., c.106
105. NAI, OPW 1/1/4/15, Public Works Letters No 15, page 178
106. *Returns of the Dates of Commissions issued by the Irish Government for the Examination of Candidates for the Office of County Surveyor in Ireland, etc* HC 1856 Vol LIII (335)
107. CC, 20 March 1838
108. *Report of the Commissioners appointed to Revise the several Laws under or by virtue of which Moneys are now raised by Grand Jury Presentment in Ireland,* Appendix B, HC 1842 Vol XXIV (386)
109. *Returns of the Dates of Commissions issued by the Irish Government for the Examination of Candidates for the Office of County Surveyor in Ireland, etc,* HC 1856 Vol LIII (335); according to the Index to the Chief Secretary's Office Registered Papers for 1838, a letter (CSORP 1838/759) was received from Mat Leahy that year in relation to a county surveyor post, but the letter itself is not extant.
110. CC, 20 March 1838
111. NAI, OPW 1/1/4/15, Public Works Letters No 15, page 23
112. NAI, CSORP 1838/77/1033, letter of 18 May 1838 from the Office of Public Works
113. NAI, OPW 1/1/4/15, Public Works Letters No 15, page 178
114. NAI, OPW 1/1/4/16, Public Works Letters No 16, page 338
115. NAI, OPW 1/1/4/17, Public Works Letters No 17, page 239
116. NAI, OPW 1/1/4/21, Public Works Letters No 21, page 127
117. NAI, OPW 1/1/4/27, Public Works Letters No 27, pages 108 and 157
118. NAI, CSORP 1845/W/1044, letter of 22 January 1845
119. NAI, CSORP 1845/W/1288, letter of 29 January 1845
120. NAI, OPW 1/1/4/32, Public Works Letters No 32, page 90
121. *Returns of the Dates of Commissions issued by the Irish Government for the Examination of Candidates for the Office of County Surveyor in Ireland, etc* HC 1856 Vol LIII (335)
122. Evidence to the *Select Committee on County & District Surveyors etc (Ireland),* HC 1857 Session 2 Vol IX (270)
123. *Third Annual Report from the Board of Public Works in Ireland,* HC 1835 Vol XXXVI (76)
124. *Report from the Select Committee on County Cess (Ireland),* HC 1836 Vol XII (527)
125. NLI, Larcom Papers, MS 7753
126. *Report of the Commissioners appointed to Revise the several Laws under or by virtue of which Moneys are now raised by Grand Jury Presentment in Ireland,* HC 1842 Vol XXIV (386)
127. *Report from the Select Committee on County and District Surveyors etc (Ireland)* HC 1857 Session 2 Vol IX (270)
128. Colm Lincoln, *Steps and Steeples, Cork at the turn of the Century,* O' Brien Press, Dublin, 1980
129. CO 48/392, letter of 2 September 1858 from Edmund Leahy to the Colonial Secretary
130. Cork grand jury records were destroyed by fires at the Courthouse in Cork in 1891 and at the Four Courts in Dublin in 1922. Copies of some of the printed schedules of presentments which were made available to individual grand jurors before each assizes have survived, but these give details of the projects which were proposed rather than those which were approved by the jury. In general, therefore, newspaper reports of the period are now the best available source of information about the activities and works programmes of the grand jury in the 1830s and 1840s. A limited number of contemporary published works provide some additional information, material published in the British Parliamentary Papers is also of value, and some relevant documentation is available at the National Archives, principally in the Chief Secretary's Office Registered Papers and in the archives of the Office of Public Works.
131. Walter Murphy, *Remarks on the Irish Grand Jury System,* Cork, 1849

CHAPTER IV

ROAD BUILDING IN CORK, 1834-46

"The valleys and mountain ranges never before traversed by the path of industry have been intersected by excellent roads; … and sounds of living industry now greet the ear of the traveller where formerly he might have been indebted to the accidental civility of the mountaineer for discovering his way through the intricacies of the scarce beaten path".

Edmund Leahy, 1844[1]

Road development before 1834

Until the second half of the seventeenth century, the condition of the Irish road network seems to have received little attention from the public authorities.[2] Thereafter, and particularly during the second half of the eighteenth century, considerable progress was made by the grand juries in building new roads and improving the condition of the existing ones, while turnpike trusts were providing a network of trunk roads linking most of the main population centres.[3] As a result, Irish roads often attracted favourable comment from English visitors like Arthur Young who found "everywhere … beautiful roads" during his 1776-79 tour of Ireland.[4] Remarks such as these, however, cannot have had any relevance to large areas of the country and to the minor roads.

The period between 1800 and 1834 was marked by further development of the main road network arising from the determination of the Post Office authorities to improve the system of mail coach roads. Competent surveyors were appointed under the Mail Coach Roads Act, 1805[5] to plan new routes and the improvement of existing routes, and loans and grants were provided by the Government to assist the grand juries in implementing these plans. Over 2,000 miles of roads were constructed or improved in this way between 1805 and 1816, with a considerable mileage also in the following ten years.[6] The County Cork grand jury participated actively in this road building programme with the aid of grants and loans amounting to more than £100,000 in the period 1810 to 1828.[7] The Cork-Mallow mail coach road was built between 1805 and 1816 and a new Cork-Youghal road was built in the 1816-20 period.[8] The mail coach road from Bandon to Clonakilty and Skibbereen was built in 1813-14 as was the road from Cork to Kinsale[9] and these were improved under an Act passed in 1822.[10] In the 1820s, the Bandon to Bantry road was rebuilt, a new road from Glengariffe to Castletownbere was completed with the aid of a Government loan of £4,860, and a new Cork-Bandon road, financed by loans of over £9,000, was built to link up with a new western road out of the city which had been completed by the city grand jury.[11] Finally, special Acts were passed in 1825 and 1826 to authorise the building of new main roads from Mallow to the Limerick county boundary and from the city to Ballyhooly on the river Blackwater.[12]

In the decade immediately before the appointment of the county surveyors, another important contribution to road building was made by a team of Government engineers who were recruited in the 1820s to carry out programmes of public works with funds provided directly by the central authorities. According to the Board of Works, the roads that resulted from these programmes were "the means of fertilising the desert and of depriving the lawless disturbers of the public peace of their place of refuge, affording them, at the same time resources for an active honest industry".[13] Richard Griffith, who conducted the programme in Cork, Kerry, Limerick and Tipperary, set himself the task of opening up "direct and easy communications" through mountainous and other districts in which, at the time, there were no roads over which wheeled carriages could pass and where it was probable that the grand juries would not themselves build good roads.[14] Between 1822 and 1837, he completed over 100 miles of roads in County Cork,[15] including the roads from Macroom to Glenflesk near Killarney (19 miles), Skibbereen to Crookhaven (23 miles),[16] Skibbereen to Bantry (8½ miles), Newmarket to Listowel (32 miles), Newmarket to Charleville (14 miles), and over 26 miles of road through the mountains on the route between Cork city and Kanturk. In addition, there were numerous smaller projects such as the roads from Union Hall to Rosscarbery, Clonakilty to Ring Pier, Timoleague to Courtmacsherry, and a new entrance into the town of Bandon, designed "to avoid a very steep and inconvenient hill" on the north of the town; this latter was completed in 1823 at the joint expense of the Government and the Duke of Devonshire.[17] The work begun by Griffith continued in the 1830s with the building of the spectacular Glengariffe-Kenmare road, which incorporated a number of tunnelled sections, and the completion by 1837 of a network of 43 miles of road in the Kingwilliamstown (or Pobal O'Keeffe) area near Newmarket.[18]

Extending and maintaining the network after 1834
Given the progress that had been made in developing and extending the road network in the 30 years before the county surveyors came into office, and the determination in the 1830s and 1840s to minimise increases in the county cess, most grand jurors were unlikely to be easily convinced by their new professional advisers that there was need to continue, or to expand, the costly roads programmes that had been in place in earlier decades.

Nevertheless, the mileage of roads being maintained by the grand juries increased rapidly after 1834: according to a report presented to Parliament in 1842, the total mileage under contract was then 38,400,[19] nearly three times the 1834 figure of 13,191 miles.[20] This increase was due mainly to the effective "taking in charge" by the grand juries of existing roads, rather than the construction of new ones.

In Cork, Patrick Leahy's first report to the grand jury in July 1834 claimed that the roads in the East Riding were, in general, only in a tolerable state of repair.[21] The main defect, he reported, was the poor shape of the roads and the lack of attention to water-tables and gripes; as a result, water did not run off quickly enough, especially on hilly sections, and there was considerable winter damage. The corresponding report from Edmund Leahy on the West Riding observed that roads generally in his area were not in good condition, and the formation was bad; materials used in the past for road surfaces had been too large and the works were done "but very middlingly and in many cases very badly".[22] In the years that followed, the mileage of roads on which maintenance and improvement work was undertaken under the general supervision of the two surveyors increased substantially and, by 1842, a total of 2,875 miles (7.5% of the national total) was being maintained, 1,904 miles in the East Riding and 971 miles in the West.[23] While no comparable official figures are available for 1834, Edmund Leahy noted that the mileage being maintained at public expense in 1844 was almost double the 1834 figure[24] and his father told the grand jury that the total mileage under contract in his area in 1844 stood at 2,300 as against 1,599 in 1834.[25] Expenditure on road maintenance in Cork increased broadly in line with the increased mileage; it was held at a level of about £20,000 from 1834 to 1839, but rose rapidly in the 1840s to nearly £30,000 a year. [26]

Like all other grand jury activity, road maintenance in the nineteenth century was carried out on a contract basis. Road contractors, often local farmers, were appointed, after a competitive tendering process, to maintain particular stretches of road for periods of three to five years but these road contractors were altogether different from other contractors, according to one of the early surveyors: "one begs payment as a favour; the other, having performed his work, requires it as a right; the latter (the ordinary contractor) is almost always a capitalist; the former (the road contractor) is mostly a pauper, or a small farmer".[27] Road contractors often submitted very low bids in an effort to gain access to what was one of the few sources of off-farm income available at the time; as Patrick Leahy put it, contracts were taken by "poor ignorant persons, having no other employment, who underbid each other"[28] and by "adventurous and speculating farmers and labourers, quite unfit for such works and who underbid competent persons".[29] Sections of road for which tenders were submitted were often chosen to suit the convenience and capacity of the contractors who provided their own horses and carts, often employed their own relatives, and did little more than spread hand-broken stone or gravel on road surfaces, allowing the materials to be worn in by the traffic. In this situation, the cost of maintenance of the contract roads was very low but the standard of the work done was uneven and not always acceptable.

The Schedules of Applications for Presentments which came before the grand jury at each assizes contained long lists, in summary form, of proposed road maintenance contracts, generally in the form of the following examples:

- For five years, 843 perches of the road from Cork to Carrigaline and Kinsale, commencing at the three roads and finger-post of Maryboro, and ending at Monees Bridge. Two tenders. Jeremiah Kelleher, contractor, 9¾d. per perch, £34 4s 11¾d. each year.

- For five years, 1,113 perches of the road from Cork to Bandon, commencing at Walsh's Cross on the new Skibbereen road, and ending at the bounds of the Liberties at Ballincollig. Five tenders. Cornelius Hayes, contractor, 1s 5d. per perch, £78 16s 9d. a year.[30]

The road sections involved were often quite short, many of them less than a mile and a half, but contracts relating to sections as long as six miles were sometimes awarded. In 1834, the annual maintenance cost in most cases was three or four pence per perch in County Cork, but up to eight pence was paid in some cases, with the higher figure applying only in the first year under some contracts. Higher figures prevailed in subsequent years, especially on more heavily trafficked routes, but expenditure per mile in the county as a whole was still little more than £10 a year.

Ensuring satisfactory completion of road contracts was a perennial problem for the surveyors, especially where contracts were awarded on the basis of unrealistic prices and, like their colleagues in other counties, the Leahys regularly complained of the difficulty of conducting road maintenance operations on this basis. As the law was interpreted, contracts had to be awarded to the lowest tenderer, regardless of his experience or capacity to perform the work. Inevitably, where reckless competition had forced prices down to a fraction of the true cost, contractors defaulted, and while the surveyors could withhold payment certificates in such cases, they could not proceed in court against a defaulter without the consent of the grand jury. In 1844, Patrick Leahy put forward an imaginative scheme for overcoming some of these difficulties and, at the same time, reducing the numbers of unemployed labourers who were being admitted to the workhouses. He proposed to the grand jury that they should seek to have the law amended to enable them to enclose some of the abandoned older roads, and much of the waste ground along the existing roads, so as to build cottages for labourers who could then be engaged to maintain the roads. This scheme, he suggested, would "provide effectually for locating a nursery of labourers along the old byways which could be at hand to repair the high roads and diminish in a great measure the weight of the Poor Laws".[31] These suggestions, however, were too far ahead of their time to have any chance of acceptance: the building of labourers' cottages by local authorities was not authorised by law until 1883 and direct labour operations on the roads were not sanctioned until 1901, a few years after the grand juries had been replaced by the county councils.

Press reports of the twice-yearly discussions by successive grand juries of road maintenance operations suggest that Edmund Leahy's performance in the West Riding gave general satisfaction and he himself asserted that he had considerable success in this sphere. He claimed particular credit, for example, for bringing the Skibbereen-Crookhaven road, built by Richard Griffith in 1822-23, up to a good standard[32] and he boasted that he had resolved a long-standing road erosion problem at the Red Strand, between Clonakilty and Rosscarbery, where up to 135 horses and heavy carts were engaged daily, at particular times of the year, in drawing away over 50,000 tons of sea sand annually for use as fertiliser.[33] While there were complaints from time to time about the failure of Patrick Leahy effectively to supervise operations in the East Riding, it appears that the standard of road maintenance and the value for money achieved in the county as a whole improved considerably between 1834 and 1844. Indeed, William Johnston, the grand jury secretary, told the Devon Commission in 1844 that there are "few counties in Ireland, if there are any, in which the working of the grand jury system exhibits so gratifying a result as in the county of Cork".[34] According to Johnston, the business was done with great regularity, and extensive additional facilities had been provided for the rural community. Between 1824 and 1843, 2,600 miles of roads had been brought within the contract repair system at an annual cost of about £9 a mile and, he went on, the grand jury was determined "to supply the wants and advance the interests of their county, by opening new lines of roads and promoting useful public works, regardless of all unwise and unfair opposition".

New roads in West Cork
Based on returns from the individual county surveyors, a report presented to Parliament in 1842 noted that a total of 1,522 miles of new roads had been built since 1834 in the entire country by the surveyors then in office, including 120 miles in the East Riding of Cork and 143 miles in the West.[35] The total for the county was higher than that of any other and it accounted for 17.3% of the national total. These statistics, coupled with the fact that annual expenditure on new road construction in County Cork virtually doubled to £15,000 between 1834 and 1837[36] indicate that the Leahys had generated relatively high levels of road building activity soon after their arrival in Cork and continued to be more active in this respect than many of their county surveyor colleagues.

By 1845, Edmund Leahy claimed credit for a higher mileage of new roads, telling the Devon Commission that he had built 200 miles between 1834 and 1845.[37] At a later stage, the mileage he claimed to have built in his 12 years in Cork had grown to more than 600[38] but this is inconsistent not only with the official statistics but also with the evidence of the secretary of the County Cork grand jury who told the Devon Commission that, between 1824 and 1843, a total of 627 miles of entirely new roads had been made in Cork.[39] The claim is inconsistent also with detailed tables published in Cork in 1849 showing that the total mileage of new roads built in the county in the 30-year period from 1820 to 1849 amounted to 824, a figure which would have included the considerable mileage of Government roads built between 1820 and 1834 and some of the new roads built under relief work schemes in 1846 and 1847.[40]

Taking account of all the available evidence, it seems reasonable to conclude that Edmund was responsible for the construction of up to 200 miles of new roads in West Cork between 1834 and 1846. His own biannual reports to the grand jury[41] provide an indication of the nature and location of his road-building activities and the first edition of the six-inch Ordnance Survey map of County Cork (published in 1845 but based on survey work carried out in 1841-42) shows many of the new roads, indicating in some cases that they were "new roads" or a "new line" and, in other cases, noting that the road shown on the map was a "new line of road in progress". As these notations suggest, some of the new roads ("the new lines") were intended to replace existing roads with severe gradients or other defects, while others were entirely new roads built in some cases on tracks used by animals and local residents but unsuitable for use by wheeled vehicles. The effects of all of this activity were summed up as follows by Leahy in 1844:

> The valleys and mountain ranges never before traversed by the path of industry have been intersected by excellent roads; large tracts that lay waste beyond the memory of man have been brought under cultivation; improvement has found its way into almost inaccessible districts which have exchanged the wildness of unreclaimed nature for the smiles of fertility; and where the marsh exhaled its' vapour and the hill was covered with its barren heath, the eye is now gladdened with the sight of pastures and cornfields. The scanty population which formerly struggled for a precarious subsistence, debarred in effect from intercourse with others, have given way to an energetic and industrious population; and sounds of living industry now greet the ear of the traveller where formerly he might have been indebted to the accidental civility of the mountaineer for discovering his way through the intricacies of the scarce beaten path.[42]

This idyllic picture of a rapidly developing rural economy – if it was ever true – was shattered only a few years after Leahy had published this overblown account of the impact of his new roads, many of which were located in the areas of West Cork in which the famine and its effects were most severe. In the main, the new roads were built in the coastal districts and more remote areas in the west and north-west of the county which had tended to be neglected by the road builders in earlier years. Almost all of the roads are still in use

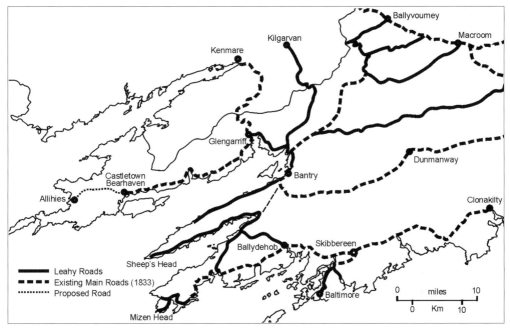

Roads in West Cork laid out by Edmund Leahy, 1834-46.

today, and many of them are among the roads most frequented by tourist traffic. The following are some examples.

- In 1839, a new Bantry–Glengariffe road, for which plans had originally been prepared in 1833, was in progress, and the 10 miles of road was almost complete in 1840 except for a causeway and bridge at Snave;[43]

- In 1839-40, new roads were being built from Bantry to Ballydehob (7 miles) and from Bantry via Durrus to Crookhaven (26½ miles) and onwards to Barleycove and the Mizen (6 miles);[44]

- In 1841, a new road from Coomhola, just off the Bantry-Glengariffe road, along the eastern side of the Borlin Valley was proposed, to open up an inaccessible district and to shorten the journey from Bantry to Killarney by 12 miles, and £3,172 was provided to build this in 1843;[45] the road climbs to over 1,200 feet above sea level before crossing the county boundary and dropping down to Kilgarvan;

- By 1842, 20 miles of a new Bantry to Crookstown road (on which work had been in progress since 1840) was open but, because of difficulties with the contractors, the 7 mile section through Coosane Gap had still to be completed;[46]

- In 1844-45, new roads were being built along Dunmanus Bay from Bantry towards Sheep's Head through a peninsula where "not one proper safe passage for a wheeled carriage" existed;[47]

- In the Skibbereen area, a 9 mile section of road along Roaring Water Bay was in progress in 1839-40 and in 1845 a new road from Skibbereen to the sea at Tragumna was under construction;[48]

- A new Macroom-Ballingeary road was under construction in 1839, leaving the Macroom-Killarney road about 5 miles from Macroom and cutting through difficult and hilly terrain; this linked up with another new system of mountain roads serving Kilgarvan and Ballyvourney;[49]

- Work was underway in 1840 on an 8 mile section of road in the Mushera Mountains on the route between Cork city and Millstreet[50] and a new road and bridge were being constructed through Anahala Bog, near Macroom, in 1842;[51]

- In the Bandon area, £806 was to be spent in 1843 on a new road from Ballinspittle and Ballinadee to Bandon, and £318 was committed for 2 miles of new road from Timoleague along the coast to Burren;[52]

- In 1844-45, new roads were being built from Bandon to Kilmacsimon Quay on the river Bandon, and from Kilbrittain to Harbour View.[53]

Perhaps the most ambitious of Edmund Leahy's road schemes was one which he prepared in 1836, with the assistance of his brother Denis, for a new road, 15½ miles long, from Dursey at the extreme west of the Beara peninsula, via Allihies to Castletownbere which itself had been linked to Bantry by a new road in the 1820s.[54] The copper mines which had been established by John Puxley at Allihies in 1812 had reached peak production of almost 7,300 tons of ore in 1835 but, with only a very steep mountain road link to the outside world, the ore was still being carried in small boats from the dangerously exposed beach at Ballydonegan Bay, through Dursey Sound, to Dunboy, near Castletownbere, for onward shipment to Swansea in a fleet of schooners built by Puxley for the purpose; supplies of all kinds, including large quantities of coal to fuel the steam engines used at the mines, and wood for pit props, had to be carried by similar means in the reverse direction.[55] Leahy's new road through the mountainous heart of the peninsula was intended to facilitate the industry by providing a direct link between the mines and Dunboy, but to achieve this, some heavy engineering, including deep cuttings, substantial lengths of retaining walls and four major bridges would have been required. In addition, there was provision for a tunnel, 300 feet long and 18 feet high, north of Allihies, where the road was to climb to about 800 feet above sea level.

In search of fame and fortune: the Leahy family of engineers, 1780-1888

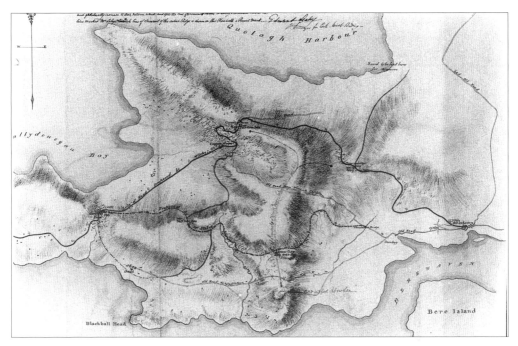

Manuscript map showing the line of the proposed Allihies-Castletownbere road as set out by Edmund Leahy in 1836 (on north of map) and the alternative line suggested by Henry Stokes, Kerry County Surveyor (OPW5HC/6/160, courtesy of the National Archives of Ireland).

Sketch of the summit tunnel on Edmund Leahy's proposed Allihies-Castletownbere road, 1836 (OPW5HC6/156, courtesy of the National Archives of Ireland).

Chapter IV — Road building in Cork, 1834-46

When Leahy's plans were submitted to the Commissioners of Public Works in January 1837 with an application from Puxley for a substantial grant to finance the new road, they noted that "the cost is considerable for a cross country road and one, judging by the map, not likely to be much of a thoroughfare". They engaged Henry Stokes, the Kerry county surveyor, to carry out an inspection and make a report on the project and, in regard to the tunnelling proposals, suggested that he should consult J B Farrell, the engineer who was then supervising the construction of the Glengariffe–Kenmare road, where tunnelling was already in progress.[56] Stokes advised that a more direct and more level line should be adopted and a grant was therefore refused for Leahy's scheme.[57] In April 1837, however, at Puxley's request, the Commissioners agreed to have the whole matter reviewed by "another engineer of confidence" and assigned Edward Russell, one of their assistant engineers, to conduct the review.[58] This led to a decision by the Commissioners in June that, while Stokes' line was best in engineering terms, "more good will be effected in opening the country and bringing out its resources by adopting the line originally recommended by Mr Leahy".[59] A recommendation was therefore made to the Treasury that a grant of £5,750 and a loan of the same amount should be sanctioned for the project[60] but the grand jury were told in June 1838 that "their Lordships do not feel justified in making so large a grant as £5,750 for any one road" but would reconsider their position if local interests were prepared to contribute.[61] At that stage, John Puxley, who had been strongly pressing the case for the new road for over two years, questioned whether the road would be of real benefit to his mines and advised that if a local subscription were to be a *sine qua non*, the project "must be dropped".[62] In the event, therefore, Leahy's spectacular new road – which if it existed today would certainly be a major tourist route – was never built.[63]

Another of Leahy's imaginative, if not particularly realistic, proposals emerged in 1840 in response to the long-standing demands from the merchants and traders of Bandon for better facilities for the import and export of goods via the Bandon river. In 1839, two new sections of road, about five miles long, were authorised by the grand jury at a total cost of £2,900 to facilitate the transport of coal and other goods to Bandon from Colliers Quay where ships of up 200 tons could berth. However, a proposal to build a further mile-long section of road from Colliers Quay eastwards along the river to deeper water was rejected by the grand jury at the Spring Assizes in 1840, notwithstanding the evidence of various witnesses about the unsuitability of the quay and the fact that a "numerous and respectably attended meeting" in Bandon had voted unanimously in favour of the new road.[64] At the following assizes, Leahy advised the grand jury that the objective could better be attained, and at less cost, by extending the navigable section of the river by one mile so as to allow ships to berth at Innishannon; goods could then be transported on the level mail coach road to Bandon.[65] He proposed deepening the river channel by five feet, banking up the sides, and constructing three weirs, each eight feet high, in the river above Innishannon so as to prevent the flow of gravel and silt to the newly deepened section of river. All of this, he suggested, could be achieved at a cost of £850, towards which the proprietor of the town of Innishannon was willing to contribute. The members of the grand jury, however, were not attracted by Leahy's novel proposal; some had doubts about the legal capacity to undertake works of this nature, others argued that the weirs and other works would seriously interfere with fish life in the river, and others still warned that raising the water level in the river above Innishannon would inevitably lead to flooding on the adjoining mail coach road. Leahy went on, however, to develop his ideas for extending the navigable section of the river Bandon and presented a more elaborate ship canal proposal (described in Chapter VI) to the merchants of the town in 1841. This, too, did not find favour and, a few years later, the earlier scheme for constructing a new road along the river to Kilmacsimon was implemented, linking up with a new quay which was being built there by the local landowner, Edward Gilman.[66]

The last of Edmund Leahy's abortive schemes was put forward in November 1843, when he told the Chief Secretary that "it is in contemplation to effect very extensive alterations" to the mail coach road between Cork and Bandon "or perhaps to construct an entirely new road for the entire distance being about 20 miles". As the Ordnance Survey six-inch maps of County Cork had not yet been published, he asked to be supplied with a trace of the relevant map sheets so as enable him "to arrive at the best determination"; he was duly supplied by Captain Thomas Larcom with "unfinished impressions" because the latter agreed that it would be wrong that "so great a line of road should be undertaken without the benefit of the Ordnance Survey".[67] In the following March, Leahy advised the grand jury that the increasing intercourse between Cork and Bandon warranted major road improvements, particularly where the road climbed at a gradient of 1:13 for over a mile at the Liberty Hill,[68] but his plans for a new road, with gradients of no more than 1:100, never materialised; generations of travellers from Cork to Bandon, using more than 10,000 vehicles per day by the 1990s, had to traverse the Liberty Hill until 1995 when a new section of road, four miles long, was completed. Given that the existing Cork–Bandon mail coach road had been constructed only about 20 years earlier, it is tempting to suggest that Leahy had no intention in 1844 of rebuilding or replacing the road and that his true purpose in obtaining privileged access to advance copies of the large-scale Ordnance Survey maps, almost two years before their publication, was to facilitate preparation of the plans for the Cork & Bandon Railway which he launched in the Summer of 1844 (see Chapter VII).

New roads in the liberties and in the East Riding
A considerable amount of the new road construction undertaken by Patrick Leahy as surveyor for the city of Cork and for the adjoining East Riding took place in and around the liberties of the city – a substantial rural area immediately outside the city boundary but which, until 1841, was part of the functional area of the city grand jury. All of the main

Chapter IV — Road building in Cork, 1834-46

approach routes to the city passed through this area and, from 1835 onwards, Leahy put forward a series of improvement projects designed to ensure, as he told the city grand jury in March 1836, that after a few years "this city will not be inferior in its entrances to any in *our* Empire".[69] Among these were:

- A short stretch of new road linking the Lough Road with Pouladuff Road, built in 1835,[70] to improve access from the city to the Kinsale road;
- An entirely new section of road from the city to Five-mile-bridge, proposed in 1835;[71]
- A new northern entrance to the city from Commons Road to York Street, plans for which were prepared in 1836;
- A new road from the city to Carrigaline, later known as the Carr's Hill road, approved in 1836;
- A new road along the Lee from the city to Glanmire (the Lower Glanmire Road), constructed between 1840 and 1842;
- The Carrigrohane "straight road", leading from the city to Ballincollig, on which work began in 1844;
- The Cork-Blarney road was to be improved in 1840-41 and a new 7 mile section of road leading to Donoughmore and Kanturk was planned;[72]
- Reflecting the suburban nature of the area, proposals for the construction of long lengths of footpaths on the roads to Bandon, Kinsale and Passage were put forward in 1840.[73]

Patrick Leahy's new road to Carrigaline introduced a novel feature to road construction in Ireland. He told the grand jury that the road was to have "a most favourable line" and moderate gradients except for half a mile near Douglas where, in order to mitigate the gradient, the new road was to "pass favourably over a branch road to the church and chapel by a viaduct". The bridge and the approach embankment, which are still in use, form one of a very small number of fly-overs of this kind built in Ireland in the nineteenth century; it became known locally as the "Bow-Wow" because of the echo which could be obtained when passing under it.[74]

One of the most important road projects undertaken by Leahy for the city grand jury was the widening and improvement of the Lower Glanmire Road. At the Spring Assizes in 1838, jurors drew attention to the fact that this section of road was the narrowest part of the route to Dublin and directed Leahy to prepare plans for a new roadway, 40 feet wide, from Fishery Quay to the quay at Glanmire, with a new river wall founded on the mud banks alongside the river.[75] A Government loan of £6,628 was sought in 1840 to finance

In search of fame and fortune: the Leahy family of engineers, 1780-1888

The bridge designed in 1836 by Patrick Leahy to carry the new Cork-Carrigaline road over an existing local road.

the scheme and work got underway quickly.[76] However, continuation of the new road through the Glanmire Valley gave rise to controversy although the scheme was designed, according to Leahy, "to benefit the community as well as to render more safe and expeditious the conveyance of Her Majesty's mails from Cork to Dublin".[77] This object was to be attained, in Leahy's words, by avoiding a steep and otherwise dangerous hill or elevation on the existing road, rising suddenly from high-water level at a rate of one foot in seven and pursuing an undulating and indirect course for three-quarters of a mile until it dropped into the valley of Glanmire again. The surface of the existing road was poor, the road was too narrow, pent in between old demesne walls, and it was overhung and darkened with aged trees, all of which, according to Leahy, created dangerous conditions for coaches and other traffic.

Leahy's plans for replacing this stretch of road by a flat new line running alongside the Lee and the Glanmire rivers were approved by the city and county grand juries in the summer of 1840[78] but when contractors were well advanced with the work a few months later, one of the landowners, Mrs Eliza MacCall,[79] sought to bring work to a stop by complaining to Dublin Castle about tree felling along the route and the fact that a minor river diversion was being effected, without proper authority, to avoid the expense of building two bridges. The Castle authorities decided, however, not to intervene and directed that Mrs MacCall should be told that "if she has any just cause of complaint

Chapter IV — Road building in Cork, 1834-46

against the proceedings of the grand jury, she may resort to the ordinary legal tribunals".[80] In July 1842, Leahy was able to tell the assizes that the new road through the Glanmire valley was open and that it constituted "one of the greatest improvements in Ireland";[81] the road was to serve as part of the main Cork-Dublin route for 150 years, carrying 10,000 vehicles per day in 1992 when it was replaced by a dual-carriageway by-pass of Glanmire.

The most enduring of Patrick Leahy's road schemes was initiated in 1844 when the city grand jury approved plans costing almost £4,000 to build a new road from the western suburbs to Carrigrohane, commencing at George IV Bridge (now O'Neill Crowley Bridge) at the end of the Western Road which itself had been constructed in the 1820s. There was considerable opposition to this new line by landowners on the northern side of the river Lee and the fact that the line had been surveyed, as a private commission, by Edmund Leahy added to the controversy. In the event, the scheme went ahead, commencing with a 2½ mile long straight section of road (known since to generations in Cork as the Carrigrohane Straight), with one branch running through Ballincollig towards Macroom and another crossing the river Lee by a skew bridge at Carrigrohane and linking up with the road to Kanturk.[82]

In the East Riding itself, some significant new road proposals were also developed under Patrick Leahy's supervision. A new road from Kinsale to Ballinspittle via Ringrone Castle was in progress in March 1836.[83] In 1842, construction of a new road from Kinsale to Innishannon was in progress as was a new Fermoy-Lismore road along the northern bank of the Blackwater.[84] Substantial improvement works on the Cork-Macroom road and a new Belgooly-Kinsale road were underway in 1845.[85] However, proposals for a new line of road from Mallow and Killavullen via Whitechurch to the city were rejected by the city grand jury in 1839 and 1840[86] and were defeated again at the barony of Cork presentment sessions in 1843.[87]

Bridges

In Cork city, some 27 bridges span the twin channels of the river Lee, the earliest of them (South Gate Bridge) dating from 1715, with several others dating from the years immediately prior to Patrick Leahy's appointment as city surveyor.[88] There is no evidence that Leahy played any part in the design, construction or significant modification of the city's stock of bridges during his years as surveyor for the city, something which is in part explained by the fact that the city streets were primarily the responsibility of Cork Corporation and the Wide Streets Commissioners.

There are no records of the construction dates of bridges generally in County Cork and relatively few bridges carry plaques indicating the date of their construction or the names of their designers or builders. Given that there are about 20,000 road bridges with spans of more than six feet in Ireland (excluding Northern Ireland),[89] one for every three miles of public road, it is likely that county Cork, with 12.5% of the country's road mileage, has about 2,500 such bridges; of these, some 2,000 are likely to be masonry arch bridges, many of them up to 200 years old, or more. However, while analysis of bridge and arch characteristics has provided a framework which can help in dating surviving bridges of uncertain origin,[90] it is still difficult in many cases to establish with certainty when a particular bridge was constructed, or which bridges may be attributed to particular engineers. If it is accepted that the Leahys built up to 350 miles of new roads in Cork, they are likely to have been responsible for at least 120 new bridges with spans of six feet or more on these roads alone, over and above essential maintenance, repair, and occasional replacement of older bridges. However, Edmund's claim that he alone "designed and made ... over 200 bridges of stone, iron and wood"[91] during his years as county surveyor is difficult to accept.

The five-arch masonry bridge carrying the main road to West Cork over the river at Bandon appears to be the only surviving major bridge which can definitely be associated with Edmund Leahy. The bridge was originally built in 1773, replacing one which had been swept away in a flood eight years earlier.[92] The presentment sessions in January 1838 accepted the need to enlarge and improve it and a local contractor was appointed by the grand jury at the Spring Assizes to carry out the work. The bridge is still in service, over 160 years later, carrying heavy traffic, and the downstream parapet carries a plaque attesting that it "enlarged and improved by county presentment A.D. 1838, Edmond (sic) Leahy esq., civ. engr., Matthew Parrett esq., archt". When account is taken of Leahy's penchant for publicity and his apparent determination to capitalise on every opportunity to advertise his engineering achievements, it seems very unlikely that he would have foregone the opportunity to have his name inscribed on other major bridges which he had

Bridge at Bandon, County Cork (Lawrence R 6706, courtesy of the National Library of Ireland).

Chapter IV — Road building in Cork, 1834-46

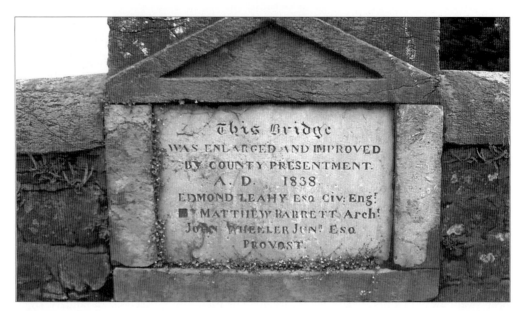

Plaque on the downstream parapet of the bridge at Bandon indicating that it was "enlarged and improved" under Edmund Leahy's direction in 1838.

built or reconstructed. Thus, even when allowance is made for the replacement of bridges and for bridge widening and parapet reconstruction during the last 150 years, the absence of major bridges carrying Leahy's name must be taken as *prima facie* evidence that, in spite of his claims to the contrary, he is not entitled to be included in any list of prominent bridge builders. It seems likely that the majority of his new bridges were relatively simple stone structures, perhaps little more than culverts, carrying minor roads over small rivers and streams, with only a very small number of more substantial structures such as Donemark Bridge, a three-arch masonry bridge built in 1839-40 to carry the new road from Bantry to Glengariffe over the river Mealagh at the falls of Donemark.[93] Some wooden bridges, long since replaced, were also built on sections of Leahy's new roads as, for example, a bridge forming part of the original causeway at Snave on the Bantry–Glengariffe road, but the location of the iron bridges for which he claimed credit remains a mystery.

From the outset, Patrick Leahy expressed concern about the condition of bridges in the East Riding. In his first report to the grand jury in 1834, he commented adversely on the design of some of the older hump-backed bridges which he had inspected, mentioning the bridges at Kanturk and at several places on the Blackwater and the Bride as examples; he believed that these bridges had "been raised an unnecessary height … after making every allowance for ordinary and extraordinary floods", and as a result "the approaches are inconvenient and their general stability is endangered".[94] In the years that followed, he

In search of fame and fortune: the Leahy family of engineers, 1780-1888

regularly told the grand jury about his concern for the condition of the bridges over the Blackwater at Mallow and west of the town; in July 1840, for example, he reported that six of these bridges had suffered badly in the unprecedented floods of the previous few winters and one of them, at Ballymaquirke near Banteer, was "tottering from the foundations".[95]

Leahy made some progress towards eliminating these defects. At the summer assizes in 1838, an estimate of £129 for the improvement and widening of Kanturk Bridge was submitted for approval, a presentment of £120 was approved for the same purpose in 1845[96] and an estimate of £180 for a replacement bridge over the Bride at Castlemore was put forward in 1838. In July 1840, a proposal to spend £599 on a new three-arch bridge at Carrigaline to replace a bridge which had been undermined by river drainage works was deferred[97] but the work was subsequently authorised and good progress was being made on the construction of the new bridge in 1844.[98] However, work on a new bridge over the river Allow near Kanturk had been delayed because the contractor was "a rash adventurer" who had incautiously undertaken the project for little more than half of what would have been a fair price.[99] In 1845, £34 was sought for underpinning Rathcool Bridge on the

Bridge over the falls at Donemark on the Bantry-Glengariffe road, constructed under Edmund Leahy's direction in 1839-40 (Lawrence R 6027, courtesy of the National Library of Ireland).

Millstreet-Mallow road[100] and small sums were allocated "to water-pave and secure the abutments" of a number of others.

One project near Glanmire with which Patrick Leahy was involved turned out to be a very controversial one and an interesting precursor of the controversy which arose 150 years later when plans were being made for the nearby downstream crossing of the river Lee. In 1842, in conjunction with his new Lower Glanmire Road scheme, he proposed the construction of a stone arch bridge to replace a wooden lifting bridge over the Glanmire River at Dunkettle, a few miles east of Cork city, on the main road to Youghal and Waterford. In January 1843, tenders were invited for the construction of the bridge (with a cost limit of £1,000 set by the grand jury) but the Harbour Commissioners and others objected on the grounds that a fixed bridge would prevent ships reaching the small quay at Glanmire to unload coal for the local mills and distillery and cargoes of corn, sand and limestone.[101] The grand jury decided in March 1844 to go ahead with the building of the stone arch bridge but their decision was quashed by the Assizes Judge later that month.[102] Forced to consider a swivel bridge, even though this would be more expensive to maintain, the grand jury, under pressure from the Judge, decided at the Spring Assizes in 1845 to implement this option at a cost of £1,215.[103] Leahy reported one year later that the bridge was almost completed.[104]

A Practical Treatise on Making and Repairing Roads
Whatever the truth of his claims about extensive new road and bridge building, Edmund Leahy obviously considered himself to be something of an expert on road engineering when, in 1844, he published *A Practical Treatise on Making and Repairing Roads*.[105] This was one of a very small number of publications by Irish engineers of the period and may be compared with Sir Henry Parnell's *A Treatise on Roads*, first published in 1833 and described by Thomas Telford as "the most valuable Treatise which has appeared in England on the history, principles, and practice of that species of national improvement".[106] Leahy's book of 306 pages was printed by George Purcell & Co of Cork, published by a reputable London publisher and "with permission, most respectfully dedicated" to General Sir John Fox Burgoyne". It was advertised for sale in the *Cork Examiner* in May 1844 at a price of 7s 6d, the equivalent of about €45 today[107] and a second edition was published in 1847, presumably in response to consumer demand. The stated object of the work was to instruct "those who may have to do with the making and repairing of roads" and to provide them with useful and practical information on their business. In addition, Leahy hoped that his treatise would be "interesting in no small degree to the public at large" although he apologised for the fact that, in some sections "it became necessary to depart in some degree from that simplicity so desirable in works of a popular nature" and to introduce "embarrassing calculations".

Leahy's book had a curious blend of subject matter. There were lengthy chapters on the history of roads, on administrative matters and on legal systems, as well as chapters on

In search of fame and fortune: the Leahy family of engineers, 1780-1888

> A PRACTICAL TREATISE
>
> ON
>
> MAKING AND REPAIRING
>
> ROADS.
>
> ILLUSTRATED BY ENGRAVINGS AND TABLES.
>
> BY
>
> EDMUND LEAHY,
> CIVIL ENGINEER.
>
> "Our present modes of conveyance, excellent as they are, both require and admit of great improvements."
> QUARTERLY REVIEW.
>
> LONDON:
> JOHN WEALE, ARCHITECTURAL LIBRARY,
> 59, HIGH HOLBORN.
> 1844.

Title page of Edmund Leahy's Treatise on Making and Repairing Roads, published in 1844.

different methods of road construction, maintenance and drainage, and sections which went into great detail about road materials, instruments and tools. He claimed that many of the deductions presented were new and differed widely "from previously received opinion" and that his work would elucidate, to a large extent, subjects "which heretofore had been either wholly overlooked or but imperfectly understood". He believed that his "humble labour", if attended to, would have the effect of introducing "more uniformity, with greater accuracy, in the design and construction of our roads" and claimed in 1858 that it had been "in part copied into the Aide-Memoire of Military Engineering edited by British Military Engineers".[108]

How much of the content of Edmund Leahy's book was, in fact, genuinely new and how much may have been based on earlier publications, or involved use without acknowledgement of the work of others, would take some time to determine. However, there is clear evidence that some passages involved plagiarism as the following comparison with sentences found in Adam Smith's classic *The Wealth of Nations* (first published in 1776)[109] demonstrates:

Chapter IV — Road building in Cork, 1834-46

ADAM SMITH	EDMUND LEAHY
The Wealth of Nations, Book V, Chapter I, Part III, Article I (page 317)	*A Treatise on Making and Repairing Roads,* (pages 14 and 15)
But what we call the cross-roads, that is, the far greater part of the roads in the country, [France] are entirely neglected, and are in many places absolutely impassable for any heavy carriage.	But the cross roads, which form by much the greater part of the roads of the country [France] are very much neglected, and are in many places quite impassable for heavy carriages.
In some places it is even dangerous to travel on horseback, and mules are the only conveyance which can safely be trusted.	In some places they are so bad that riding is unsafe with any but the horses or mules accustomed to their rugged and rutty surfaces.
The proud minister of an ostentatious court may frequently take pleasure in executing a work of splendour and magnificence, such as a great highway …	A Minister, high in office, may take pleasure in executing a work of splendour and magnificence, such as a great highway,
But to execute a great number of little works, in which nothing that can be done can make any great appearance, or excite the smallest degree of admiration in any traveller, and which, in short, have nothing to recommend them but their extreme utility, is a business which appears in every respect too mean and paltry to merit the attention of so great a magistrate.	but to execute a great number of comparatively small works, which excite but little or no admiration, in the traveller, and in short have nothing to recommend them but their utility, is a business, which however important to the interests of society, appears unworthy the attention of one placed in an office of high authority.
Under such an administration, therefore, such works are almost always entirely neglected.	Under such regulations therefore these works are almost always neglected.

In search of fame and fortune: the Leahy family of engineers, 1780-1888

Despite whatever faults it may have had, an interesting tribute to Leahy's book appeared in a letter to the *Irish Builder and Engineer* in 1906 from the county surveyor of Gloucestershire who described Leahy as an "eminent Irishman" and "the most important Irishman who had written on the subject" of road maintenance; the book, he went on, contains "by far the best information of any book I have seen, and what no other work contains".[110] Later still, in 1922, Sidney and Beatrice Webb, the much-respected writers on English local government, thought fit to cite Leahy's book in their discussion of road pavements in London.[111]

REFERENCES

1. Edmund Leahy, *A Practical Treatise on Making and Repairing Roads*, John Weale, Architectural Library, London, 1844, page 55
2. J H Andrews, "Road Planning in Ireland before the Railway Age", *Irish Geography*, Vol 5, No 1, 1964, pages 17-41; for a first-rate account of road development in Cork and Kerry between 1700 and 1830, see David Dickson, *Old World Colony: Cork and South Munster 1630-1830*, Cork University Press, 2005, pages 427-433; see also Colin Rynne, "Connecting Cork" in John Crowley and others (eds) *Atlas of Cork City*, Cork University Press, 2005.
3. Peter O Keeffe, *Ireland's Principal Roads 1608-1898*, National Roads Authority, Dublin, 2003; David Broderick, *The First Toll-Roads: Ireland's Turnpike roads 1729-1858*, The Collins Press, Cork, 2002; see also Peter J O' Keeffe, *The Development of Ireland's Road Network*, paper presented to ICEI, 1973, and *Roads in Ireland: Past and Ahead*, An Foras Forbartha, 1985.
4. Arthur Young, *A Tour in Ireland 1776-1779*, 2 vols, London, 1780
5. 45 Geo III c. 43
6. J H Andrews, "Road Planning in Ireland before the Railway Age", *Irish Geography*, Vol 5, No 1, 1964; for a full picture of the mail coach routes in 1832, see J H Andrews, Map No 69, in *A New History of Ireland*, Vol IX, Maps, Genealogies, Lists, Oxford, 1984
7. *Return of all Sums of Money voted or applied either by way of Grant or Loan in Aid of Public Works in Ireland since the Union*, HC 1839 (540) XLIV
8. *Accounts of All Sums advanced by the Public to Grand Juries in Ireland, during each of the last Ten Years for Mail Coach Roads and Prisons*, HC 1829 Vol XXII (327); NAI, OP 494/25
9. NAI, OP 423/22, 403/11, 423/24
10. An Act for more effectually repairing the Road leading from the City of Cork to the Town of Skibbereen in the County of Cork, and a Branch therefrom communicating with the Town of Kinsale in the said County, 3 Geo. IV, c.cviii
11. *Accounts of All Sums advanced by the Public to Grand Juries in Ireland, during each of the last Ten Years for Mail Coach Roads and Prisons*, HC 1829 Vol XXII (327)
12. 6 Geo.IV, c.xcvi and 7 Geo IV, xxvi
13. *Third Annual Report from the Board of Public Works in Ireland*, HC 1835 Vol XXXVI (76)
14. *Report on the Roads made at the Public Expense in the Southern Districts of Ireland by Richard Griffith, Civil Engineer*, HC 1831 Vol XII (119)
15. Peter J O' Keeffe, "Richard Griffith: Planner and Builder of Roads", in *Richard Griffith Centenary Symposium Papers*, RDS, 1980, page 57; *Report on the Southern District in Ireland*, HC 1829 Vol XXII (153)
16. For an account of the construction of the Skibbereen–Crookhaven road, see Patrick Hickey, *Famine in West Cork: The Mizen Peninsula, Land and People, 1800 – 1852*, Mercier Press, Cork, 2002, pages 51-56.
17. *Report on the Roads made at the Public Expense in the Southern Districts of Ireland by Richard Griffith, Civil Engineer*, HC 1831 Vol XII (119)
18. *Report upon the experimental improvements on the Crown Estates at Pobble O'Keeffe in the County of Cork for 1837*, HC 1837-1838 Vol XLVI (69)
19. *Report of the Commissioners appointed to revise the several Laws under or by virtue of which Monies are now raised by Grand Jury Presentment in Ireland*, HC 1842 Vol XXIV (386); the total has been adjusted to include the mileage in County Tyrone
20. *Report from Select Committee on County and District Surveyors etc* (Ireland) HC 1857 Session 2 Vol IX (270)

Chapter IV — Road building in Cork, 1834-46

21. CC, 29 July 1834
22. ibid
23. *Report of the Commissioners appointed to revise the several Laws under or by virtue of which Monies are now raised by Grand Jury Presentment in Ireland*, Appendix B, Queries transmitted to the County Surveyors, with their Answers, HC 1842 Vol XXIV (386)
24. Edmund Leahy, *A Practical Treatise on Making and Repairing Roads*, London, 1844, page 57
25. *Cork Southern Reporter*, 16 March 1844
26. Abstracts of Grand Jury Presentments were submitted to Parliament each year; see, for example, HC 1835 Vol XXXVIII (220), 1836 Vol XLVII (119), 1837 Vol LI (110), 1836-37 Vol XLVI (207), 1839 Vol XLVII (654), 1840 Vol XLVIII (41), 1841 Vol XXVII (143), 1842 Vol XXXVIII (90), 1843 Vol L (146), 1844 Vol XLIII (194)
27. John Neville, "Grand Jury Laws and County Public Works, Ireland", *Dublin University Magazine*, March 1846, Vol XXVII, No CLIX; Neville, who was a son of the architect of the same name who had carried out some building work at New Birmingham, became Louth County Surveyor in 1840 and served until 1886.
28. NLI, William Smith O' Brien Papers, Ms 436, letter No 1540 of 7 April 1846 from Patrick Leahy, referred to again in Chapter VIII below
29. CC, 28 July 1840
30. CC, 3 March 1840, list of presentments submitted to Cork city grand jury
31. *Cork Southern Reporter*, 16 March 1844
32. CO 48/392, letter of 2 September 1858; OPW General Letter Books (NAI OPW 1/7/2/4 and 1/7/2/5) show that Leahy acted as agent for OPW in 1834-36 in preparing a specification for repair and maintenance works and inspecting and certifying the performance of the contractor who was paid a total of £765 for his work.
33. Edmund Leahy, *A Practical Treatise on Making and Repairing Roads*, London, 1844, pages 161-164
34. *Supplement to the Minutes of Evidence taken before the Commissioners appointed to Inquire into the Occupation of Land in Ireland*, HC 1845 Vol XXI (657), Statement of William Johnston
35. *Report of the Commissioners appointed to revise the several Laws under or by virtue of which Monies are now raised by Grand Jury Presentment in Ireland*, Appendix B, HC 1842 Vol XXIV (386)
36. Abstracts of Grand Jury Presentments were submitted to Parliament each year; see, for example, HC 1835 Vol XXXVIII (220), 1836 Vol XLVII (119), 1837 Vol LI (110), 1836-37 Vol XLVI (207), 1839 Vol XLVII (654), 1840 Vol XLVIII (41), 1841 Vol XXVII (143), 1842 Vol XXXVIII (90), 1843 Vol L (146), 1844 Vol XLIII (194)
37. *Evidence taken before the Commissioners appointed to Inquire into the Occupation of Land in Ireland*, HC 1845 Vol XXI (657); one year earlier, in his *Practical Treatise on Making and Repairing Roads*, he had stated that "upwards of 300 miles of new road in the County of Cork have been designed and executed under our direction" (page 61) but this may have been intended to include roads designed and built by his father in the East Riding.
38. CO 48/392, letter of 2 September 1858
39. *Supplement to the Minutes of Evidence taken before the Commissioners appointed to Inquire into the Occupation of Land in Ireland*, HC 1845 Vol XXI (657), Statement of William Johnston
40. Walter Murphy, *Remarks on the Irish Grand Jury System*, Cork, 1849
41. In most years between 1835 and 1846, the full text of the surveyors' biannual reports may be found in the *Cork Constitution* and *Cork Southern Reporter*, generally in mid- March and late July.
42. Edmund Leahy, *A Practical Treatise on Making and Repairing Roads*, London, 1844, pages 55 - 56
43. CC, 18 July 1839, 28 July 1840
44. CC, 18 July 1839, 28 July 1840
45. CC, 6 March 1841, 16 March 1843
46. CC, 10 March and 16 July 1842.
47. *Cork Southern Reporter*, 14 March 1844, 13 March 1845, and 29 July 1845
48. CC, 18 July 1839, 28 July 1840, 13 March 1845
49. CC, 18 July 1839
50. CC, 28 July 1840
51. CC, 16 July 1842
52. CC, 16 March 1843
53. *Cork Southern Reporter*, 14 March 1844, 13 March 1845, and 29 July 1845
54. NAI, OPW5HC/6/156, Map and Section of the Proposed New Road from Castletown through Teernahillan Gap to Durzey Sound, by E Leahy MSCEI, Civil Engineer, 1836; see also OPW5HC/6/157, 158 and 159
55. R A Williams, *The Berehaven Copper Mines*, The Northern Mine Research Society, Sheffield, 1991, pages 39, 56 and 101; Lady Georgina Chatterton, *Rambles in the South of Ireland during the Year 1838*, Vol I, London 1838, page 73

In search of fame and fortune: the Leahy family of engineers, 1780-1888

56. NAI, OPW 1/1/4/11, Public Works Letters No 11, pages 355 and 356
57. NAI, OPW 1/1/4/12, Public Works Letters No 12, pages 17, 19 and 62
58. ibid, pages 187, 196 and 219; NAI, OPW5HC6/160, Map … to accompany report made to Commissioners of Public Works by E Russell, June 1837
59. NAI, OPW 1/1/4/12, Public Works Letters No 12, page 306
60. NAI, OPW 1/1/4/13, Public Works Letters No 13, page 301
61. NAI, OPW 1/1/4/15, Public Works Letters No 15, page 163
62. ibid, pages 204 and 308
63. Daphne du Maurier's novel *Hungry Hill*, first published by Victor Gollancz in 1943 and based loosely on the Puxley family and their involvement in the Allihies mines, attributes to Copper Jack Brodrick (John Puxley) the thought that while "there was constant talk of a new road being built,… there the matter ended, like everything else in the country, and never a penny would come from the Government for the improvement of the roads".
64. CC, 5 March 1840
65. CC, 28 July 1840
66. *Cork Southern Reporter*, 23 July 1845
67. NAI,CSORP 184/W/15624, letter of 11 November 1843 from Edmund Leahy and subsequent correspondence
68. *Cork Southern Reporter*, 16 March 1844
69. CC, 12 March 1836
70. NAI, OPW5HC/6/385, Map and Section of an Intended Line of Road from Cork to Kinsale, between Pouladuff Road and the Lough Road, 1835
71. CC, 21 March 1835
72. CC, 5 March 1840, 9 March 1841
73. CC, 3 March 1840
74. CC, 12 March 1836; Con Foley, *A History of Douglas*, second revised edition, Cork, 1991
75. CC, 17 March 1838; *Cork Southern Reporter*, 17 March 1838; for Leahy's drawings, at a scale of 5 inches to 40 perches, showing the new road extending east as far as Dunkettle Bridge, see NAI OPW 5HC/6/168.
76. *Ninth Annual Report from the Board of Public Works in Ireland*, HC 1841 Vol XII (252)
77. NAI, CSORP 1840/A/ 13564, letter of 17 October 1840 from Patrick Leahy
78. CC, 30 July 1840
79. Mrs McCall was the owner of the rented house at Glenville occupied by the Leahys.
80. NAI, CSORP 1840/A/ 13564, note of 20 October on letter of 17 October 1840 from Patrick Leahy
81. CC, 16 July 1842
82. CC, 12 March 1846; *Cork Southern Reporter*, 21 March and 27 July 1844; the bridge was designed and built by Leahy's successor, Sir John Benson, whose short paper on the subject to the Institution of Civil Engineers of Ireland in November 1849 was published in *TICEI*, Vol IV, Part I.
83. CC, 10 March 1836
84. CC, 16 July 1842
85. *Cork Southern Reporter*, 29 July 1845
86. CC, 12 March 1839, 20 July 1839, 3 March 1840
87. CC, 10 June 1843
88. Antoin O Callaghan, *Of Timber, Iron and Stone: A Journey through time on the Bridges of Cork*, Cork, 1991; Malachy Walsh, "Cork Bridges", *The Engineers' Journal*, August, 1981
89. *Reports on inspection, assessment and rehabilitation of masonry arch bridges and of concrete bridges*, Department of the Environment, Dublin, 1988 and 1990
90. Peter O Keeffe and Tom Simmington, *Irish Stone Bridges: history and heritage*, Irish Academic Press, 1991
91. CO 48/392, letter of 2 September 1858
92. D J O Donoghue, *History of Bandon*, Cork Historical Guides Committee, 1970, page 14
93. CC, 18 July 1839
94. CC, 29 July 1834
95. CC, 27 July 1840
96. Schedules of Applications for Presentments, Summer 1838 and Summer 1845; *Cork Southern Reporter*, 29 July 1845
97. CC, 30 July 1840
98. *Cork Southern Reporter*, 16 March 1844
99. ibid

Chapter IV — Road building in Cork, 1834-46

100. ibid
101. CC,19 and 26 January, 7,9,16 and 23 March 1843
102. CC, 14 March 1844; *Cork Southern Reporter*, 30 March 1844
103. CC, 18, 20 and 22 March 1845
104. CC, 12 March 1846
105. Edmund Leahy, *A Practical Treatise on Making and Repairing Roads*, John Weale, Architectural Library, London,1844; books on a wide variety of subjects bearing the title *"A Practical Treatise"* were common in the first half of the nineteenth century.
106. Quoted in the Preface to *A Treatise on Roads* by Sir Henry Parnell (second edition), London 1838
107. CE, 17 and 20 May 1844
108. CO 48/392, letter of 21 August 1858
109. Adam Smith, *The Wealth of Nations Books IV-V*, Penguin Classics, 1999, page 317
110. *The Irish Builder and Engineer*, 27 January 1906, page 50
111. Sidney and Beatrice Webb, *English Local Government: Statutory Authorities for Special Purposes*, Longmans, Green and Co, London, 1922, page 295

CHAPTER V

PUBLIC BULDINGS IN CORK

"extensive architectural works seldom occur or come under the charge of the surveyor".

Thomas Jackson Woodhouse, 1836.[1]

Engineers "know but little of architecture"
The first Antrim county surveyor, Thomas Jackson Woodhouse, told a Parliamentary Committee in 1836 that "extensive architectural works seldom occur or come under the charge of the surveyor" and when the need arose, as in the case of hospitals, gaols and courthouses, "other parties would be called in" because the county surveyor could not be expected to undertake such work.[2] Woodhouse agreed that a man could be a very good engineer and a very bad architect and he offered the view that engineers "know but little of architecture". A small number of the early surveyors (including Charles Lanyon who was to succeed Woodhouse in Antrim, and John Benson who succeeded Patrick Leahy in Cork) did, in fact, distinguish themselves as architects, both on foot of work which they undertook for their grand juries and commissions from private clients. Generally, however, the grand juries, as well as the authorities in Dublin, seem to have had little confidence in the design skills of the average county surveyor and, for this and other reasons, the design and construction of public buildings cannot generally be listed among the more significant achievements of the early surveyors.

A "general asylum for insane people"
Edmund Leahy's claim in 1858 that he had designed and built "one general asylum for insane people"[3] was completely without foundation. In the 1820s and early 1830s, under an extensive building programme undertaken by the Commissioners for the Erection of Lunatic Asylums, nine new district asylums were provided throughout the country from designs by Francis Johnston and William Johnston Murray.[4] A much older asylum for lunatics from the city and county was already in existence in Cork when the Leahys took up duty there but the two grand juries resisted pressure from the authorities in Dublin to

replace this by a new district asylum, arguing that the existing institution could be made "perfectly sufficient" by the expenditure of a relatively small sum on the construction of additional buildings on the existing site.[5]

Whether or not advice from the Leahys influenced the grand juries' approach is not clear but, in any event, this approach was not accepted by the Government. An Act was eventually passed in August 1845 to establish a new asylum in Cork[6] - the largest in a second series of asylums to be provided by the Office of Public Works and handed over on completion to the grand juries for operation and maintenance. The commission for the design of the new building was not awarded until September 1846, after the Leahys had left their county surveyor posts, and it went to a little-known Cork architect, William Atkins, whose plans were approved in December of that year.[7] County surveyors in the other six areas where new asylums were provided in the 1840s were also passed over in favour of consultant architects, but at Ballinasloe, where only an extension to the existing building was required, the commission went to the surveyor for the eastern division of the county, James Forth Kempster.

Harbours, docks and piers

Edmund's statement in 1858 that he had built "harbours, docks and piers" at various locations on the Cork coast also seems to have had no basis in fact. The record shows that piers, landing places and minor coast roads were constructed at State expense all along the Cork coast by the Commissioners of Irish Fisheries in the 1820s, leading to increased activity in the fishing industry and introducing "other proofs of civilization to the sequestered districts" of West Cork.[8] New or extended piers were built, for example, at Castletownbere, Bantry, Baltimore, Glandore, Ring (near Clonakilty), and Courtmacsherry, usually at a cost of £500 to £800[9], but most of these did not become the responsibility of the grand jury until 1854.[10] At Glandore, however, Edmund was engaged in a personal capacity by the Commissioners of Public Works in January 1840 to inspect and report to them at two-monthly intervals on the progress of the works which were being carried out for them by a local contractor, and he was paid a fee of £3 for each inspection.[11] Apart from assignments such as this, and routine maintenance and improvement works undertaken on an agency basis for the Commissioners, there is no evidence that the Leahys were directly responsible for any work on piers and landing places during their years in Cork and there are no references in their biannual reports to the grand jury to proposals for works of this kind.

Bridewells and sessions houses

The claim by Edmund Leahy in 1858 that he had "designed and made …15 local prisons and sessions houses" appears to have been a gross overstatement of his work in this field. Only a few years before the Leahys' arrival in Cork, the Inspectors General of Prisons had reported that "the bridewell system of the county of Cork is unequalled".[12] With the aid of Government loans of £277 in each case,[13] new gaols had been built in 1825-26 at Midleton, Clonakilty, Macroom, Kanturk, Skibbereen, Bantry and Mallow, all according to the same plan and connected to the local sessions houses. The buildings were "on a handsome scale and formed a very conspicuous object in their towns, calculated to excite a feeling of respect connected with the administration of justice at the quarter sessions".[14] Gaols or bridewells, to accommodate those convicted of petty offences and those jailed for the recovery of debt, had been built at Mitchelstown, Millstreet, Dunmanway and Bandon. The latter cost about £650[15] and had "separate yards, dayrooms and cells (eight in number) for males and females, with a spacious insulating passage surrounded by a high exterior wall".[16] Bridewells had also been provided at Fermoy, Rosscarbery, Cobh, Youghal, Charleville and Kinsale by 1830, although some were considered defective by 1832.[17] In 1835, the Inspectors General were again able to report in glowing terms on the condition of the County Cork bridewells[18] but Edmund Leahy took a different view, noting in his first report to the grand jury in July 1834 that many of the sessions houses and bridewells in the West Riding were in bad condition, and adding that this was inexcusable since the buildings were so recent.[19]

In addition to building a substantial number of bridewells in the 1820s, the Cork grand jury built new sessions houses (local courthouses) in Midleton, Bantry, Clonakilty, Skibbereen, Kanturk, Mallow and Macroom in the same period, with the aid of advances of £462 in each case from the Government.[20] In all, over £26,000 was spent on the building of bridewells and sessions houses in the county in the twenty years ending in 1843[21] but most of that expenditure had been incurred before 1834 when the Leahys took up duty as county surveyors. Improvement and maintenance works at the existing buildings were certainly undertaken by the two surveyors and, while he complained that there were few tenderers for the work, Edmund was able to report by 1842 that the courthouses generally were in good repair except the building at Bandon[22] where repairs were said to be essential one year later.[23] However, his report about the good condition of some of the other courthouses should, perhaps, be viewed in the light of his attitude to the grand jury's suggestions that he should transfer his own office to the building at Bantry: his life would be in danger, he told the jury in March 1845, if he had to work in such a damp and unwholesome building.[24]

Some new buildings were provided by the Leahys. Proposals from Edmund for a bridewell at Skull, estimated to cost £200, were approved in March 1845 but a proposal to provide a similar building at Ballineen was rejected.[25] At the same Assizes, a proposal by Patrick to build a new bridewell at Kinsale at a cost of £499 was approved[26] and the building was almost complete a year later. However, although a scheme to provide a new building at Cobh, at a cost of £470, was also approved in 1845, work had still to commence in March 1846 because a site could not be acquired.[27] In general, therefore, the evidence suggests that the Leahys were responsible for little more than routine maintenance work on the county's extensive stock of bridewells and sessions houses and that they contributed no more that two or three of the total of some thirty-five buildings which were required for the administration of justice in the county.

There is evidence also that the Commissioners of Public Works and their architects were not particularly impressed by Patrick Leahy's efforts in the field of building design. In April 1843, after the Under-Secretary had directed that a new sessions house and bridewell should be provided at Youghal, Leahy's plans were rejected by the Commissioners; they went on to tell him that they were sure that "a person of your qualifications can have no difficulty in effecting the required alterations" and instructed that the object should be to give sufficient accommodation, without resorting to a great outlay, combining, as much as possible, economy of arrangement with simplicity of design".[28] Two months later, Leahy was told that his next set of revised plans was a big improvement but still needed "much consideration and improvement in several particulars", and he was given detailed recommendations and notes drawing attention to serious errors and inconsistency in the drawings and numerous unsatisfactory aspects of the design.[29] After a third set of plans had been rejected because they did not embody all of the recommendations already made, the fourth and final set was sent to the Lord Lieutenant for approval towards the end of July.[30] That, however, was not the end of the saga: when the plans were presented to the grand jury in March 1844, with a cost estimate of £3,000, the jury decided that the new buildings were not necessary and they confirmed this decision a year later, even though a construction contract had been entered into by Leahy at that stage.[31]

The city and county gaols

Before the Leahys' arrival in Cork, the massive limestone buildings of the county gaol on the city's Western Road (only the portico and outer wall of which now remain) were improved and extended, a new city gaol had been constructed and later enlarged, and a house of correction was provided; loans of more than £55,000 to finance these undertakings were provided by the Government between 1819 and 1829.[32] All of them were designed by the brothers James and George Richard Pain which makes a nonsense of Edmund's claim in 1858 that he had built "one general county prison" during his time as county surveyor. Patrick, rather than Edmund, was asked by the grand jury to inspect the county gaol in 1835 but reported that only a relatively small outlay was needed to carry out essential repairs.[33] Edmund, in 1842, reported to the grand jury that he had erected a flour mill at the county gaol, powered by a tread-wheel, but in spite of the savings which he anticipated, the mill was not being used by the prison authorities.[34] That seems to have been his only involvement with the gaol during his years in Cork.

The Cork Courthouse

When Patrick Leahy took up duty in 1834, he noted that work on the new Cork Courthouse was "under the supervision of a gentleman of high eminence and celebrity" (a reference to George Richard Pain),[35] and one year later, the building was expected to be ready for the Summer Assizes having cost over £23,000.[36] However, as early as January 1837, serious dissatisfaction arose about the condition of the courthouse and the facilities available there for the conduct of county business, with judges and lawyers complaining that, like all of Pain's buildings, it was not properly ventilated, that there were excessive draughts of cold air and that the acoustics were poor. The city and county grand juries decided against asking Pain "who committed the blunders" to set matters right but, significantly, did not entrust either of the Leahys with responsibility for drawing up proposals for remedial works; instead, they decided to advertise for a competent expert from Dublin to advise as to what should be done[37] and subsequently engaged the prominent architect, Richard Morrison, for the purpose. Both Morrison and Pain attended a meeting of the city grand jury in March 1837 at which there was a long and acrimonious debate on the matter during which Pain attempted to convince the jury that the defects arose from alterations in his original plans which had been made by Morrison's son, William Vitruvius Morrison, and the Cork architect, Sir Thomas Deane.[38] Neither of the Leahys was invited to offer an opinion at that stage on the remedial measures proposed by Morrison and by Pain but they do seem to have undertaken some relatively minor works at the courthouse in subsequent years; in 1841, for example, Patrick told the city grand jury that works to make the roof secure had been completed[39] and, in 1845, Edmund reported that he had made alterations to stop the dangerous draughts, which, apparently Morrison had not succeeded in eliminating.[40]

Although he had never built a county gaol or courthouse, Patrick Leahy offered himself as an expert witness at an inquiry conducted by a Committee of the Privy Council in June 1837 into the hotly contested question of whether Thurles or Nenagh should be selected as the headquarters town of the separate grand jury which was about to be established for the north riding of Tipperary.[41] In support of the case for Thurles, he told the inquiry that, based on estimates prepared by himself and his son, Matthew, the cost of making the town's existing courthouse and bridewell suitable for use as a county courthouse and county gaol would be less than £7,000. This was disputed by the Inspector of Prisons who put the cost of a suitable gaol at Thurles at £14,000, and by the architect, John B Keane, who estimated that the cost of a gaol and a county courthouse would be £15,000.[42] Judge Perrin, Master of the Rolls, who presided at the inquiry, thereupon remarked that it was quite plain that Leahy did not understand the building of gaols or courthouses and advised the counsel who represented Nenagh that there was no need to continue cross-examining him;[43] whatever about his competence as a surveyor and engineer, this was a fitting verdict on Leahy's pretensions to proficiency in architecture.

Chapter V — Public buildings in Cork

REFERENCES

1. *Report from the Select Committee on County Cess (Ireland)*, HC 1836 Vol XII (527)
2. ibid
3. CO 48/392, letter of 2 September 1858 from Edmund Leahy to the Colonial Secretary
4. Frederick O'Dwyer, "The Architecture of the Board of Public Works 1831-1923", in *Public Works, The Architecture of the Office of Public Works, 1831-1987*, Architectural Association of Ireland, 1987
5. David M Nolan, *The County Cork Grand Jury, 1836-1899*, MA Thesis, UCC, 1974; *Copies of Correspondence between Her Majesty's Government in Ireland and the Grand Juries assembled at Summer Sessions 1844, on the subject of providing additional accommodation for Pauper Lunatics*, HC 1844 Vol XLIII (603); CC, 14 March 1844
6. 8 & 9 Vict., c. 107
7. *Report of the Commissioners for Inquiring into the Erection of District Lunatic Asylums in Ireland*, HC 1856 Vol LIII (9); the asylum at Cork was completed in 1853 and was in use until 1991
8. *Tenth Report of the Commissioners of the Irish Fisheries*, HC 1829 Vol XIII (329), Appendix No 10
9. *First and Second Reports of the Commissioners of Inquiry into the state of the Irish Fisheries*, HC 1837 Vol XXII (77) (82)
10. Grand Jury (Ireland) Act,1853, 16 & 17 Vict., c. 136
11. NAI, OPW 1/1/4/19, Public Works Letters No 19, page 185
12. *Eighth Report of the Inspectors General on the General State of the Prisons of Ireland*, HC 1830 Vol XXIV (48)
13. *Return of all Sums of Money voted or applied either by way of Grant or Loan in Aid of Public Works in Ireland since the Union*, HC 1839 Vol XLIV (540)
14. *Fifth Report of the Inspectors General on the General State of the Prisons of Ireland*, HC 1827 Vol XI (471)
15. *Accounts of all Sums advanced by the Public to Grand Juries in Ireland during each of the last Ten years for Mail Coach Roads and Prisons*, HC 1829 Vol XXII (327)
16. *Seventh Report of the Inspectors General on the General State of the Prisons of Ireland*, HC 1829 Vol XIII (10)
17. *Tenth Report of the Inspectors General on the General State of the Prisons of Ireland*, HC 1831-32 Vol XXII (152)
18. *Thirteenth Report of the Inspectors General on the General State of the Prisons of Ireland*, HC 1835 Vol XXXVI (114)
19. CC, 29 July 1834
20. *Return of all Sums of Money voted or applied either by way of Grant or Loan in Aid of Public Works in Ireland since the Union*, HC 1839 Vol XLIV (540)
21. *Evidence taken before Her Majesty's Commissioners of Inquiry into the state of the Law and Practice in respect to the Occupation of Land in Ireland*, Statement of William Johnston, Secretary of the Grand Jury of County Cork, HC 1845 Vol XXI (657)
22. CC, 16 July 1842, half-yearly report from Edmund Leahy
23. CC, 29 July 1843, half-yearly report from Edmund Leahy
24. *Cork Southern Reporter*, 13 March 1845
25. *Cork Southern Reporter*, 29 July 1845
26. *Cork Southern Reporter*, 13 March 1845
27. CC, 12 March 1846
28. NAI, OPW 1/1/4/28, Public Works Letters No 28, page 74
29. ibid, page 231
30. ibid, pages 270, 294 and 315
31. NAI, CSORP 1843/S/1090, 3174, 3966, etc; CSORP 1844/S/4328; CC 29 July 1843 and 14 March 1844, half-yearly reports by Patrick Leahy; *Cork Southern Reporter*, 13 March 1845
32. *Accounts of all sums advanced ... between 1819 and 1829 for public works in Ireland and for the employment of the Poor*, HC 1829 Vol XXII (317)
33. CC, 21 March 1835
34. CC, 10 March 1842, half-yearly report from Edmund Leahy
35. Report to the Grand Jury, CC, 29 July 1834
36. CC, 21 March 1835
37. CC, 31 January 1837
38. CC, 16 March 1837
39. CC, 6 March 1841
40. *Cork Southern Reporter*, 13 March 1845, half-yearly report from Edmund Leahy

41. NAI, PCO MB 8, Privy Council Minute Book; NAI, Council Office Papers, Box 44, report of the Committee of the Privy Council which sat on 16 June 1837; see also Donal A Murphy, *The Two Tipperarys*, Relay Publications, Nenagh, 1994
42. After the Privy Council ruled in December 1837 that Nenagh should be the new county town, Keane was commissioned to design a new County Gaol and a new County Court House there; the buildings were completed by 1843.
43. *Clonmel Advertiser*, 21 June 1837

CHAPTER VI

PRIVATE PRACTICE, 1834-46

"Private practice helps to get a better class of man to undertake the duties of county surveyor".

Charles Lanyon, 1857[1]

"many advantages connected with a private practice"
While county surveyors were forbidden by the Grand Jury (Ireland) Act 1833 to receive payments from contractors to the grand jury, there was nothing in that Act, or in the 1836 Act which replaced it, to prevent them from engaging in private practice or, indeed, from accepting fees for assignments obtained from other grand juries or public bodies. The majority of the early surveyors appear to have taken advantage of this liberal regime and a number of them became better known for their private work than for their public duties. Most of the surveyors confined their work for private clients to projects in the field of engineering, but a significant number of them also engaged so extensively in architectural work as to cause continuing controversy among the membership of the Institute of the Architects of Ireland. Charles Lanyon, for example, who took up duty as county surveyor in Antrim in 1836 (after two years in County Kildare) was one of the most important and prolific architects in Ireland in the 1850s and 1860s[2] and was responsible for many of Belfast's prominent public buildings, including the gaol, courthouse, asylum and university,[3] as well as a large number of churches and country houses. John Fraser, the surveyor for County Down, freely admitted in 1841 that he earned as much from his private engineering practice as he was paid for his official duties.[4] And Sir John Benson, who succeeded Patrick Leahy as county surveyor in Cork, had an extensive private practice as an architect, both in the Cork area and elsewhere in Ireland, providing the buildings for the Great Exhibition in Cork in 1852 and for Dublin's Industrial Exhibition in the following year.

The fact that so many of the surveyors engaged extensively in private practice in the early years is not surprising when account is taken of their dissatisfaction with their salary level and with the lack of any provision for payment of travelling and subsistence expenses. In addition, allowance must be made for the fact that there was still a scarcity of private

Chapter VI — Private Practice, 1834-1846

engineering practices in the Ireland of the 1830s and 1840s, especially outside the capital. One indication of this is the high proportion of the membership of the Civil Engineer's Society of Ireland (later the Institution of Engineers of Ireland) accounted for by the county surveyors and other holders of public office. Of the 36 founder members of the Society in 1835, no less than 31 were county surveyors, with four others employed elsewhere in the public service. By 1849, despite the expansion in public works programmes and the railway boom, at least 50 of the 85 members of the Institution were persons who were employed, or were shortly to be employed, in public positions.

It was not until August 1861, when increased salaries for county surveyors were authorised, that the law was amended to allow a grand jury to direct their county surveyor not to engage in private professional practice.[5] This change was made on the advice of a Select Committee of the House of Commons which had given "anxious attention" to the issue in 1857.[6] By then, the fact that the official salary was, in several cases, only a small part of the income of the surveyor was attracting critical media and other comment and a *Dublin Evening Post* editorial went so far as to suggest that the reason the office was anxiously sought after was "because it is generally the means of obtaining a very large amount of private practice".[7] However, the surveyors themselves strongly defended the status quo in their evidence to the Select Committee[8]. Charles Lanyon, for example, saw "many advantages connected with a private practice" and argued that it helped "to get a better class of man to undertake the duties of county surveyor". Henry Brett, who served successively in Counties Laois, Mayo, Waterford and Wicklow, thought that "an efficient officer is decidedly improved by allowing him to have private practice, provided it is not to a very great extent". And Alexander Tate, a Dublin District Surveyor, claimed that "where private practice is enjoyed to a large extent, you will find concurrently with it county works efficiently and satisfactorily managed".

Prior to 1861, the law as well as being silent on the question of private practice, actually encouraged to some extent the employment of county surveyors on public projects outside the area of responsibility of the grand juries. For example, the Commissioners of Public Works were empowered in 1834 to employ the surveyors to repair the mail coach roads and the extensive network of other roads built by various government agencies in the 1820s and early 1830s, with the cost of the work being recouped from the grand jury. But the Commissioners decided that no additional payments should be made to the surveyors themselves for this work and the 1836 Select Committee on County Cess saw no reason for any change in the position.[9] Payments for expenses were, however, allowed where new works were involved. Edmund Leahy, for example, received about £6 from the Commissioners in 1835 on foot of his supervision and inspection of work on the Skibbereen-Crookhaven road[10] and in 1836 his father earned £21 for a survey of a proposed new road in County Wexford.[11] Some of the other surveyors earned far more substantial amounts for the inspection and superintendence of public works on behalf of the Commissioners: Charles Lanyon, in Antrim, earned £341 in this way between 1834 and 1839; Henry Brett in Mayo earned £261 in the same period; while in Galway, Henry Clements earned £328; and in Kerry, Henry Stokes received £414.[12] In addition, a few of the surveyors profited handsomely from commissions to design major public buildings. However, surveyors generally, with the possible exception of Charles Lanyon, are unlikely to have earned as much from private practice, on a continuing basis, as did the Leahys whose total income from this source must have far exceeded their official salaries.

"I do all the private business I can get"
In response to an official enquiry in 1841 about the extent of his private practice, Patrick Leahy said that he could do very little private business because of the extent to which his time was devoted to the duties of his onerous posts as county and city surveyor and claimed that his only private business since coming into office related to three law suits.[13] Edmund dealt with the same enquiry by stating that he had held no other office since his appointment as county surveyor, but freely admitted: "I do all the private business I can get, which is very little". In fact, during the whole period of their service as county surveyors in Cork, Patrick and Edmund Leahy carried on a thriving civil engineering practice, operating from offices at 20 South Mall, Cork, under the title "Messrs P & E Leahy", Civil Engineers. They were involved in civil engineering work of all kinds throughout Munster during the 1830s and early 1840s and occasionally took on construction contracts. Matthew, when he qualified as an engineer, was associated with the family firm and Denis was actively involved at intervals, particularly in the 1840s.

The building at South Mall, Cork where the offices of Messrs P & E Leahy, Civil Engineers were located in the 1840s.

Chapter VI — Private Practice, 1834-1846

The extent to which the Leahys engaged in private practice gave rise to criticism. For example, at the city assizes in March 1838 several of the grand jurors condemned Patrick Leahy's neglect of the roads and attributed this to his involvement in other activities.[14] Leahy responded that "it could be nothing to anybody what his property was, or what other situations he held, provided he kept the roads in order and gave satisfaction". A Cork magistrate, in sworn evidence to the Devon Commission in 1844, claimed that "the taxation of the county will always be increasing, so long as the engineer of the county has the privilege or the power of a private office".[15] He went on to give an instance of the kind of abuse that was occurring:

> You want a road to a certain place, and I want a road to a certain place. We differ as to the line of road to be taken. If you want your line to be advocated by the county engineer, you will employ that person to survey it in his private capacity; and therefore, when I go with my road before the cess-payers, and you go with your road, the magistrates and the cess-payers naturally turn to the county engineer - "Sir, which is the best line of road between these two gentlemen?" So long as the county engineer is allowed to carry on business on his private account, it is in his interest to be continually making roads and bridges, and thereby increasing the tax.

Patrick Leahy hotly denied these allegations and told the Commissioners that he could mention circumstances that would "deteriorate the importance to be attached to the evidence", although he refrained from mentioning what circumstances he had in mind.[16] He argued that the private business of a county engineer could not have anything to do with the costs falling on the ratepayers and insisted that "I never received payment from private individuals for surveying a road which afterwards became a public work since I became a county engineer". Edmund Leahy also refuted the charge that he had been receiving fees for surveys of new roads to be made by the grand jury, although he added that "it would be only just if it were true".[17] He claimed that he had "altogether declined such surveys" (presumably within his own West Riding) and had earned only about £2 a year, on average, for road surveys and plans between 1834 and 1844. The Leahys were not, however, as innocent as they claimed to be and their fee for surveying a new line of road was publicly known to be £6 per mile.[18] They attempted to fudge the issue by drawing a distinction between their personal activities and those of "Messrs P & E Leahy" and, of course, the fact that the public works business of the grand jury was divided between them allowed them to channel assignments to one another (on paper, if not in reality) as the circumstances required. Even in the case of one of the most substantial and high-profile road projects which was recommended to the grand jury by Patrick Leahy in 1844 – the construction of a new road running westwards from Cork city to Carrigrohane and Ballincollig – Edmund not only admitted that it was he who had been employed to survey the line, but stated that he was very proud to have done so.[19]

In search of fame and fortune: the Leahy family of engineers, 1780-1888

The family relationship between the two surveyors was a major source of criticism from an early stage: according to one observer, in works affecting both ridings, there was only "one mind to bear upon the new work, the father seldom wishing to differ from the son".[20] In March 1839, the county grand jury found it necessary to order Patrick Leahy to give his services solely to the East Riding but he ignored this and had to be reprimanded again in July at the Summer Assizes.[21] Five years later, a grand juror complained that "if a road were to be brought before the East Riding, it was privately surveyed by the officer for the West Riding" and *vice versa*,[22] thus enabling the surveyors to increase their earnings considerably. On that occasion, the grand jury considered whether the surveyors' salaries should be withheld if this happened again, but drew back from taking firm action: instead, they simply recommended that the surveyors should no longer give their services for private surveys of works which might be brought before the presentment sessions.[23]

Very little of the work undertaken by the Leahys for private clients involved architectural projects - indeed, it seems likely that they knew "but little of architecture", as was true of the majority of engineers, according to the Antrim surveyor, Thomas Jackson Woodhouse.[24] There is no record of any building of significance designed by them, either in Cork city or county or elsewhere. The major projects arising in the city during the Leahys' years in Cork were carried out either by the brothers James and George Richard Pain or by the Deane brothers - Sir Thomas and Kearns Deane - assisted by Benjamin Woodward. It was Sir Thomas Deane who was selected by the Commissioners of Public Works in December 1845 to prepare the plans for the new Queen's College in Cork even though the 1845 Act, under which the college was established, enjoined the Commissioners to consider first the possibility of employing the county surveyor for the purpose. The Antrim surveyor, Charles Lanyon, was, in fact, commissioned to design the Belfast college but the Leahys could hardly have been seriously considered for the corresponding assignment in Cork, in view of their lack of experience in architecture and the difficulties which had arisen by that time in their relationship with officialdom.

Churches
Notwithstanding the family connection with the Catholic Church, the Leahys do not appear to have had any involvement in the building or refurbishment of catholic churches during their years in Cork. Patrick Leahy, however, was employed for a few years by the Ecclesiastical Commissioners for Ireland, a body established by statute in 1833 to take over responsibility from the Board of First Fruits for the maintenance and improvement of the property of the established Church of Ireland.[25] The Board had financed the construction of a large number of new churches between 1810 and 1833, with the aid of grants of over £1 million from the Government, but maintenance of existing churches seems to have been neglected: "churches became in almost every instance out of repair" and in some cases "verged towards total ruin".[26] To deal with the large number of applications made by

incumbents for grants to repair their churches, to build new churches, and to build or repair glebe houses, the Commissioners appointed a number of Provincial Architects to inspect and report on the nature and extent of the repairs needed to the different church buildings, to recommend those which were in greatest need of repair and to provide detailed estimates of the cost of the works.[27]

Patrick Leahy was employed as Provincial Architect for Munster in 1835 and 1836 in what might be seen as an ecumenical gesture towards one whose eldest son had already been ordained priest in the Catholic Church. Leahy earned about £200 a year in fees from the assignment in those years[28] and since his fees were calculated at 4% of the cost of the works for which he was responsible,[29] he must have been involved in an annual programme of works valued at about £5,000. In the absence of records of the activities of the Ecclesiastical Commissioners, it is impossible to determine whether Leahy was responsible for the design of any of the seven new churches built in the province in the period 1833-43 at a cost of £500-£600 each.[30] The available evidence, however, suggests that he was engaged primarily on church reconstruction and repair projects. The fact that his rate of fee at 4% was less than the rates paid to other architects engaged by the Commissioners (James Pain, Alexander Deane, Joseph Welland and Frederick Darley got from 5% to 9%) implies that the projects in which he was engaged did not involve design or significant new work. An advertisement of his in the *Cork Constitution* in July 1836 sought tenders for the repair of some 30 churches in the Dioceses of Cork, Cloyne and Limerick in accordance with specifications which were available for inspection in his Cork offices[31] but, without the original records, it is not possible to distinguish which projects may have been completed by 1837 when Leahy "declined the continuance of the office because of its heavy duties".[32] He was succeeded as Provincial Architect by James Pain, one of the most distinguished Cork architects of the early nineteenth century.

Poor Law Valuation

The Poor Relief (Ireland) Act, 1838[33] set out to establish a system "for the more effectual relief of the destitute poor" with relief to be provided only in workhouses operated by poor law guardians who were to be elected in 130 separate poor law unions. To finance the system, arrangements had to be made quickly for the valuation of all land and buildings in each union as the basis for a new poor rate. The Poor Law Commissioners, who were responsible at national level for putting the system into place, issued instructions in March 1839 suggesting that "professional or paid valuators" would be necessary in most cases, that the valuation should be made "by persons practically conversant with the letting value of land and houses in each union", and that the guardians in each area should advertise for tenders from persons who were duly qualified.[34] There is no evidence that Patrick, Edmund or Matthew Leahy sought to be appointed as valuator for any of the unions in the Cork area. However, Denis Leahy, describing himself as a professional valuator and civil engineer, submitted tenders to the guardians in Tipperary and in Thurles and was successful in both cases, being among the earliest of the poor law valuers to be appointed in the country. There was no suggestion that local politics or any improper considerations influenced these particular appointments although one of the Assistant Poor Law Commissioners reported in 1844 that, while "appointments were, in general, very honestly made in the north ... I should hardly be disposed to make so favourable a statement of the state of things in the south".[35]

In Tipperary, Leahy's appointment was made by majority vote of the guardians at a meeting on 25 May 1839, in preference to ten other applicants, one of whom had sent in a proposal for valuing the union at the same rate.[36] His proposed valuations for 185,562 acres of land and other property in the union were submitted to the guardians in August 1840 and, after objections and various submissions from ratepayers had been considered, the task was declared to be complete at the end of September. It appears that Leahy's work gave general satisfaction at the time but the individual valuations may have been up to 25% below prevailing letting values.[37] Some years later, the Senior Assistant Poor Law Commissioner was critical of the performance of surveyors generally in making the valuations: "surveyors we find are not much accustomed to value as some other men; they are more surveyors than valuers".[38]

When the appointment of a valuator for the 125,139 acres of land and other property in the Thurles union came before the guardians there on 22 June 1839,[39] there were ten tenders for consideration. Initially, the guardians seemed to have doubts about Denis Leahy's capacity to carry out the work satisfactorily in view of his prior commitments in Tipperary but he assured them that he would personally inspect and value all of the property in the union, that he would not need an assistant, and that he would have the valuation completed by 25 March 1840. On the basis of these assurances, he was appointed as valuator by majority vote. His fee was to be £1 19s per 1,000 acres and he was allowed 12 months to complete the assignment, with his father and his brother Edmund acting as securities for timely and proper completion. In the event, the complete valuation of the union was not available until 1841. It was thought to be a very low one, perhaps 25% below letting values as in Tipperary.

Building the workhouses
In addition to the establishment of a new valuation and rating system, a crash programme of workhouse construction was necessary to enable the poor law system to be brought into operation. Irish architects and engineers had expected to earn substantial fees from the design of the workhouses but they were outraged to learn that all of the 130 new buildings were to be designed by George Wilkinson, a 25-year-old English architect who was employed directly by the Poor Law Commissioners for the purpose.[40] The exclusion of Irish architects from the workhouse programme led them to form the Institute of the Architects of Ireland

to protect their future interests. But for others, including engineers and contractors, there were still opportunities to be grasped as the building programme got under way.

By May 1839, Wilkinson had prepared a series of standard plans for workhouses capable of accommodating from 300 to 1,300 inmates[41] and, following a competitive tendering process, 64 workhouses had been contracted for one year later.[42] Three of these were to be built by Denis Leahy: under a contract dated 12 August 1839 he was to build a workhouse for 700 inmates in Tipperary at a cost of £6,240; in March 1840 he contracted to erect a building for 1,000 inmates at Nenagh at a price of £8,320; and in July 1840 he was appointed to be the contractor at Thurles where places for 700 paupers were to be provided at a cost of £5,840.[43] A commentator in the *Tipperary Free Press*, noting the Nenagh appointment, remarked that "from the ability he is said to possess, I have no doubt but he will give ... every satisfaction".[44] Meanwhile, Patrick and Edmund Leahy (operating as Messrs Leahy) won the contracts for two further workhouses. At Lismore, they were to provide a building for 500 inmates at a cost of £5,500, and at Mallow accommodation was to be constructed for 700 inmates at a cost of £6,000.[45] It seems that the Leahys were the only county surveyors who were awarded workhouse contracts, although Henry Stokes, their Kerry colleague, submitted an unsuccessful tender for the buildings at Tralee.[46]

Construction work at Tipperary, Nenagh, Thurles, Lismore and Mallow seems to have been carried out reasonably satisfactorily by the Leahys, despite the unusually severe and protracted winter of 1840-41, and a strike by masons and labourers at Nenagh which led to threats to the life of Denis Leahy and his steward. The Poor Law Commissioners were able to declare all five of the workhouses fit for the reception of paupers at various dates between June 1841 and April 1842.[47] A report submitted to Parliament in 1845 found that construction costs at workhouses generally were about 12.5% in excess of "the proper costs", bearing in mind prices paid for other construction work in the areas concerned; there was no specific criticism of the cost of the buildings erected by the Leahys although their building at Lismore had cost almost 30% more than the contract price.[48] Another report to Parliament in 1844 indicated that there were no serious complaints from the guardians about the quality of the workmanship in the Leahys' buildings although the walls were said to be damp in Nenagh and the well had failed, leading to the withholding of part of the contract price and threats of legal proceedings.[49] However, entries in the minute book of the Thurles guardians suggest that there were some serious defects in their workhouse: the boilers in the kitchen were considered dangerous, the chimney flues were not brick-lined as they should have been, the roof was leaking, the doors defective, and the cesspools and drains badly constructed.[50]

Messrs Leahy contracted on 30 June 1841 to provide another 700-place workhouse at Listowel at a cost of £5,980.[51] In April 1842, with 81 of the 130 workhouses already completed, and another 30 well advanced, the Poor Law Commissioners were advised by

In search of fame and fortune: the Leahy family of engineers, 1780-1888

Lismore workhouse constructed by Messrs Leahy in 1840-42 (Lismore Union Workhouse, General view from the North-East, R Armstrong, 1842, PD 1956TX, courtesy of the National Library of Ireland).

their architect that "less satisfactory progress has been made at Listowel than ought to have been".[52] The works were "proceeding very unsatisfactorily" according to the architect, and the Leahys had been warned "of the intention of forthwith carrying on these works by the Commissioners unless proper steps shall be immediately taken by the contractors for completing the building". But none of this seemed to have had any effect on the Leahys' performance and eventually it was necessary for the Commissioners to take over the works and to complete the building themselves by direct labour.[53] As a result, the workhouse at Listowel was not opened until February 1845, making it one of the last of the workhouses to be brought into use.[54]

Tralee Ship Canal

A quay at Blennerville, built about the middle of the seventeenth century, still served as the port of Tralee in the first half of the nineteenth century and a flourishing import and export business was conducted there by coasters trading with ports in Britain and Ireland.[55] With agricultural prosperity in the area, expansion and commercial growth in the town itself, and the silting up of the quay and channel at Blennerville, interest developed in the idea of providing a ship canal capable of bringing ships of 200-300 tons from Tralee Bay into a basin nearer the town centre. In 1825, preliminary plans for such a canal were prepared for a local group by Richard Griffith,[56] and it was he who prepared the later plan and cost estimate of £8,335 on foot of which the project was authorised by Parliament in July 1828.[57] Harbour Commissioners were appointed to carry out the scheme which was to be financed by local contributions and harbour dues, borrowing from the Exchequer Loan Commissioners, and an Exchequer grant. The canal was to be about $1^1/_2$ miles long, terminating in a basin 400 feet long, 150 feet wide, and 15 feet deep, and it was to be completed within eight years of the passing of the Act.

Construction work was assigned to local contractors and had begun by 1832.[58] Some three years later, work was stopped with £8,000 already spent and the realisation that local springs would not provide sufficient water for an impounded canal, with only one lock at the entrance, as envisaged by Griffith. The plans were then revised to provide for a tidal canal, with greater depth and width, and with one pair of gates to retain the water between tides. However, there was difficulty in raising the funds required to resume work and the partly completed canal was virtually abandoned, becoming a ditch of stagnant water, the subject of great local controversy and of repeated representations to the Government by Daniel O'Connell, among others. In April 1840, the Commissioners of Public Works took possession of the works and, shortly afterwards, invited tenders for completion of the project.[59] When the tenders were opened after the closing date of 10 June, the Commissioners were faced with a dilemma on finding that the lowest tender was submitted by Matthew Leahy who was then only 22 years old and had no experience of comparable works.[60] As Patrick Leahy, along with Edmund, had been named as Matthew's securities to the extent of £500 each for proper completion of the contract, the Commissioners wrote to Patrick indicating that they were unwilling to accept the tender but went on to say that –

> if you still think it advisable that your son should engage in such a contract, they will consider the tender and if accepted, the security offered will be, in the usual course, referred to their solicitor for his approval, when evidence of property must be given by the sureties independent of any life interest from salaries of Office".[61]

The Leahys, collectively, appear to have been determined to secure the Tralee contract – and may well have submitted the tender in Matthew's name because of the fear that the simultaneous involvement of the other three members of the family in the construction of six workhouses would militate against their prospects of winning it. But fresh difficulties arose when it emerged that none of them held property which would be acceptable as security: all they could offer was a property in Thurles belonging to Denis, but this was rejected as it was held only on a yearly tenancy.[62] Finally, after special pleas by Patrick, the Commissioners relaxed their requirements for securities, telling him that they were "inclined to believe that you and your son will proceed in a proper manner with the work".[63]

The contract with Matthew Leahy, valued at £8,520, was signed in September 1840 and Joseph Coneys, who had successfully overseen the construction of the masonry abutments and central pier for the suspension bridge at Kenmare, was transferred to Tralee at the same time to act as resident engineer.[64] Coneys soon began to send a series of adverse reports to the Commissioners' headquarters in Dublin; in October, he reported with alarm that statements had appeared in the Kerry newspapers to the effect that Leahy did not intend to commence the works until the following Spring, and in early November, he was complaining that Leahy had yet to meet him on site in Tralee.[65] From March 1841 onwards, after Coneys had reported that progress on the works was unsatisfactory,[66] a continuous stream of letters was sent to Leahy complaining of the inadequate progress, directing him to rectify defective work identified by the resident engineer, ordering him to take down and rebuild 150 feet of the basin wall which had been built in water with no proper foundations, and instructing him to dismiss his foreman who had used abusive language to Coneys.[67] He was told in May that "nothing can be more unsatisfactory than the manner in which the works are carrying on at Tralee" but, to make matters worse, the labourers went on strike in June because of unpaid wages.[68]

Having been warned by the Commissioners in August that, if he did not have at least 200 men employed within a week, they would step in and take over the works themselves,[69] Leahy seems to have made somewhat better progress so much so that it was suggested in a local newspaper in October that "the long and shamefully neglected work" would be nearing completion by the following spring.[70] However, by January 1842, the Commissioners were complaining that there had been virtually no progress since the previous October and that workers' strikes had again been caused by the low rate of wages and irregular payments.[71] In the following month, formal notice of intention to enter on and take possession of the works was served on Leahy because of the dilatory manner in which he was proceeding and because it was clear that he did not have access to sufficient working capital to enable payments to workmen and suppliers to be made on time.[72] But, in April, after Leahy had established that he then had sufficient credit at the local bank to meet his commitments, he was allowed to resume the works, although the specialised task of constructing and installing a cast-iron swivel bridge over the canal was excluded from his contract and assigned instead to the prominent Dublin engineering firm of J & R Mallet.[73]

Work continued throughout 1842, with up to 450 men engaged at one stage, but difficulties continued to arise: piling was found to be necessary, at extra cost, for the abutments of the swivel bridge, and combinations among the tradesmen forced some workers to leave.[74] Nevertheless, the Commissioners of Public Works considered the project to be well advanced by the end of the year[75] and the *Kerry Evening Post* also commented favourably:

Chapter VI — Private Practice, 1834-1846

Swivel Bridge at Blennerville, Tralee Ship Canal, constructed by J & R Mallet of Dublin after it had been excluded from the Leahys' contract with the Commissioners of Public Works (Lawrence, R 6663, courtesy of the National Library of Ireland).

We are glad to be able to announce that the drawbridge across the canal is fast progressing towards completion. The contractor, Mr Leahy, has, during each night of the past fortnight, had the full complement of men employed on the work by torchlight.

These works, when seen by night, present a startling appearance when beheld by a person not previously aware of their existence. The first sign of them are the pinewood torches flitting hither and thither - now gleaming full on the bronzed countenance of the mason, now giving to the view the labourer bending under the ponderous pod or moving about under one of the supporting beams. And when the beholder approaches within earshot - the busy hum of many voices, the rattling of bars, the knocking of hammers on the stonework, the rushing, fill the air with a medley of sounds. At close quarters, the flickering and uncertain lights at one time bring into view the bustling activity of the workmen, and at another moment leave them in undefined obscurity, displaying instead the experienced artisan carefully setting the last brought up stone, with other figures passing and repassing every instant like shadows.

The completion and opening of the bridge will be of the utmost convenience to the inhabitants of Blennerville and its vicinity, and to the carmen engaged in carrying corn to, and coals from Blennerville quay, as the road by the canal has become almost impassable, particularly since the recent rains. The state of things has been occasioned by the unfinished condition in which it was left, and the heavy loads carted on it since the commencement of the drawbridge.[76]

Progress continued to be made in the early months of 1843, when steam-driven pumps were being used to remove water from the canal cutting so that dredging of the bed to the required depth could proceed. By early June, however, the Commissioners of Public Works wrote that they had "learned with great regret" that Leahy had again suspended the works and they were receiving complaints from stone cutters and others that he had left the country, leaving behind large arrears of unpaid wages and other debts.[77] Patrick Leahy, as surety, was forced to step in at that stage, telling the Commissioners on 7 June that he had not heard from Matthew for some time but that his other two sons, Denis and Edmund, would make arrangements to complete the works.[78] Almost £6,000 had already been paid to Matthew under the contact and the resident engineer estimated that another £2,000 would need to be spent by the Leahys to complete the project.[79]

The indications are that Edmund Leahy took no part in the subsequent efforts to salvage something from what promised to be a costly venture for the Leahys who were facing severe penalties for failure to complete the contract as well as having to meet whatever costs would be incurred by the Commissioners of Public Works if they were to take over the works. Denis, however, stepped into Matthew's shoes and struggled on for almost a year, but his own shortage of capital and his failure to secure payment for works which were held not to be in conformity with the contract led to increasing concern about lack of progress and a growing volume of complaints about unpaid wages and other unpaid bills. In September, he was told by Jacob Owen, the Commissioners' Engineer and Architect, that "it has rarely occurred that works carrying on under the direction of this Board have been attended with difficulties similar to those on the Tralee Canal" and he was warned to complete all outstanding work without further delay.[80]

By early 1844, work had again come to a standstill and, after another intervention by Patrick Leahy, the Commissioners agreed in April to take the contract off the Leahys' hands and to complete the work themselves.[81] Coneys, the resident engineer, then took charge of the works and quickly began operations on a direct labour basis.[82] This was welcomed locally, one newspaper commenting that "if the Board of Works had adopted this course before, the canal would now be available for the use of the merchants of Tralee, and not an eyesore and disgrace as it has been for the last dozen years".[83] The Board, however, refused to accept any criticism for having assigned a substantial construction contract to the inexperienced Matthew Leahy, insisting that, when they accepted his tender he was "a respectable trustworthy person" and they had hoped "from his respectability and station that the works would have been executed without the annoyance and delay that has been experienced".[84]

When the Leahys finally left Tralee in April 1844, a considerable amount of work had still to be done: the excavation of a lay-by above the lock to allow outward-bound ships to await the tide; the construction of the banks extending beyond the lock and out into the bay; completion of the swivel bridge and basin; paving of the sides of the canal; and the excavation of the last 100 yards of the canal at the basin end. In addition, it was decided that a second pair of gates and a sea lock would have to be installed to enable vessels to enter and leave while maintaining a fixed water level in the canal and basin.[85] All of this work went ahead during 1844 and 1845 but, ironically, there was a further setback in October 1845 when Coneys left his position as superintendent of the works to take up a better-paid position with the Leahys who, by then, were heavily involved with the Cork & Bandon and other railway projects.[86]

Tralee ship canal was finally completed by the Commissioners of Public Works, under Frederick Villiers Clarendon as resident engineer, and the first ships used it on 13 April 1846.[87] The whole project had cost almost £12,500, roughly half of which had been paid under Matthew Leahy's contract,[88] but it was heavily burdened by debt and was virtually obsolete by the time it became available for use. When first proposed in the 1820s, the canal had been intended for sailing ships of 200-300 tons but, by 1846, larger steamships were in regular use in the coastal trade. Some steamships did use the canal in the 1850s but, in the 1860s, with still larger ships and the silting up of the canal, ships were obliged to anchor off Fenit while cargo was unloaded into lighters and brought by canal into Tralee. The extension of the railway from Killarney to Tralee in 1859 and the completion in 1887 of a new deep-water port at Fenit, with its own railway spur, brought the canal's short period of relative prosperity to an end although it did not finally close to shipping until 1940.

Bandon Navigation

While his brother Matthew was still involved as contractor on the Tralee ship canal project, Edmund Leahy was engaged to prepare proposals for a somewhat similar but longer canal to serve the town of Bandon. The assignment arose from a public meeting in the town on 17 December 1841 which had considered three options (described as a common road, a rail or train road, and a ship canal) for improving access to Collier's Quay, the town's traditional port which was some five miles away by road.[89] The first two of these options were rejected on cost grounds and because of the inconvenience of repeated loading and unloading of goods. The meeting decided in favour of a proposal to develop a ship canal which was felt to be most suitable for exports of cattle and corn, and for imports of coal. Besides, it would enable sea sand for farmers to be brought by lighters right into the town, thus cutting out a journey of up to 20 miles by horse and cart; enormous quantities of this sand were used for manuring small land-holdings at the time, and great effort was involved in its collection and transport inland. In reaching its decision, the 1841 meeting took account of the success of the canals in the Liverpool and Manchester areas and noted that the canal which was then being built at Tralee was expected to be of great benefit to the town. But the decisive influence seems to have been the advice of Edmund Leahy who, according to a letter read to the meeting, strongly favoured a five-mile extension of the navigation into the town of Bandon: he claimed that this could be achieved at no very great cost and would bring major benefits not only to Bandon but to West Cork generally and a great part of Kerry. In his view, it would be more advantageous to bring goods by ship direct to the town than to use a small port at an inconvenient part of the river. As a bonus, he claimed that the canal scheme could provide water power of up to 1,000 horse power for the operation of industrial machinery in the town.

Following the 1841 meeting, the General Committee for the Improvement of the Navigation of the Bandon River was formed and it submitted a report a year later.[90] The Committee engaged Leahy to prepare plans and costings for the proposed ship canal and he lost no time in completing the assignment, even though the Committee's limited funds did not enable them to award him "an adequate compensation". According to a report of 25 October 1842 from Leahy, his scheme involved deepening the river between Shippool and Downdaniel to admit vessels of 14 feet draught at ordinary spring tides, removing the bridge at Innishannon, and constructing a large embankment or dam across the valley at Downdaniel, with a new rock-cut waste-water channel on the northern side. A flight of four locks was to be set into the embankment, each 180 feet long and 45 feet wide, and giving a lift of 32 feet in all to bring the water level up to the level at the bridge in Bandon; the locks would "correspond in almost the nicest details" with the flight of locks at Fort Augustus, where the Caledonian Canal (completed by

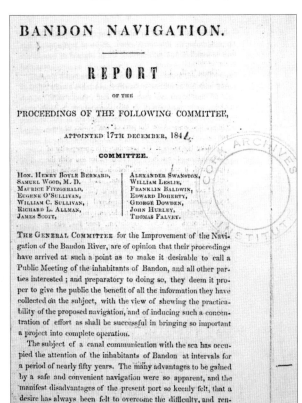

Title page of the Report of the General Committee for the Improvement of the Navigation of the Bandon River, 1842 (Cork Archives Institute U 140/D).

Chapter VI — Private Practice, 1834-1846

Thomas Telford in 1822) entered Lough Ness. There were to be three miles of new public road between Innishannon and Bandon, crossing the river on top of the embankment and carried by an arched bridge of 50-foot span across the waste-water channel, and by a cast-iron swivel bridge across the locks. The valley as far west as Bandon was to be flooded behind the dam (affecting 96 acres), with the river channel widened and deepened to provide a navigation channel 110 feet wide. Quays and wharves were to be constructed near the bridge in Bandon, enabling steamships drawing 14 feet of water to berth virtually in the town itself.[91]

When consulted about Leahy's proposals, the Commissioners of Public Works initially expressed serious reservations in a letter to Hon H B Bernard: "the difficulties in the way of continuing the Navigation up to Bandon are quite out of proportion to the prospective advantages" and, besides, the estimated income from charges was overstated and the estimated cost was "very greatly understated".[92] Nevertheless, in deference to pressure from the Bandon area, they agreed in August to direct their chief engineer, Jacob Owen, to visit the area and to report on Leahy's plans.[93] Owen suggested that, if the scheme were to go ahead, it should be modified in a number of ways to avoid flooding such a large area of land and to give "the permanency that is necessary in works of this kind".[94] He estimated that a properly designed scheme would cost nearly £72,000 (instead of the £50,000 estimated by Leahy) and advised that this would "far exceed what either the present trade, or any prospective increase, may reasonably justify". A tramway to the river at Shippool, made with second-hand rails, would fully meet the needs of the town in Owen's view, but Leahy rejected this option out of hand; he argued that the capital and maintenance costs of a tramway would be excessive and that it would not give any of the advantages which would accrue from a ship canal. In addition, it would "retard the advancement of the town, or perhaps entirely destroy whatever trade it possesses at present by driving it into another channel, tending to the establishment of a rival town, with rival interests, in that part of the river where the shipping could load and discharge their cargoes".[95] Leahy was so confident of the viability of the canal that he undertook to construct it himself for £51,300, exclusive of land acquisition, with £15,000 to be paid in cash and the balance in 5% debentures.

The Navigation Committee accepted Leahy's proposals which, they believed, would enable Bandon to attain "an eminent position among the towns of the British Empire" and they authorised Leahy to make the necessary preliminary arrangements to obtain an Act of Parliament in the 1843 session to authorise the scheme. In inserting the statutory notices, Leahy exceeded his instructions by introducing proposals for a piped water supply and a corn market among the objects of the proposed legislation but was overruled in this respect by the committee. The canal scheme itself was approved by acclamation at a public meeting in Bandon on 23 December 1842 as a "practicable scheme of unquestionable value", one of its supporters noting that China was "well conditioned and happy" because it was "intersected by canals which allowed produce to be widely distributed".⁹⁶ Some further steps were then taken to advance the scheme, but trade and industry were extremely depressed at the time, in Britain as well as in Ireland, and for this and other reasons, the promoters failed to raise the funds required to bring it to fruition. Less than two years later, Leahy had completely revised his views on the relative merits of railways and canals and was actively engaged in promoting a railway rather than a canal to serve Bandon. Viscount Bernard, who had been a supporter of the original canal scheme, attempted unsuccessfully to revive it in January 1846 under the famine relief works programme.⁹⁷

Cork Harbour

Because he had been "a pupil of Mr Nimmo" who, in 1815, had prepared a report on *The Means of Improving the River and Harbour of Cork*,⁹⁸ Edmund Leahy felt obliged to intervene when the matter came up again for consideration before the Tidal Harbour Commissioners in Cork in September 1845.⁹⁹ In his evidence to the Commissioners, he strongly opposed the construction of embankments and the reclamation of slob-lands between the city and Blackrock, which had been advocated by Nimmo and had already been partly carried out. He argued that the 156 acres involved, covered by over three feet of water at high tide, stored 500,000 tons of water which acted as a powerful scouring agent and helped to maintain an adequate depth of water from the city to Passage West. He was critical of the fact that some 170 acres had been enclosed by Belvelly Bridge and argued that other reclamation work in the Douglas area had been "very injurious to the navigation". Leahy's ideas did not, however, carry any weight with those responsible for the development of the port of Cork where reclamation and canalisation has continued almost to the present day. Neither did the authorities take up his suggestions that the Glanmire, Carrigaline and Douglas rivers could be of great service to the district, if properly cleaned, or that the Bandon River, if dredged and properly buoyed, would also be "very important".

REFERENCES

1. *Evidence to the Select Committee on County & District Surveyors etc (Ireland)*, HC 1857 Session 2 Vol IX (270)
2. Frederick O'Dwyer, "The Architecture of the Board of Works 1831-1923", *in Public Works, The Architecture of the Office of Public Works 1831-1987*, Architectural Association of Ireland, 1987, page 13
3. Maurice Craig, *The Architecture of Ireland from the earliest times to 1980*, (revised edition) Batsford, London, 1989, page 298
4. *Report of the Commissioners appointed to revise the Several Laws under or by virtue of which moneys are now raised by Grand Jury Presentment in Ireland*, Appendix B, HC 1842 Vol XXIV (386)
5. An Act to enable Grand Juries in Ireland to increase the Remuneration of County Surveyors and for other purposes, 24 & 25 Vict., c.63, section 6
6. *Select Committee on County & District Surveyors etc (Ireland)*, HC 1857 Session 2 Vol IX (270)
7. *Dublin Evening Post*, editorial, 22 June 1861
8. Evidence to the *Select Committee on County & District Surveyors etc (Ireland)*, HC 1857 Session 2 Vol IX (270)
9. *Report from the Select Committee on County Cess (Ireland)*, HC 1836 Vol XII (527)
10. ibid, Appendix 11

Chapter VI — Private Practice, 1834-1846

11. *A Return of the Number of County Surveyors, and of their Deputies or Clerks, 1834 -1839 ... and of the Sums paid by the Commissioners of Public Works to each County Surveyor etc.*, HC 1840 Vol XLVIII (291)
12. ibid
13. *Report of the Commissioners appointed to revise the Several Laws under or by virtue of which Moneys are now raised by Grand Jury Presentment in Ireland*, Appendix B, HC 1842 Vol XXIV (386)
14. CC, 17 March 1838
15. Evidence of Robert A Rogers, Esq (No 746) in *Evidence taken before the Commissioners appointed to Inquire into the Occupation of Land in Ireland*, HC 1845 Vol XXI (657)
16. *Supplement to Minutes of Evidence taken before the Commissioners appointed to Inquire into the Occupation of Land in Ireland*, HC 1845 Vol XXI (657)
17. ibid
18. Evidence of Robert A Rogers, Esq (No. 746) in *Evidence taken before the Commissioners appointed to Inquire into the Occupation of Land in Ireland*, HC 1845 Vol XXI (657)
19. *Cork Southern Reporter*, 30 July 1844
20. Evidence of Robert A Rogers, Esq (No. 746) in *Evidence taken before the Commissioners appointed to Inquire into the Occupation of Land in Ireland*, HC 1845 Vol XXI (657)
21. CC 16 March and 18 July, 1839
22. *Cork Constitution*, 30 July 1844
23. *Cork Constitution*, 1 August 1844
24. *Report from the Select Committee on County Cess (Ireland)*, HC 1836 Vol XII (527)
25. An Act to alter and amend the laws relating to the Temporalities of the Church in Ireland, 3 & 4 Wm. IV, c 37
26. *Annual Report of the Ecclesiastical Commissioners for Ireland to the Lord Lieutenant 1836*, HC 1837-1838 Vol XXVII (53)
27. *Annual Reports of the Ecclesiastical Commissioners for Ireland, 1834 and 1835*, HC 1835 Vol XXII (113), 1836 Vol XXV (130)
28. *Report of the Commissioners appointed to Revise the Several laws under or by virtue of which moneys are now raised by Grand Jury Presentment in Ireland*, Appendix B, HC 1842 Vol XXIV (386)
29. *Returns of the Expenses of the Ecclesiastical Commissioners for Ireland for each of the last six years*, HC 1844 Vol XLIII (319)
30. *Return of Numbers of Churches and Chapels built, rebuilt or enlarged in each Diocese of Ireland since September 1833*, HC 1844 Vol XLIII (279); there is no record of the activities of the Ecclesiastical Commissioners in the archives of the Representative Church Body - such records may have been deposited in the Public Record Office of Ireland as civil records and destroyed in the 1922 fire.
31. CC, 21 July 1836
32. *Report of the Commissioners appointed to Revise the Several laws under or by virtue of which moneys are now raised by Grand Jury Presentment in Ireland*, Appendix B, HC 1842 Vol XXIV (386)
33. An Act for the more effectual Relief of the Destitute Poor in Ireland, 1 & 2 Vict., c 56
34. *Fifth Annual Report of the Poor Law Commissioners, 1839*, Appendix No 7 and No 8
35. Evidence of Edward Senior to the *Select Committee on Townland Valuation of Ireland*, HC 1844 Vol VII (513)
36. *Valuations for Poor Rates, Ireland, Local Reports, Second Series, Part I, Report on Tipperary Union*, HC 1841 Vol XXII (308)
37. *Reports relative to the Valuations for Poor Rates and to the Registered Elective Franchise in Ireland*, HC 1841 Vol XXII (326), Second Appendix
38. Evidence of Edward Gulson to the *Select Committee on Townland Valuation of Ireland*, HC 1844 Vol VII (513)
39. *Reports relative to the Valuations for Poor Rates and to the Registered Elective Franchise in Ireland*, HC 1841 Vol XXIII (326), Second Appendix
40. Frederick O'Dwyer, "The Foundation and Early Years of the RIAI", in *150 Years of Architecture in Ireland*, RIAI, 1989, page 19
41. *Fifth Annual Report of the Poor Law Commissioners, 1839*
42. *Sixth Annual Report of the Poor Law Commissioners, 1840*, Appendix B, No 20
43. *Sixth Annual Report of the Poor Law Commissioners, 1840*, Appendix C, No 4; *Seventh Annual Report, 1841*, Appendix D, No 1
44. *Tipperary Free Press*, 1 April 1840
45. *Sixth Annual Report of the Poor Law Commissioners, 1840*, Appendix C, No 4; *Seventh Annual Report, 1841*, Appendix D, No 1; there is no indication as to whether the Leahys may have submitted tenders for the construction of any of the other eleven workhouses built to serve Cork city and county.

In search of fame and fortune: the Leahy family of engineers, 1780-1888

46. *Report of the Commissioner for inquiring into the Execution of the Contracts for Certain Union Workhouses in Ireland*, HC 1844 Vol XXX (562)
47. *Eighth Annual Report of the Poor Law Commissioners, 1842*, Appendix E, No 10; the minute book of the Thurles Board of Guardians, however, records that Denis Leahy did not give up possession of the Thurles workhouse until 24 October 1842 – see Anne Lanigan, "The Workhouse Child in Thurles, 1840-1880", in *Thurles: The Cathedral Town, Essays in honour of Archbishop Thomas Morris*, Geography Publications, Dublin, 1989.
48. *Report of the Commissioner appointed to enquire into the Execution of the Contracts for Certain Union Workhouses in Ireland, with Copy of Treasury Minute thereon, 20 March 1845* HC 1845 Vol XXVI (170)
49. *Report of the Commissioner for inquiring into the Execution of the Contracts for Certain Union Workhouses in Ireland*, HC 1844 Vol XXX (562)
50. Anne Lanigan, "The Workhouse Child in Thurles, 1840-1880", in *Thurles: The Cathedral Town, Essays in honour of Archbishop Thomas Morris*, Geography Publications, Dublin, 1989, page 57
51. *Seventh Annual Report of the Poor Law Commissioners, 1841*, Appendix D, No 1, Appendix E, No 13
52. *Eighth Annual Report of the Poor Law Commissioners, 1842*, Appendix D, No 1
53. *Ninth Annual Report of the Poor Law Commissioners, 1843*, Appendix B, No 11
54. *Return of the Number of Inmates in Irish Workhouses*, HC 1846 (297) XXXVI
55. Liam Kelly, "A History of the Port of Tralee", in *Blennerville: Gateway to Tralee's Past*, Tralee, 1989, page 205; Russell McMorran and Maurice O'Keeffe, *A Pictorial History of Tralee*, Tralee, 2005, pages 69 and 87
56. Plans deposited at the House of Lords Record Office, London
57. An Act for Making and Maintaining a Navigable Cut or Canal … and for otherwise improving the said Harbour of Tralee, 9 Geo. IV, c.cxviii
58. Liam Kelly, "A History of the Port of Tralee", in *Blennerville: Gateway to Tralee's Past*, Tralee, 1989, page 218
59. NAI, OPW 1/1/4/20, Public Works Letters No 20, pages 98, 118 and 188
60. NAI, OPW 1/1/4/20, Public Works Letters No 20, page 251
61. ibid, pages 207 and 251
62. ibid, pages 119 and 139
63. ibid, pages 154 and 155
64. ibid, page 221
65. ibid, pages 271 and 315
66. NAI, OPW 1/1/4/22, Public Works Letters No 22, page 356
67. NAI, OPW 1/1/4/23, Public Works Letters No 23, pages 34, 72, 151, 177 and 196
68. ibid, pages 92 and 215
69. ibid, page 352
70. *Connaught Journal*, 8 October 1840, reprinted from the *Kerry Examiner*
71. NAI, OPW 1/1/4/24, Public Works Letters No 24, pages 326, 352, 358 and 393
72. NAI, OPW 1/1/4/25, Public Works Letters No 25, pages 30 and 51
73. ibid, pages 70, 155 and 242
74. NAI, OPW 1/1/4/25, Public Works Letters No 26, pages 11, 254 and 336
75. *Eleventh Annual Report of the Board of Works in Ireland*, HC 1843 Vol XXVII (467)
76. *Kerry Evening Post*, 19 November 1842
77. NAI, OPW 1/1/4/28, Public Works Letters No 28, pages 202 and 241
78. ibid, pages 227 and 228
79. ibid, page 234
80. ibid, page 406
81. NAI, OPW 1/1/4/30, Public Works Letters No 30, page 121
82. NAI, OPW 8/354, note of 2 April 1844
83. *Kerry Evening Post*, quoted in the *Cork Examiner*, 17 May 1844
84. NAI, OPW 1/1/4/29, Public Works Letters No 29, page 155
85. *Thirteenth Annual of the Board of Works in Ireland*, HC 1845 Vol XXVI (640)
86. NAI, OPW 1/1/4/32, Public Works Letters No 32, pages 349 and 350
87. *Kerry Evening Post*, 15 April 1846
88. *Returns of the several Sums of Money advanced by the Treasury for the Improvement of Tralee Harbour and for opening a Canal there, etc*, HC 1852-53 Vol XCIV (781)
89. CC, 21 December 1841

Chapter VI — Private Practice, 1834-1846

90. *Bandon Navigation: Report of the Proceedings of the Committee appointed 17 December 1841*, Cork Archives Institute, U 140/D
91. Leahy's plans were sent to the Commissioners of Public Works in June 1842 but were returned to Hon H B Bernard in October 1842 (NAI, OPW 1/1/4/26 Public Works Letters No 26, page 400); it seems unlikely that they have survived.
92. NAI, OPW 1/1/4/26 Public Works Letters No 26, page 48
93. ibid, pages 184 and 215
94. Report of 15 October 1842, quoted in the Report of the Navigation Committee
95. Report of 25 October 1842, quoted in the Report of the Navigation Committee
96. CE, 26 December 1842
97. NAI, OPW 1/1/4/33, Public Works Letters No 33, page 158
98. Report by Alexander Nimmo, signed and dated 11 September 1815 and printed by Edwards & Savage, Cork
99. *Second Report of the Commissioners appointed to Inquire into the State and Condition of the Tidal and Other Harbours, Shores and Navigable Rivers of Great Britain and Ireland,* Appendix B, HC 1846 Vol XVIII (692)

CHAPTER VII

RAILWAY ENGINEERING, 1836-46

The early railway engineers
The civil engineering profession in Britain and Ireland gained status and experience because of the involvement of its members in the major programmes of canal, road and harbour construction which were carried out in the early part of the nineteenth century. But experienced and competent engineers were still relatively few in number when railway construction began in Britain in the 1820s and in Ireland ten years later. In many ways, it was the railways which made the profession, forcing it to react urgently to the demands for new skills, and more scientific methods of design and construction, which were thrown up by the frantic rush to develop the network. The profession responded well to this demand: those who turned to railway engineering included some of the leaders of the profession and men who, without training or experience or academic background, were prepared to grapple with new problems and pioneer the development of a revolutionary new form of transportation. By the mid-1840s, "railways were the cry of the hour, and engineers the want of the day".[2] The situation was well described by Sir John Hawkshaw, writing in 1871:

> In less than one generation, there arose a demand for men of constructive skill and ability such as had never been known before … In this, as in other cases, the occasion created the men, as great occasions do … Men were wanted to design these works and to execute them, and engineers and contractors sprang, as it were, from the earth on whose surface they were going to make such impression. There was no time for preliminary education. The demand arose, and men who felt or thought themselves capable to undertake the duties stepped forward from every rank. It was by a process of what Mr Darwin has termed "natural selection", more than in any other way, that they were found …[3]

It is against this background that the involvement of the Leahys in railway engineering must be examined.

Chapter VII — Railway Engineering, 1836-46

Cork & Passage Railway

Patrick Leahy's association with the abortive Limerick & Waterford Railway project in 1825-26 and his involvement, with Denis and Edmund, in some of Nimmo's English railway projects in 1830-32 have been discussed in Chapters I and II. In May 1834, when the Leahys took up their positions in Cork, Ireland's first railway – the Dublin & Kingstown – designed originally by Nimmo, was well on the way to completion. Only two years later, when the little railway mania developed, Patrick and Edmund launched their own first venture into railway engineering and promotion.

A large group of the city's merchants and traders were present in Cork's Imperial Hotel in May 1836 when maps and plans and a draft prospectus for a Cork to Passage railway were presented by Patrick, assisted by Edmund.[4] Leahy senior explained that the plans had been drawn up on his own initiative and without remuneration or reward of any kind, and that a complete survey of the proposed route had been carried out at his firm's expense. The line was to run from a terminus at the South Terrace in Cork by an inland route to Passage, but Leahy suggested that the terminus could easily be moved to Anglesey Street, nearer the city centre, and at Passage the line could be extended to the quayside. Piers were to be erected at Passage and at Cobh to allow for a steamer service between the two places. The line selected was said to be "a most eligible one and exceedingly level". According to Edmund, the excavations "would, on an average, be found less than on any other line perhaps in existence" and the Commissioners of Public Works "had spoken very favourably of it". The whole venture would cost no more than £60,000, based on "a very lengthy but minute estimate", and the project would certainly turn out to be "a most profitable speculation".

With the remarkable early success of the Dublin & Kingstown Railway, promotion of a line to serve the corresponding suburbs in the Cork area was a logical development. Such a line could be expected to attract much of the heavy passenger traffic carried by road and by paddle-steamers to Passage, Cobh, and other settlements and resorts in the lower harbour. In addition, Passage, with its new dockyard and direct access to deep water, had obvious potential for major development as steamship services expanded. These factors had led a group of Dublin businessmen, connected with the Dublin & Kingstown line, to form the Cork & Passage Railway Company which had been launched a few days before the meeting arranged by the Leahys to promote their scheme. Nevertheless, the Leahys succeeded initially in gaining support: an influential committee was formed to carry their proposals into effect and £15,000 was quickly subscribed towards the cost. But it was the Dublin company, with a scheme planned by Charles Vignoles, which went on to obtain the approval of Parliament in 1837 for a line to Passage.[5] Because of subsequent financial and other difficulties, the railway to Passage did not reach construction stage until the mid-1840s.

The Leahys' 1836 plans for the Passage line - if detailed plans ever existed - are no longer available but, bearing in mind the terrain involved and the distance of little more than six miles, the line is unlikely to have involved any novel engineering features. Provision was made, however, for a tunnel, 80 yards long, near the city terminus, between Langford Row and the Evergreen area, whereas eight years later, Edmund Leahy was ruling out tunnelling on other lines as "an objectionable expedient" which should be avoided at all costs.[6]

Bringing turf by rail to Cork city

In September 1838, Patrick Leahy, on behalf of himself and some other interests in the city, proposed to the Commissioners of Public Works that they should make a loan of £5,000 available to him for the construction of a railway "from the bogs into Cork at Blackpool" so as to provide a good supply of turf for the city. He and Edmund offered themselves as security for the loan, but were told that security must be given in land, stock or other real property, none of which the Leahys were able to offer. Quite apart from this, the Commissioners were sceptical, if not incredulous, about Leahy's scheme which seemed to be a step backwards in design terms. They challenged his claim that he could build a single line of wooden railway, plated with wrought iron, for £500 a mile, pointing out that "it would probably cost that sum alone for the purchase and laying down of the rails after the ground is formed". Are there no cuttings and fillings, bridges etc, they asked, and how was the necessary land and right of way to be obtained without an Act of Parliament? But instead of withdrawing his application for support for what can only be described as a half-baked scheme, Leahy wrote again to the Commissioners in November challenging their narrow interpretation of the legislation under which they were empowered to make loans and grants for productive and other purposes and suggesting new forms of security for the loan. However, after he was told firmly that prospective future profits could not be accepted as security, and that a personal security would not be accepted for more that £1,000,[7] he seems to have dropped the proposal.

Cork & Bandon Railway

With the failure of their efforts to build a Cork-Passage line, and the cyclical downturn in railway investment generally, the Leahys made no attempt to re-enter the railway engineering field until 1844 when railway promotion, speculation and investment increased sharply in Ireland as well as in Britain. In the summer of that year, they launched the project for which they are best known in the history of railway engineering - the Cork & Bandon Railway. Edmund Leahy was remarkably successful in promoting that line and it seems clear that he, rather than his father, was the originator of the scheme, as he claimed in 1845.[8] It was Leahy "who originally got the company up", according to one of the directors,[9] and they themselves were "mere cyphers" in the early stages, according to another of them.[10] The Cork & Bandon line was authorised by Parliament in July 1845[11] and construction work

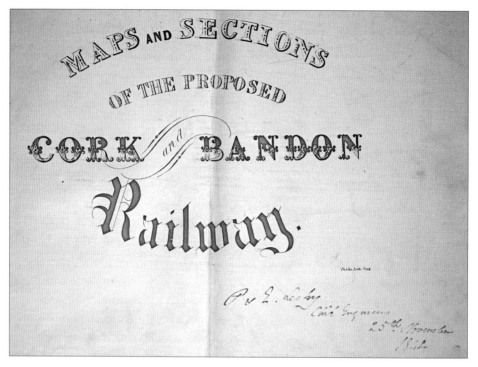

Cover page of the book of plans for the Cork & Bandon Railway lodged by the Leahys at the House of Lords, London, in November 1844.

began in September of that year. By then, Edmund Leahy's railway interests, and those of his family, had expanded enormously but his competence in railway engineering and, indeed, his ability to plan and manage any large-scale project, was brought seriously into question by his performance on the Bandon line in the following six months.

Cork-Cobh-Youghal[12]
Although his role in relation to the Bandon line was a secondary one, it was Patrick Leahy who took the lead in promoting a Cork, Youghal and Cobh Railway Company in 1844.[13] Under his original scheme, the line was to run from Cork to Passage before crossing the harbour by means of a pontoon bridge, and going on to serve Cobh and Youghal. Leahy's plans were available for public inspection in December 1844 at the offices of the company's solicitor in Cork.[14] However, the necessary parliamentary notices were not served before 30 November 1844[15] and the plans were not submitted to the Board of Trade in time to allow a Bill to authorise the proposed line to come before Parliament in the 1845 session.

Beginning on 30 December 1844,[16] a series of public meetings was held in Cork and Youghal to promote the building of a railway between the two places. The discussions were complicated by proposals to take the line through Passage, proposals to serve Cobh and

Fermoy by branch lines, and proposals to build a trunk line from Cork to Waterford, serving both Midleton and Youghal. Patrick, Edmund and Matthew Leahy attended these various meetings and were able to report in February 1845 that they had surveyed every possible route for a line to link Cork, Cobh, Fermoy and Youghal and to present various options for public consideration.[17] But there was little support for their proposal to take the line via Passage and Cobh, even with the modification that a steam ferry would be substituted for the pontoon bridge by which they had earlier proposed to cross the lower harbour.[18] The idea of providing a wooden railway did not find favour either; Patrick and Edmund Leahy, among others, were highly critical of it. The most widely supported scheme involved a direct link between Cork and Youghal, built on conventional lines, with the possibility of branch lines to serve Fermoy and Cobh.[19] Controversy about the route continued throughout 1845 and, by October, no less than five separate companies had been established to build lines to serve Youghal and other East Cork towns. One of these described its scheme as a continuation of the Cork & Bandon line running via Passage and Cobh (linked by a steam ferry) and on into east Cork.[20] The company concerned was promoted largely by the Leahys, and was generally known as the Cork, Fermoy and Cobh Railway[21] but the scheme never advanced beyond the preliminary stages.

Schemes linked with the Cork & Bandon line

The railway mania of 1845 and the successful promotion of the Cork & Bandon Company made it possible for the Leahys not only to revive their 1836 scheme for a railway between Cork and Passage but also to bring forward three other schemes linked to the Bandon line. As a result, four companies closely associated with the Cork & Bandon had been formed by October 1845. These were:

- the *Bandon & Bantry Railway Company*, which was to extend the Cork & Bandon line to Bantry, with branches to Clonakilty and Skibbereen;

- the *Cork, Bandon & Kinsale Junction Railway Company*, which was to build a branch line from the Cork & Bandon line near Crossbarry to Kinsale;

- the *Cork, Macroom & Killarney Railway Company*, whose line would diverge from the Cork & Bandon line at a junction at Ballyphehane, about a mile from the city, and go on to serve Macroom, Millstreet, Mallow (by a branch line), Killarney and, later, Tralee and Valentia;

- the *Cork & Passage Railway Company*, which was to share the city terminus with the Cork & Bandon and run from the same junction near Ballyphehane by an inland route to the deep water of Cork Harbour.

Chapter VII — Railway Engineering, 1836-46

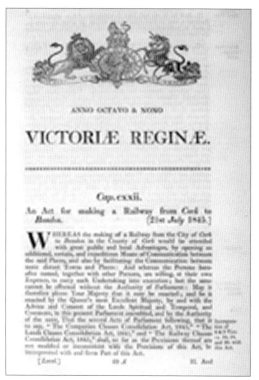

First page of the Cork & Bandon Railway Act, 1845.

All of these schemes were based on hasty surveys by the Leahys who were appointed (or appointed themselves) as consulting engineers to the companies involved. The Bills necessary to authorise the schemes came before Parliament early in 1846 but none of them became law. The Cork & Passage and the Cork, Bandon & Kinsale Junction Bills were rejected by a House of Lords Committee on 5 March, 1846, primarily because the plans were defective and not in compliance with Standing Orders. Lord Redesdale, a member of the Committee, was said to have "convicted (Edmund) Leahy to his face ... of incompetence and misrepresentation".[22] Confidence in Leahy must have suffered very seriously as a result of this episode. The *Irish Railway Gazette* added fuel to the flames by noting that, while most of the Irish Railway Bills had passed Standing Orders, "in one case, which we need not particularise, the Bill would have passed had it not been for gross and palpable incapacity on the part of the engineer who had ample time to complete his surveys and perfect his plans and sections, but shamefully failed in his duty".[23] The *Gazette* did not mention Leahy by name but its criticism could only have been directed at him. The Bandon & Bantry and Cork, Macroom & Killarney Bills survived the initial challenges in Parliament but, with a change in the investment climate, the shareholders of both companies decided to proceed no further and withdrew their Bills in June 1846.[24] That marked the end of the Leahys' inglorious involvement in railway engineering in West Cork.

Schemes outside County Cork

Not content to engineer and promote an extensive network of railways radiating from Cork city, the Leahys brought forward a number of other substantial proposals in 1845. Edmund travelled the length and breadth of Munster and south Leinster in a headlong rush to promote these schemes and, to cope with the extra work, the family maintained offices at 28 Westmoreland Street, Dublin (where their brother-in-law Patrick Ryan, a solicitor, also had an office) as well as the offices in Cork. Their major projects included:

- *the Great Munster Railway*, running from a point near Mountrath, on the Great Southern & Western Railway, to Nenagh, Limerick and Tralee, with a branch to Killarney;

- *the Cork, Limerick & Galway Railway*, running from Charleville, on the Great Southern & Western's Dublin-Cork line, to Limerick, Ennis and Galway;

- *the Clonmel & Kilkenny and Kilkenny, Carlow & Wicklow Railway*, running from Clonmel to Kilkenny, Carlow, Aughrim and Wicklow.

When added to their six schemes in County Cork, these proposals brought the Leahys' planned railway mileage to more than 500 by the end of 1845. Edmund's associates in the promotion of the Bandon line assisted on the legal, administrative and financial side in promoting the wider network of schemes and, on the engineering side, his father Patrick and brother Matthew played significant supporting roles. In addition, Denis who had been based in County Tipperary and had played no part in getting the 1844-45 schemes off the ground, became heavily involved in railway work in 1845-46, acting in some cases as an

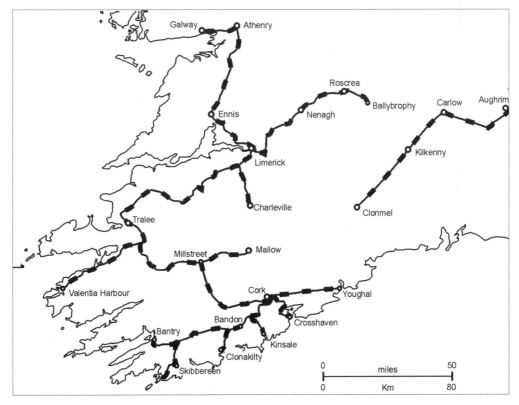

Railways promoted by the Leahys, 1844-46.

independent promoter and, in other cases, on behalf of Messrs Leahy. In his own right, he was named as engineer to the Cork & Waterford Railway Company which published a prospectus in March 1845 proposing a line 100 miles long to serve Midleton, Youghal, Dungarvan, Tramore, and Waterford.[25] In addition, it was Denis who deposited the plans for all of the Leahys' lines at the Board of Trade in London in November 1845[26] and he seems to have remained on in London throughout the early months of 1846 to promote the interests of the various new companies.[27]

Working from the Leahys' offices in Cork, Matthew was involved from the beginning in the firm's railway projects and in some others; he was a member of a deputation from the Waterford & Kilkenny Railway which met the Earl of Dalhousie at the Board of Trade in London on 3 December 1845 and was described as Acting Engineer to the Kilkenny Junction Railway when a deputation from that company was received on the same day at the Board of Trade[28]. He assisted in the surveys for the Bandon line and it was he who supervised the lockspitting of the route in July 1845.[29] Later that year, he was a regular visitor to the Vulcan Foundry at Warrington where the Cork & Bandon's locomotives were being built.[30]

The bubble bursts

Whatever about Edmund Leahy's competence as a railway engineer, there can be little doubt about his imagination, drive and capacity for hard work. His schedule in the summer and autumn of 1845 must have been a daunting one, arranging surveys, supervising the preparation of plans, and developing support for his schemes by personal contact with persons of influence throughout the south of Ireland. He addressed numerous public meetings in support of his various projects, with occasional back-up from his father and brothers. Reports of his speeches suggest that he spoke eloquently, forcefully and persuasively. Besides, he had a flair for advancing plausible but often exaggerated claims about the need for railways in different areas and the economic development that was certain to follow. He told a meeting in Newcastlewest that the Great Munster Railway would bring enormous prosperity to the town because it was situated on the largest coalfield in Great Britain.[31] A meeting in Rathkeale next day was assured that "a vein of the richest and finest marble in Ireland" would be exploited when the railway came through.[32] Those interested in the Clonmel & Kilkenny line were guaranteed that it would lead to early development of the Slieveardagh coalfield on the Tipperary-Kilkenny border.[33]

In the rush to promote railways in 1845-46, there was generally no time for proper surveys, a fact painfully shown up in many cases when companies later reached construction stage. The Leahys' surveys seem to have been particularly hasty and inadequate, their plans were often based on the first edition of the six-inch Ordnance Survey maps and other available maps rather than on field work, and Edmund was accused at least once of plagiarising plans made by other engineers: for example, a commentator on his proposed line from Kilkenny to Carlow noted that it "coincides with singular accuracy, with that submitted to the gentry of this county in November last by Mr Walker, our County Surveyor, to whom it must be gratifying that a hasty survey, to meet the demands of a moment, should be followed so exactly by that preparing for Parliament".[34] Objections on engineering grounds to the lines chosen by the Leahys were casually dismissed. For example, when asked how the Great Munster line would cross Barna Hill, near Abbeyfeale, to enter County Kerry, Edmund Leahy assured his audience that his "present knowledge and practical experience in the construction of railways and in the use of locomotive engines" enabled him to say that difficulties "which to unprofessional persons seemed so formidable were, in reality, no difficulty at all".[35]

Totally carried away by ambition and the desire to maximise their profits, the Leahys appear to have attempted far more in 1845-46 than any men in their positions could reasonably have hoped to achieve. In his haste to promote railway schemes, Edmund virtually abandoned his duties on the Cork & Bandon line for long periods, even though it was the only one of his projects to reach construction stage and, as such, would seem to have merited his special attention. As a result, the works on that line were left in the hands of inexperienced engineers and contractors, working from hopelessly inadequate plans. When matters started to go wrong, as they inevitably did, Edmund and his family were too far away, or too heavily involved in other work, to take corrective action, even if they had the ability to do so – and the predictable outcome was dismissal from his post as Chief

The Illustrated London News engraving of "The Earl of Bandon cutting the first turf for the Cork & Bandon Railway" at Kilpatrick, between Innishannon and Bandon (ILN, Vol VII, No 178, 27 September 1845).

Engineer with the Cork & Bandon Company in July 1846. All of the other railway projects promoted by the Leahys, and from which they had received substantial fees for their engineering work, had crashed out of existence by then, with heavy losses for those who had advanced the initial capital.

E Leahy CE v Cork & Bandon Railway Company
Leahy's dismissal by the Cork & Bandon led to lengthy legal proceedings which began in Dublin before the Lord Chief Baron and a jury on 15 October 1847, with no less than three Queen's Counsel representing each side. His claim for £8,000 in damages was strongly contested by the railway company and the case attracted considerable publicity.[36] The claim was based on a contract under which Leahy was to receive £5,000 for providing the engineering services necessary to complete a double line from Cork to Bandon but under which he had received only £500 before his dismissal. He was a "well-known and eminent engineer", according to his counsel, he had laboured incessantly in the interests of the company, given its affairs his undivided attention, and had been dismissed without reasonable cause. The directors had constantly expressed the highest opinion of him and the speeches when the first sod was cut were "pregnant with approbation" of his skill. It was dishonest of the company to assert some months later that there was a lack of skill or intelligence in the laying out of the line: respectable builders had contracted to build it and good progress was being made until land acquisition and financial difficulties were encountered. Leahy's absence in London in the spring of 1846 was necessary to further the interests of the company and the associated companies, and it was the directors who were responsible for slowing down the works to take account of their limited resources.

Evidence in support of Leahy's claims was given by the former Locomotive Superintendent of the Birmingham & Gloucester Railway and by Henry Dubbs, manager at the Tayleur engineering works, who spoke of the stage which had been reached in the construction of locomotives for the Bandon line, and confirmed that Leahy and his brother Matthew had regularly visited the foundry to inspect the work in progress. John Valentine, an engineer who advocated wooden railways, confirmed that he had been in communication with Leahy about the use of Payne's process to prepare the beech sleepers for the line. A number of Leahy's assistant engineers testified to the progress which had been made in detailed design work on the line, and some of the contractors gave evidence about the state of the works on sections where the necessary land had been acquired.

The presentation of Leahy's case occupied three full days. According to the *Irish Railway Gazette*,[37] the facts elicited were "of a painful character" and they must indeed have come as a shock to many of the shareholders who, until then, had been given no indication of the severity of the crisis through which the company had passed in 1846. But worse was to come when, on the fourth day, counsel for the railway company began the defence of his clients with a slashing all-day attack on Leahy, who was accused of being incompetent, incapable and unreliable. He had made the company the laughing stock of every scientific mind and some of his proposals would have been perfectly ruinous for the company. He seemed to have "the greatest possible taste in constructing things to make the wondrous gaze" and, in the case of a proposed wooden viaduct at Chetwynd, "must have lost his senses" by devising something which would be an enormous trap for human life. He had made "monstrous blunders" in adopting "a peculiarly novel form" for the rails, and even a tyro in the profession should have known that beech sleepers could not stand the weather. He had "humbugged" the directors into the "imprudent making of premature contracts before they had the money to meet them" and some of his reports on the state of the works constituted "positive misrepresentation" and were "without a particle of truth". He had neglected the business of the company because of his extensive involvement in the promotion of other railways and it was "absolutely miraculous that the company had not long since fallen to pieces in consequence of Mr Leahy and his line". As it was, his blundering had cost the company over £10,000 - "and had almost swamped it".

Evidence for the railway company was given by several of its directors and officers. In addition, architects, engineers and builders took the stand to give evidence about the state of the works when Leahy was dismissed, stressing the inferior quality of the workmanship and materials, and the problems that had arisen in relation to rails, sleepers and working drawings. Alexander Deane, a Cork architect, had been engaged to carry out a detailed inspection of bridge foundations and masonry with a view to giving evidence for the company but had died before the court case began[38]. Sir John MacNeill, the most prominent Irish railway engineer of the day, testified that neither beech nor yellow pine (which Leahy planned to use) was suitable for use as sleepers, and insisted that any competent engineer ought to know this. Sir John went on to condemn Leahy's proposal for light rails and longitudinal sleepers with triangular sections; he himself would not use such a system and he knew of no justification for its use on a normal line, where the rails would wobble and loosen.

After the case had been at hearing for 13 days, the Lord Chief Baron began his summing up on Friday, 12 November, and went on until 3 pm next day. Then, after a three-hour adjournment, the jury returned their verdict. They awarded Leahy damages of £325 (his claim was for £8,000) with only six pence towards his costs. The *Irish Railway Gazette* noted that the costs to both parties of the hearing "must be frightful" and that while Leahy got the verdict, he would be considerably out of pocket.[39] The lawyers were the only ones who gained in the long run: according to the *Gazette*, the case had dragged along with "a degree of procrastination" that must have proven to be "peculiarly refreshing" to them, and they had gained "a rich harvest".[40]

Chapter VII — Railway Engineering, 1836-46

Blundering buffoonery?

Edmund Leahy's knowledge and practical experience of railway engineering, such as it was, had been acquired before coming to Cork in 1834 as a 20-year-old county surveyor and it must have been very limited indeed. Although he still clearly lacked experience when he promoted the Bandon line ten years later, Viscount Bernard was the only one connected with the railway company who seemed to be prepared to question Leahy's competence; speaking at a public meeting in Bandon in October 1845 in regard to the Bandon & Bantry line, Lord Bernard said that " highly as I value the services of Mr Leahy … able as I believe him to be, still youth must be allowed for" and, in so great an undertaking, a more experienced and better known engineer such as Sir John Rennie was required.[41] All of the others involved seem to have been prepared to overlook Leahy's lack of experience, influenced perhaps by his skill and persuasiveness in marketing his projects and the attractiveness of the returns which he forecast for potential investors.

In Ireland, as well as in Britain, most railway companies preferred to entrust their schemes to a relatively small number of established engineers. These included Robert Stephenson, Sir John Rennie, and Isambard K Brunel in Britain and Sir John MacNeill and Charles Vignoles in Ireland. At the height of the mania in 1845, MacNeill was consulting engineer to no less than 37 railway companies, while Vignoles was associated with 22 others.[42] The demands made on these leading engineers were phenomenal. Long hours were spent in travel by carriage or on horseback, with only brief snatches of rest between assignments. Days were taken up in all weathers walking and surveying possible routes, with reports and plans to be written up at night in the frugal comfort of some rural hostelry. Public meetings to promote new schemes had to be addressed and provisional committees and directors often demanded first-hand progress reports. Inevitably, a good deal of work had to be delegated to unqualified assistants, so much so that in some cases the consultant's association with a particular scheme was purely a nominal one. But even this had to be accepted by the railway companies if the prestige that followed association with one of the leading engineers was to be retained. For most companies, the alternative of engaging a less experienced or local engineer was to be avoided at all costs, in the interests of maintaining the confidence of the public, enhancing the attractiveness of their schemes for investors, and increasing the prospects of gaining approval in Parliament.

Although engineering experience was at a premium in 1845, only a few of the Irish railway companies established at that time appear to have been seriously affected by careless or incompetent engineering.[43] The Cork & Bandon is one of the best examples of a company which suffered heavily in this way. In other cases, the active involvement of an experienced consulting engineer compensated for the inadequacies of a less experienced local engineer, but Charles Vignoles, although nominally acting as consulting engineer, was not sufficiently involved to save the Bandon line: to quote Joseph Lee, Vignoles' "perfunctory supervision of the Cork & Bandon did nothing to rescue the concern from the blundering buffoonery of Edward (sic) Leahy".[44] This may seem a harsh verdict on both men but there is ample evidence to support it. Vignoles' journal[45] shows that he paid only one flying visit to Cork on 5 November 1844 when he was "engaged incessantly the whole day with Messrs Leahy in examining the several lines of railway between Cork and Bandon, especially as to the possibility of taking in Kinsale ... and on the whole was satisfied that the line of Messrs Leahy ought to be adopted". Despite Leahy's frequent assertions to the contrary, it seems clear that Vignoles could not have played a major role in preparing the parliamentary plans – in fact, he saw these plans for the first time when they were brought to London on 30 November 1844 for lodgement, before close of business that day, with the Clerks of the Houses of Parliament[46] - and he was not involved at all in preparing the working drawings or in supervising the construction work. Thus, the engineering of the line was left almost entirely to Edmund Leahy, his father and his brothers and, when construction of the line began, their lack of skill and relevant experience was cruelly exposed. The situation was compounded by their carelessness and negligence. If Edmund had been prepared to concentrate his undoubted energies on the works underway on the Bandon line, it is just possible that he would have managed to complete what would then have been an unique line, with bogie locomotives, bogie passenger carriages, large wooden viaducts, an unusual form of permanent way and several other novel and interesting features, all of which were provided for in his plans. But Leahy's pursuit of fame and fortune and his apparent determination to extend his railway interests throughout Munster dictated that this was not to be.

Edmund Leahy seems to have had an almost pathological aversion to following the norms of established English railway engineering practice; instead, his engineering incorporated features of American railroad practice and technology of which he could have had no direct experience. A possible explanation for this is suggested by the fact that Leahy's work was similar in many ways to design features of the Birmingham & Gloucester line, constructed between 1836 and 1841 in the English midlands, a line to which he frequently referred when aspects of his plans were called into question. The Birmingham & Gloucester's engineer was Captain William Scarth Moorsom (1804-1863), one of the few English engineers who favoured American railroad methods. Following his discharge from the army in 1832, Moorsom practiced as a civil engineer and was responsible for the design of many railways in Britain and some in Ireland.[47] His Irish projects included the Waterford & Kilkenny, which has been well described as the "outstanding engineering fiasco" of the early Irish railway era.[48] "On no railway in either country, for the extent of it, has there been from first to last, a greater number of gross blunders committed - and extravagant blunders too", was how the *Irish Railway Gazette* saw the situation on that line in 1849.[49] It was Moorsom who ordered for the Birmingham & Gloucester Railway the American-built Norris bogie locomotives on which Leahy's designs for the Cork & Bandon were based. On that line, too, Moorsom had used beech sleepers, as Leahy proposed to do on the Bandon line, and built a number of timber lattice girder bridges in the American style, very similar to some of those proposed by Leahy.[50] Between 1846 and 1850, Moorsom built a large viaduct of this kind for the Waterford & Kilkenny Railway, with a 200-foot span over the river Nore.[51] Yet another unusual component of Moorsom's engineering which featured in Leahy's plans was the use of light iron bridge-rails on longitudinal sleepers with triangular cross-section: although these were not used on the Birmingham & Gloucester, they were introduced by Moorsom, with costly and damaging effects, on the Waterford & Kilkenny line after that company's initial flirtation with the idea of using wooden rails.[52] While it may be that these similarities between the engineering of Moorsom and Leahy were no more than coincidence, it is tempting to suggest that Leahy was influenced or advised by Moorsom in his first major venture into railway engineering. The fact that Matthew Leahy was a member of a deputation to the Board of Trade on behalf of Moorsom's Waterford & Kilkenny line in December 1845 suggests that such an association existed.[53]

Leahy's pauline conversion to support for railway development appears to have been a rather opportunistic one. His subsequent frenetic efforts at railway promotion, beginning with the Cork & Bandon in September 1844, were in marked contrast to his attitude to steam locomotives and railways as expressed in *A Practical Treatise on the Making and Repairing of Roads*, published only a few months earlier. In its limited treatment of the subject, Leahy's textbook appeared to suggest that railways, in the conventional sense, using a separate right of way and steam locomotives, would not be viable in most situations. In fact he suggested that outside the large cities, railways would be "advantageously adapted to great roads". What he had in mind was that —

> On such roads when they are sufficiently broad railways might be placed on each side of the road. On these, small wagons for heavy burdens linked together might be drawn, and carriages for swift conveyance might be received in cradles suited to railways: and thus light carriages may be transported on railways without any alteration in their present construction, so that coaches and chaises may travel for any convenient distance upon any public railway and may turn off instantly to a common road. [54]

Leahy's views on the viability of railway development — even in relatively remote areas — seem to have changed dramatically once the railway mania arrived or, perhaps, it was a case of discounting the practicalities in the interests of maximising his earnings in the short term. He was not, of course, the only engineer or railway promoter who advanced extravagant claims about the ease with which new lines could be constructed and the profits they would instantly generate. In fact, as one of the earliest railway historians pointed out, "engineers promised all and everything" in favour of particular lines: "they plunge through the bowels of mountains; they undertake to drain lakes; they bridge valleys with viaducts; their steepest gradients are gentle undulations; their curves are lines of beauty; they interrupt no traffic; and they touch no prejudice".[55]

It is part of the normal duty of an engineer to design a project and to see that it is properly carried out. Under both headings, Edmund Leahy must be adjudged to have failed miserably in his railway engineering. In the context of railway history, he can be remembered only as one of those who flitted across the stage for a brief period in the mid-1840s, while the boom raged. He appeared to be omnipotent for a few months, trading astutely on the credibility of the public whose imagination had been stimulated to an extraordinary extent by the prospect of rapid railway development and the high rates of return which were forecast. Intoxicated by his early success, he resented any effort to question his schemes, depending on his skill in argument and his persuasive abilities to cover up his basic lack of experience. Many of his schemes involved no more than wild, reckless speculative projects, hastily surveyed but imaginatively packaged, and sold effectively to gullible provisional committees and investors. Even before the bubble was pricked and burst in 1846, Leahy's deviousness, economy with the truth, and lack of judgment, responsibility and skill had already been exposed. The crash of 1846 finally put an end to his ambition to be the Railway King of the south of Ireland: he failed to see even one of his many projects through to completion or, indeed, to complete even one mile of railway, before leaving Ireland a year later.

A vision ultimately realised

In justice to Edmund and Patrick Leahy, however, it must be recognised that the West Cork railway network which they envisaged and designed – if only in a preliminary way – between 1844 and 1846 was eventually completed and largely followed the route alignments they had proposed in those years. After serious financial and land acquisition problems, the Cork & Bandon line opened in December 1851, with conventional permanent way and vehicles and some minor deviations from the original route. This was followed in 1863 by a line to Kinsale and in 1866 by a line to Macroom, both branching off the Cork & Bandon line at the locations which Edmund Leahy had proposed. The main line itself was extended westwards, as Leahy had also proposed, reaching Dunmanway in 1866, Skibbereen in 1877, Bantry in 1881 and Clonakilty in 1886. Thus, within Edmund Leahy's own lifetime, the original Cork & Bandon line had become the central element in a network of over 100 miles of railway.[56]

Further extensions to Baltimore and Courtmacsherry, which were not included in the Leahys' original plans, were completed in the early 1890s. By then, most of the original companies had amalgamated to form the Cork, Bandon & South Coast Railway Company whose substantial network remained unconnected with the rest of the Irish railway system until the completion of the Cork City Railways in 1912. But while the railways of West

Chapter VII — Railway Engineering, 1836-46

Cork went on to become an important element in the economic and social life of the area until the closure of the entire system on 31 March 1961, Edmund Leahy's predictions at the ceremony to mark the turning of the first sod on the Bandon line in September 1845[57] failed to materialise. The untold wealth which he foresaw for West Cork was illusory and the unfortunate investors in his original Cork & Bandon scheme soon found – as did their counterparts in some of the other early railway schemes – that his guarantee that the railway would be one of the best and most profitable in the United Kingdom was worthless.

REFERENCES

1. *Personal Recollections of English Engineers, by a Civil Engineer*, London, 1868
2. ibid
3. John Hawkshaw, letter of 11 December 1871, quoted in Arthur Helps, *The Life and Labours of Mr Brassey*, London, 1872
4. CC, 7 May 1836
5. Cork & Passage Railway Act 1837, 7 Wm. IV & 1 Vict., c. cviii
6. IRG Vol I, No 48, 29 September 1845
7. NAI, OPW 1/1/4/16, Public Works Letters No 16, pages 130 and 319
8. CE 22 October 1845
9. CC 15 February 1849
10. IRG Vol III, No 53, 1 November 1847
11. An Act for making a Railway from Cork to Bandon, 8 & 9 Vict., c.122
12. The modern name *Cobh* is substituted throughout this chapter for *Cove*, which was the name used in the 1830s and 1840s.
13. IRG Vol I, No 9, 30 December 1844
14. ibid
15. IRG Vol I, No 10, 6 January 1845
16. CC 31 December 1844
17. IRG Vol I, No 17, 24 February 1845
18. CE, 28 and 31 December 1844
19. IRG Vol I, No 17, 24 November 1845
20. IRG Vol I, No 49, 6 October 1845
21. The company was also known as the Midleton, Cove and Passage, the Midleton & Fermoy, and the Carrigaline, Midleton & Fermoy.
22. IRG Vol III, No 53, 1 November 1847
23. IRG Vol II, No 24, 13 April 1846
24. IRG Vol II, No 32, 8 June 1846; RT Vol IX, Nos 23 and 24, 6 and 13 June 1846
25. Prospectus of the Cork & Waterford Railway Company, Cork Archives Institute
26. IRG Vol II, No 6, 8 December 1845
27. IRG Vol III, No 52, 25 October 1847
28. *The Times*, Court Circular, 4 December 1845
29. IRG Vol III, No 53, 1 November 1847
30. IRG Vol III, No 52, 25 October 1847
31. IRG Vol I, No 35, 30 June 1845; the Munster coal field, occupying "considerable portions of the counties of Clare, Limerick, Cork and Kerry" had already been described by Sir Robert Kane as "the most extensive development of the coal strata in the British Empire" in *The Industrial Resources of Ireland*, Dublin, 1844, page 11.
32. ibid
33. IRG Vol I, No 50, 13 October 1845
34. ibid
35. IRG Vol I, No 35, 30 June 1845
36. IRG Vol III, Nos 52, 53, 54, 55, 25 October and 1, 8 and 15 November 1847

37. IRG Vol III, No 53, 1 November 1847
38. Alexander Sharpe Deane and Kearns Deane, both of whom died early in 1847, were brothers of Sir Thomas Deane (1792-1871) one of the most prominent Irish architects of the early Victorian period.
39. IRG Vol III, No 55, 15 November 1847
40. IRG Vol III, No 53, 1 November 1847
41. IRG Vol I No 52, 27 October 1845
42. IRG Vol II, No 11, 12 January 1846
43. Joseph Lee, "The Construction Costs of Early Irish Railways 1830 -1853", *Business History*, Vol 9, 1967, pages 98 - 103
44. ibid, page 102
45. BL, Ms 34,530 Vol III, Journal of Charles Blacker Vignoles, 1842-1845, entry for 5 November 1844
46. BL, Ms 34,530 Vol III, Journal of Charles Blacker Vignoles, 1842-1845, entry for 30 November 1844
47. ODNB 2004-05, biographical note by Mike Chrimes
48. Joseph Lee, "The Construction Costs of Early Irish Railways 1830 -1853", *Business History*, Vol. 9, 1967, page 102
49. IRG Vol V, No 6, 5 February 1849
50. J G James, "The Evolution of Wooden Bridge Trusses to 1850", *Journal of the Institute of Wood Science*, Vol 9, No 3, June 1982
51. P J Flanagan, "The Nore Viaduct, Thomastown", *Journal of the Irish Railway Record Society*, Vol 9, No 52, June 1970, page 236
52. D B Mc Neill, "Waterford & Central Ireland Railway", *Journal of the Irish Railway Record Society*, Vol 13, No 74, October 1977, page 114
53. *The Times*, Court Circular, 4 December 1845
54. Edmund Leahy, *A Practical Treatise on the Making and Repairing of Roads*, John Weale, Architectural Library, London, 1844, page 85
55. John Francis, *A History of the English Railway*, London, 1851
56. For the most recent and most comprehensive account of the development of the railways of West Cork, see Ernie Shepherd, *The Cork Bandon & South Coast Railway*, Midland Publishing, Leicester, 2005.
57. IRG, Vol 1, No 47, 22 September 1845

CHAPTER VIII

LOSS OF COUNTY SURVEYOR POSTS

"the law having invested the grand jury with the power of dismissing any person holding that office, His Excellency can only proceed to appoint your successor as soon as circumstances will permit".

Chief Secretary's Office, 1846[1]

Exciting the enmity of grand jurors

Despite occasional disputes, and complaints about lack of attention to their official duties, the Leahys seem to have had a generally good working relationship with the county grand jury and with the authorities in Dublin Castle, at least until 1844. In 1836, the Commissioners of Public Works reported that the Leahys had given great satisfaction as county surveyors, although they were careful to add "like many others".[2] In 1858, Sir John Fox Burgoyne attested that Edmund had performed his duties "to the satisfaction of the county and of the Commissioners of Public Works".[3] Edmund himself claimed that he had discharged his functions "without complaint of any sort and to the full satisfaction of the Government" between 1834 and 1846,[4] that he had repeatedly "received the expressed thanks of Grand Juries and Road Sessions"[5] and that he had enjoyed "as much of public approval during that time as fell to the lot of most if not any other such officer in Ireland".[6] Both Patrick and Edmund were formally thanked by the county grand jury in 1836 "for the zealous and able manner in which they have performed their duties"[7] and there were similar resolutions in some subsequent years. At the Summer Assizes in 1844, the grand jury adopted a resolution supporting Edmund's claim to be appointed "Principal Surveyor and Engineer for the County of Dublin" because of the efficient and satisfactory manner in which his duties in Cork had been carried out.[8]

There is ample evidence, however, that the Leahys' relationship with the county grand jury was not entirely a harmonious one and that matters deteriorated as the years went by. The extent of their private practice caused problems throughout their entire period in Cork and there were disputes with the grand jury, or with individual grand jurors, about particular projects and on procedural matters. But, according to Edmund, occasional exhibitions of hostility were inevitable because "county engineers have too much responsibility placed upon them – should they perform their duties independently and properly, they are sure to create much individual opposition … and this opposition finds easy opportunities of exhibiting itself, and often in a very troublesome manner".[9] More specifically, he alleged that he was occasionally pressed by grand jury members to certify works which were not properly executed and that his refusal to do so "in the independent discharge of the duties of the county surveyor … could excite the enmity of grand jurors".[10] Patrick Leahy, although he was more frequently accused of having certified faulty work, also found himself in difficulty with the city and county grand juries from time to time when he refused to grant the certificate without which a road contractor could not claim payment. On one such occasion in 1840, when some of the city grand jurors alleged that his specifications for road maintenance were too exacting, he replied that, for £1,000, he could not improve on the form he had been using, adding that it had been highly approved of by the Office of Public Works and that copies had been sought by his colleagues in neighbouring counties.[11]

Patrick Leahy and the city grand jury
From the beginning, Patrick Leahy's relationship with the city grand jury was an uneasy one. On learning of his appointment, many of the jurors took the view that one man could not properly discharge the duties of the city and county posts and that a separate surveyor should have been appointed for the city;[12] that view was expressed again at intervals in subsequent years and it was largely for this reason that the city grand jury refused initially to pay Leahy the maximum salary allowed by law. There were complaints at the city assizes about Leahy's lack of attention to his duties, allegations that he was neglecting to visit and inspect road works, suggestions that his certificates of completion could not be relied on, and claims that the roads were in far better condition before his appointment.[13] At one stage, he was directed to submit monthly reports on his activities to a committee of the grand jury, but he failed to do so, although he insisted that he visited all roads in the area at least once a month.[14] For his part, Leahy was not slow to complain about his employers. In January 1839, he was writing to the authorities in Dublin Castle about the propensity of the grand jury "to depart from the spirit and the letter of the law."[15] Later that year, he was again in touch with the Under-Secretary requesting rulings from the Attorney General on his duties in relation to roads, dilapidated buildings, the lunatic asylum, and so on; in addition, he questioned the legality of the grand jury's procedure, the times fixed for their meetings, their practice of adjourning from day to day, and other minor matters.[16] Leahy was also in dispute with the county grand jury in 1839 about the additional expense which was arising in cases where he stepped in to take over the work of a defaulting contractor; in this case, the Attorney General ruled that Leahy should keep his expenditure within the limit of the approved presentment.[17]

In search of fame and fortune: the Leahy family of engineers, 1780-1888

Legislation which came into effect in November 1841 transferred responsibility for the liberties of Cork from the city to the county grand jury and, since the paving and repairing of the roads in the city itself was the responsibility of Cork Corporation and the Wide Streets Commissioners, the city grand jury took the view that Leahy's duties in relation to their much reduced functional area were "very trifling" and did not warrant the continuation of his office as city surveyor. At the summer assizes in 1842, the jury therefore decided to dismiss him from that position and sent a memorial to the Lord Lieutenant asking him to confirm their decision so as to ensure that the surveyor could not claim the salary of £300 a year "for which the city received no benefit".[18] Leahy responded by asking the authorities in Dublin to take into serious consideration the great loss which would accrue to himself and his family if the grand jury decision were allowed to stand and argued that if it were left to grand juries to dismiss a surveyor, "that species of jobbing so much condemned in Ireland heretofore would be acted on again and again".[19] The Attorney General, when consulted, advised that the grand jury did indeed have the power to dismiss the surveyor and either to reappoint him at a salary appropriate to his more limited duties, or to make a new appointment.[20]

Leahy refused to accept that he had been lawfully dismissed and, at the next city assizes in March 1843, applied as usual for payment of his annual salary and expenses, even though the total cost of the works he had certified in the previous period amounted only to £270.[21] The jury agreed to stand by their decision to dismiss him and refused to recognise him as their officer. By the time of the Summer Assizes in July, matters were becoming more serious, with Leahy claiming £525 on foot of arrears of salary and expenses and the grand jury directing that counsel should be briefed to deal with the action which Leahy was expected to take in the courts.[22] Two years later, at the city assizes in July 1845, Leahy was reported to be claiming salary arrears of £1,250[23] but it appears that the threatened court proceedings did not go ahead and that he had to resign himself to the loss of the city surveyor post.

Edmund's resignation

Newspaper reports of discussions at meetings of the county grand jury in the late 1830s and early 1840s suggest that grand jurors generally were impressed by Edmund Leahy's road building activity in the West Riding and by the standard of road maintenance there which, in the opinion of many grand jurors, was far better than the results achieved by his father in the East Riding. While grand jurors were critical of the extent to which Edmund engaged in private practice, they were seldom in a position to point to specific deficiencies in his performance in the years up to 1844. In later life, he claimed that his resignation from his position as county surveyor was "of his own free will and accord"[24] and that it arose because "other professional engagements became so important and numerous".[25] But while it is true that Leahy's career as county surveyor was ended by resignation, it is equally true that the resignation was forced on him and that he avoided dismissal only by the narrowest of margins.

Grand juries in the 1830s and early 1840s regularly dismissed their county surveyors for relatively trivial reasons or sought their removal to other counties. The Leahys were exceptionally fortunate, therefore, that there was a lack of firmness on the part of the Cork grand jury in their dealings with them for while it was apparent from early on that the surveyors were not being entirely open and honest about their business affairs, the grand jury took no action for years. However, when the Leahys became heavily involved in railway promotion, in addition to their earlier private practice activities, their difficulties with the grand jury were exacerbated. From the summer of 1844 onwards, Edmund, in particular, devoted a great deal of his time to railway work, initially on the Cork & Bandon line and later on the associated lines in other parts of County Cork and lines further afield. He and his father virtually abdicated their duties as county surveyors during this period and severely tested the tolerance of the grand jury.

Between April and July 1845, when the Cork & Bandon Railway Bill was before Parliament, Edmund Leahy, as the promoter of the company and its engineer, was obliged to spend long periods in London. Relying on a provision in grand jury law which allowed a county surveyor to appoint a deputy "in case of indisposition or other unavoidable cause",[26] he appointed his brother Matthew to act for him at the presentment sessions preceding the 1845 Spring Assizes and he obtained the approval of the Office of Public Works for this arrangement, claiming that his absence was unavoidable if he was to comply with the Speaker's Warrant to attend the parliamentary committees dealing with Railway Bills. However, when the grand jury convened in March 1845 it soon became clear that some members were not at all happy with this arrangement and objected in principle to the fact that their surveyors were "mixed up with other speculations".[27] A motion demanding that the two men should in future give their whole and undivided attention to their county surveyor duties was proposed by one juror, another moved that their salaries for the half year should be withheld, and yet another proposed that the two surveyors should be censured for engaging some of their assistant surveyors on railway work. Eventually, a compromise motion, proposed by the foreman, and applying to Edmund Leahy only, was agreed; this criticised his absence from the presentment sessions and his use of assistant surveyors on railway work, cautioned against any repetition of these practices and strongly advised him against "in any way suffering other business or occupation to interfere in the slightest degree" with his duties as county surveyor.[28]

Leahy, of course, ignored the warnings of the grand jury and spent the bulk of his time in the following months carrying out surveys and engaging in promotional work for additional railway schemes. Thus, when the cesspayers met for the road sessions in Clonakilty on 4 June, they adopted a resolution which was severely critical of the surveyor, charging him with "gross neglect of his entire duty" for the previous two months, and alleging that he had omitted to inspect the public works, to keep a record of inspection visits, and to open an office in Bantry, as he had earlier undertaken to do.[29] When asked by the Dublin Castle authorities to respond to these charges, Leahy provided a robust defence in July 1845.[30] Complaining that he was "being assailed in terms not bespeaking much of measured phrases", he claimed that the views expressed by the cesspayers were those of a minority: they were not shared by the cesspayers of the baronies of Ibane & Barryroe and Kinalmeaky who, at their meeting in Bandon on 9 June, had expressed views "of an opposite nature" and he was sure that every other sessions with which he had any dealings would also dissent from every syllable of the criticism. He denied that he had neglected his duties, asserting that he had travelled over 1,000 miles through West Cork during April and May inspecting the public works, and insisted that he had been legally and satisfactorily represented by his brother at the presentment sessions while he himself was in London promoting a railway which would be of immense importance to the West Riding, and meeting Sir Robert Peel, the Prime Minister, with a large delegation of the nobility and gentry of West Cork, to press for the establishment of a naval and packet station in Bantry Bay.

The Dublin authorities decided to opt out of the dispute between Leahy and the cesspayers: they referred the whole matter back to the grand jury indicating that "the Lord Lieutenant, after inquiry, does not consider that his interference in the conduct of Mr Leahy is called for and any neglect on the part of that officer should be brought under the notice of the grand jury at the ensuing assizes".[31] At the assizes in August, the jury did indeed take up the matter again when Daniel Connor of Manch House criticised the fact that Leahy had been represented by a deputy during his absence in London and alleged that, while mercy had been shown to him on previous occasions, Leahy had, since the last assizes, "in the most open and notorious manner" neglected his official duties. According to the *Cork Southern Reporter*, a rather warm discussion ensued but the grand jury were clearly reluctant to take decisive action against Leahy and allowed him to escape again with a vote of censure; the majority agreed with the view of the foreman that some leniency was due to Leahy because of the success of the Cork & Bandon Railway Company, whose Bill had become law on 21 July. Many of the jurors themselves were, of course, associated in one way or another with the company.[32]

For his part, Leahy was not concerned by the grand jury's censure, telling them that he intended to resign his office anyway. But he deferred his resignation because, according to himself, many of the most influential members of the grand jury induced him to continue until the works which were in progress were more advanced.[33] In the months that followed, Leahy devoted very little time to the county works, bringing in an English engineer, Samuel H Kettlewell, to assist him and giving the impression locally that Kettlewell had been officially appointed as deputy county surveyor. Thus, freed of his work for the grand jury, Leahy went on to develop his railway interests still further. By October 1845, in addition to his association with no less than 500 miles of proposed railways, he claimed to be connected with a company which was planning to launch a steamer service from Bantry to St John's, Newfoundland.[34] Some months later, he was again publicly proclaiming that he no longer had any real interest in the post of county surveyor and that he was retaining it only so that the public interest would not suffer by his sudden withdrawal.[35] By then, his railway commitments were so extensive and so potentially profitable that he had effectively abandoned his post with the grand jury, leading inevitably to further complaints to the Dublin authorities.[36]

Even without the onerous duties of a county surveyor, any engineer would have found it difficult to cope with the number of railway projects initiated by Leahy in 1845-46 and to give each of them the personal attention that its directors and shareholders felt entitled to demand. With the commencement of the new session of Parliament in January 1846, Leahy's position became quite impossible since it was essential for railway engineers to remain in London for long periods in order to be available to give evidence before the committees which both Houses of Parliament had set up to examine Railway Bills. He therefore sought permission from the Lord Lieutenant to appoint Kettlewell to carry out his functions as county surveyor:

> I beg leave to acquaint you that my evidence is required before Parliament at the coming Sessions in reference to the contemplated Railway from here to Bantry, as well as other Public Works of a similar nature through this County.
> I beg leave also to state that I have done all in my power to promote their success under the conviction that I could not better serve the interests of the County of which I am Engineer than by contributing my best endeavours towards carrying forward public works tending to develop the resources of the country and of especial importance at the present time as promising to afford large employment to the labouring classes and so to provide in some measure against the danger of a Season of Scarcity.
> The Promoters of these works have subscribed upwards of £60,000 as the Parliamentary Deposits required previous to seeking the necessary Acts for authorising their construction, and my evidence is indispensably necessary before Parliament to the success of these important measures. May I therefore hope His Excellency will permit Mr Samuel H Kettlewell to act as Deputy County Surveyor during my temporary absence in London to attend upon the Committees of the next Session of Parliament. Mr. Kettlewell was educated in Woolwich Military College as an Engineer Officer. He obtained his Commission as a Lieutenant in the Artillery, in which Corps he served from the year 1829 to 1837, serving during that time at home and in the West Indies, in which latter place he acted as private Secretary to the Governor of Honduras, Colonel McDonald, and finally owing to ill health retired from the service.[37]

Leahy's letter was referred to the Commissioners of Public Works who, notwithstanding their approval of a similar arrangement in 1845, told him that "whatever may be due to you for promoting railways, it cannot be received as a reason for having your duty performed by a deputy".[38] The Commissioners took the view that it would be wrong to accede to Leahy's request because of the serious conflict of interests which would arise, and set out their position extensively in a letter to the Under-Secretary, Richard Pennefather:

> I am directed to state for the information of His Excellency, that this Board consider that in principle it would be very unadvisable that he should countenance the employment of County Surveyors on Railroads in their own Counties. It is well known by an inspection of the Plans recently lodged for Parliament that the attention of the County Surveyor is most urgently required to the manner in which the crossings of the Roads by Railways is proposed to be effected.
>
> The Legislature has made provisions as to the heights and spans of all Bridges that are to cross Public Roads, but there does not appear to be any person appointed to watch over the interests of the Public. This ought to be fixed upon the County Surveyor, as the responsible officer of the County, and who ought to watch with a jealous eye the movements of Railway Companies, so far as these works are likely to injure or interfere with the Roads, Bridges, or Public Buildings within his district. In many cases, the crossings are marked to be on a level with the Road, and in other cases important Roads are marked as Byeroads in order that the Bridges may be built of smaller dimensions than they ought to be. The County Surveyors, as the officers of the Grand Juries, who are the Trustees of the Roads, should pay particular attention to these points.
>
> The Board would not feel themselves justified in recommending His Excellency to allow of a County Surveyor having his duty done by Deputy to enable him to absent himself for an indefinite period on private professional business ... It does not appear to the Board that the cause of absence put forward by Mr Leahy is such as the Act contemplates to justify the appointment of a Deputy.
>
> The Board are most unwilling to put obstacles in the way of the legitimate employment of County Surveyors in private business, being fully aware of the low rate of remuneration they receive for their services as County Officers, but they are of the opinion that the interests of the Counties will not be properly guarded, having reference to the injury that their Public Works may receive from the construction of Railway Works passing over or near them, if the County Surveyors are not in their own Counties unconnected with the management of the latter.[39]

Chapter VIII — Loss of County Surveyor posts

Leahy's request to appoint a deputy was therefore rejected both by the Lord Lieutenant and by the grand jury.[40] In addition, he was told that the Lord Lieutenant could not sanction his involvement in any way with railways passing through County Cork while he continued to act as county surveyor. He was forced therefore to resign his post 21 January 1846, bitterly attacking Daniel Connor, his main opponent on the grand jury.[41]

Dismissal of Patrick Leahy

Just as his son had done, Patrick Leahy also attempted, after the event, to create the impression that he had voluntarily given up his post as county surveyor but in his case, too, the facts were somewhat different. At the Spring Assizes in 1843, serious dissatisfaction was expressed by several grand jurors at the condition of the roads in the East Riding and Leahy was heavily criticised for not devoting more of his own time to the work and for relying too heavily on assistants who were not as active as they should have been. When Leahy foolishly joined in the criticism of two named assistants, the foreman sharply pointed out that he had never previously advised the jury of the deficiencies of these assistants, whom he had appointed himself, and suggested that the men should not be condemned without a hearing. The debate ended with the passing of a vote of censure on Leahy and a statement by the foreman that the jury were being as lenient as they could be and that if the surveyor was unable to perform his duties satisfactorily, he should adopt the honourable course and resign. There was no leniency, however, for Regan, one of the unfortunate assistants, who was summarily dismissed.[42] Leahy's public criticism of his assistants, and his attempt to pass the blame for their deficiencies to the Office of Public Works who had examined them, brought him a sharp rebuke in a letter from that Office which pointed out that it was he who had to carry full responsibility for the selection, appointment and supervision of the assistants.[43]

During the Summer Assizes in July 1844 nearly a full day was devoted by the grand jury to a further debate on Leahy's negligent performance of his roads duties. Specific objections were again raised to the fact that he kept a private office and it was said to be "a perfect absurdity" that his son and his assistants were allowed to survey proposed road schemes in his area while he, as county surveyor, was expected to give impartial advice to the jury on these schemes. There was criticism also of the fact that Leahy had been receiving annual allowances from the city and county grand juries to cover the cost of employing clerks, whereas in fact his office was staffed only by a young apprentice who had been there for three years without receiving any payment. At the end of a long debate, during which Leahy was accused of making a fool of his employers, the jury divided evenly on a compromise motion to fine him £50 by way of deduction from his salary; the foreman's casting vote was given in Leahy's favour and, once again, he escaped with a caution and a direction to be more diligent in the future.[44]

In search of fame and fortune: the Leahy family of engineers, 1780-1888

In March 1845, following suggestions by a number of grand jurors that Leahy's salary should be withheld on the grounds that he had certified faulty road work, the jury agreed that payment of the salary would be suspended pending the outcome of a report which they sent to the Lord Lieutenant.[45] Four months later, when the jury convened again for the Summer Assizes, Leahy was absent on the opening day because of the death of his wife, but his bi-annual report, which was read by one of his assistants, insisted that the roads generally in his area were then in better condition than at any time since his appointment. Whether this view was accepted by the grand jury is not clear, but they decided in any event to restore his salary, one of his supporters on the jury asking rhetorically "how was he to continue to support his family after having sustained a severe domestic affliction".[46]

When the Spring Assizes opened in Cork on 12 March 1846, Leahy submitted his 24th half-yearly report which presented a comforting picture of the state of the roads and of the progress of the county works. However, in possible anticipation of difficulties which were to come, he had written a few days earlier to the Commissioners of Public Works stating his desire to "take up some line of employment connected with the improvement of the country", rather than remain as county surveyor.[47] As the grand jury continued their discussions of fiscal matters, Leahy found himself in serious trouble when a questionable transaction involving the repair of a breach in a sea wall near Youghal was brought to their attention. A presentment for £60 for this work had been approved in 1843 and Leahy, on behalf of the grand jury, let the work on contract for less than £50. He swore at the Spring Assizes in 1844 that the work had been completed satisfactorily and drew down the full £60 but none of this was paid to the contractor who subsequently obtained a court decree ordering Leahy to pay over the amount due to him.[48] The matter was brought up at the Assizes on 16 March 1846 by Daniel Connor – the same man who, according to Leahy, had preferred the vexatious charges which led to Edmund's resignation two months earlier. "This gentleman", he told the Under-Secretary, "not satisfied to stop here, comes from the West Riding where he resides into the East Riding, and at its eastern extremity at Youghal, finds out that I had not paid a contractor for building a sand quay on the coast which had been washed away by the tide, and that I was open to a serious charge for not paying for this dilapidated work rather than refund the amount back to the public purse, the contractor having wholly refused to make good the defect".[49]

Leahy's explanation of his conduct failed to satisfy the grand jury and they rejected his offer to resign at the next Summer Assizes; having lost confidence in him, they decided there and then to dismiss him and to report this to Dublin Castle.[50] Leahy professed to be unconcerned about the loss of his post, telling the grand jury that he could do better without it "for his son and himself had made £15,000 by the railways".[51] Nevertheless, he attempted to forestall the decision of the jury by writing to Dublin Castle to advise that the Lord Lieutenant need not be put to the trouble of enquiring into the cause of his dismissal; he suggested that a sudden withdrawal from the post of county surveyor in a county as extensive as Cork would cause great public loss, as well as disappointment for many contractors, and confusion and inconvenience of the kind which had already occurred in the West Riding where the vacancy caused by Edmund's resignation had yet to be filled.[52] For these reasons, Leahy again proposed that he should be allowed to resign at the following summer assizes "as soon as I can finish the works now in progress and settle the outstanding accounts". In addition, recalling his association with public works in Ireland since 1808, he offered to accept some other office "connected with the improvement of the country and the employment of the people at this period of distress". The Chief Secretary's Office, however, was not prepared to entertain Leahy's pleas and advised him on 19 March 1846 that "the law having invested the grand jury with the power of dismissing any person holding that office, His Excellency can only proceed to appoint your successor as soon as circumstances will permit".[53]

In spite of his dismissal, and the circumstances which had led to it, Leahy was bold enough to offer advice and assistance a few weeks later to William Smith O'Brien, the Young Ireland MP, following reports in the press that O'Brien was planning to introduce a Bill to amend the Grand Jury Acts.[54] The main problem, according to Leahy, was that a Grand Jury "perhaps chosen from their rank in society rather than for their skill and judgement in public works" could not deal with the needs of a great county like Cork during a three or four day session. Besides "our present grand juries are now turned Engineers and Surveyors, by taking all of the business out of the hands of the Surveyor or Surveyors, and pronouncing on the works themselves" – the first step towards "renewing the jobbings of the olden times". O'Brien's reaction to this warning from an official who had himself effectively been dismissed for fraud is not recorded.

Aftermath

As a general rule, where a vacancy had arisen in the ranks of the county surveyors in the years up to 1846 due to the death, resignation or dismissal of an incumbent, it was filled by the Lord Lieutenant within a matter of days by the appointment of the person whose name was next on the list of qualified candidates maintained by the Commissioners of Public Works. However, after Edmund Leahy's resignation in January 1846, it emerged that none of the small number of candidates whose names were then on the list were available to fill the vacancy. Following the dismissal of Patrick Leahy two months later, the Lord Lieutenant directed that new county surveyor examinations should be held, but the Commissioners told him that "in consequence of the heavy pressure of business" caused by their involvement in the planning and administration of famine relief schemes, the holding of examinations at that time would cause serious inconvenience. They decided therefore to "send for the gentlemen they think likely to be suitable" and, four days later, after what can only have been a cursory examination, they recommended that John Benson (1812-74), a Sligo-born architect and engineer, should be appointed as county surveyor for the two ridings of County Cork "under the emergency of the case" and pending the selection of another qualified candidate.[55] As this would have been an unrealistic arrangement, even in the short term, Benson was appointed in early April to serve in the West Riding only.[56] Shortly afterwards, William Augustus Treacy (1818-68), a surveyor and engineer who had been in private practice in Cork for some years, was recommended for appointment to the second post, but since he was "too intimately connected" with the East Riding, Benson was switched from the West Riding and Tracey became surveyor for that riding instead.[57]

The departure of the Leahys from their county surveyor posts came at a particularly inauspicious time. It had been evident from the autumn of 1845 that the failure of the potato crop in many districts throughout the country would lead to serious difficulties in 1846, before that year's crop became available, and that relief measures would have to be organised to provide employment and income in distressed areas. To meet this situation, two Acts were passed by Parliament in early March 1846[58] to allow relief measures to be put in place quickly. One of these allowed the Lord Lieutenant, on the application of the local magistrates and cess-payers of any barony, to make grants of half the cost of local improvement works; the other provided for the convening of special baronial presentment sessions to arrange for additional public works to be carried out in the distressed areas at the ultimate cost of the cess-payers. The County Cork grand jury decided to hold the special sessions for all 23 baronies between 13 April and 13 May 1846 and arranged that extraordinary county sessions to approve presentments would take place on 13 May. Although a high proportion of the applications coming before the special sessions were rejected, there was still a very large number of projects to be examined and reported on by the two new surveyors, Benson and Treacy, whose appointments had been announced only on 11 April.[59] Both men had to tell the county sessions on 13 May that they had been unable to give sufficient attention to the examination of the individual projects[60] and this, coupled with the reluctance of the grand jurors to spend money on works which would have to be funded from the county cess, led to the rejection of more than half the schemes put forward by the baronies.

In spite of all the difficulties, some relief works began under the supervision of Benson and Treacy relatively soon after the special county sessions and considerable numbers were employed before the Government directed that all works were to be closed down by 15 August. Both surveyors, with augmented engineering organisations, were heavily involved again in relief works in 1846-47, and came in for strong criticism at times on account of the design, execution and cost of various schemes. While one can only speculate on how the Leahys would have fared during this difficult period, had they remained in office, it is certain that their sudden departure contributed in no small way to the difficulties experienced in getting practical and cost-effective relief measures off the ground in 1846 in the distressed areas of Cork.

Chapter VIII — Loss of County Surveyor posts

REFERENCES

1. NAI, CSORP 1846 W 5222,
2. NAI, CSORP 1836/1579, letter of 25 May 1836
3. CO 48/392, letter of 2 September 1858
4. CO 48/322, Memorial of 2 June 1851
5. NAI, CSORP 1846 W 4862, letter of 21 January 1846
6. NAI, CSORP 1845 W 9768, letter of 14 July 1845
7. Advertisements to this effect, likely to have been inserted by the Leahys themselves, appeared in the *Cork Constitution* and *Cork Southern Reporter*, 24 and 26 March 1836.
8. NAI, CSORP 1845 W 9768, letter of 14 July 1845; the county surveyor system was extended to Dublin in 1844, but there was no office of "Principal Surveyor".
9. NAI, CSORP 1845 W 9768, letter of 14 July 1845
10. NAI, CSORP 1846 W 4862
11. CC, 7 March 1840
12. CC, 24 July 1834
13. CC, 17 and 20 March 1838
14. CC, 17 March 1838
15. NAI, CSORP 1839/77/95, letter of 19 January 1839
16. NAI, OP 1839/217
17. NAI, CSORP, 1839/77/6519, letter of 24 July 1839 from Richard Cotter, Secretary of the County Cork Grand Jury
18. NAI, CSORP 1842/W/10580, memorial dated 30 July 1842 from Samuel Lane, Grand Jury Foreman
19. NAI, CSORP 1842/W/ 10805, memorial from Patrick Leahy
20. NAI, CSORP 1842/W/11294
21. CC, 16 March 1843
22. CC, 29 July 1843
23. *Cork Southern Reporter*, 29 July and 12 August 1845
24. CO 48/32, memorial of 2 June 1851
25. CO 48/392, letter of 21 August 1851
26. Grand Jury (Ireland) Act,1836, 6 & 7 Wm. IV, c.116, section 40
27. CC, 15 March 1845
28. CC, 18 March 1845; *Cork Southern* Reporter, 15 March 1844; NAI, CSORP 1845/W/3974
29. NAI, CSORP 1845 W 9768
30. NAI, CSORP 1845 W 9768, letter of 14 July 1845 from Edmund Leahy
31. ibid
32. NAI, CSORP 1846/ W/ 1400; CC, 7 August, 1845; *Cork Southern Reporter,* 7 August 1845
33. NAI, CSORP 1846/W/ 4862
34. IRG Vol I, No 52, 27 October 1845
35. IRG Vol II, No 13, 26 January 1846
36. NAI, CSORP 1846/W/524, 1018
37. NAI, CSORP 1846/W/ 292, letter of 5 January 1846
38. NAI, OPW 1/1/4/33, Public Works Letters No 33, pages 98 and 108
39. NAI, CSORP 1846/ W/ 676, letter of 12 January 1846
40. ibid; see also IRG Vol II No 13, 26 January 1846 and CC, 27 January 1846
41. NAI, CSORP 1846/ W/ 1174, 1400
42. CC, 18 and 21 March 1843
43. NAI, OPW 1/1/9, letter of 24 March 1843 from OPW to Patrick Leahy
44. *Cork Southern Reporter*, 30 July 1844; CC, 27 and 30 July 1844
45. CC, 18, 22 and 25 March 1845
46. *Cork Southern Reporter*, 7 August 1845
47. NAI, OPW 1/1/4/33, Public Works letters No 33, pages 316 and 333
48. CC, 14 and 17 March 1846
49. NAI, CSORP 1846 W 5222, letter of 17 March 1846
50. CC, 17 March 1846; NAI, CSORP 1846/W/5222

51. CC, 17 March 1846
52. NAI, CSORP 1846 W 5222
53. ibid
54. NLI, William Smith O'Brien papers, MS 436, letter No 1540 of 7 April 1846 from P Leahy to W S O' Brien
55. NAI, CSORP 1846/W/5786, letter of 24 March 1846 from Jacob Owen; NAI, OPW 1/1/4/33, Public Works Letters No 33, pages 352 and 394
56. NAI, CSORP 1846/W/6654; Benson left the post of county surveyor in 1855 but continued to act as engineer to the Cork Harbour Commissioners, architect to Cork Corporation and city surveyor; he was knighted in March 1853 at the opening of the Great Exhibition in Dublin.
57. NAI, CSORP 1846 W 6976, 7108; NAI, OPW 1/1/4/33, Public Works Letters No 33, page 394; Treacy had some involvement in railway schemes in 1844-46, and had prepared plans in 1844 to bring the Cork & Bandon Railway nearer to Kinsale, in opposition Leahy's chosen route; Treacy transferred to the East Riding of Cork in March 1855 and subsequently served as county surveyor in Mayo and in Tyrone.
58. 9 Vict., c.1 and 9 Vict., c.2
59. CC, 11 April 1846
60. CC, 14 May 1846

THE LEAHYS ABROAD
1846-88

CHAPTER IX

AT THE CAPE OF GOOD HOPE, 1849-51

"My life has been remarkably full of difficulties of all sorts and perhaps for that reason I am not so easily discouraged or frightened by misfortune".

Edmund Leahy[1]

A pyrrhic victory

Edmund, Denis and Matthew Leahy were in Dublin in October and November 1847 while Edmund's case against the Cork & Bandon Railway Company was at hearing but their father, Patrick, did not attend the trial. Charles McAuliffe Keller, who had been an apprentice in Patrick's office from 1840 to 1846, told the Court that Leahy senior had not been in the office since February 1846 and he believed that he was in France at the time of the court case.[2] Leahy's wife, Margaret, had died at their residence in Cork on 28 July 1845,[3] her death notice in the *Cork Examiner* describing her as a person "endowed with a rare meekness of disposition" who had "closed her unobtrusive life as she spent it, in the fulfilment of the duties dignifying the affectionate mother, the devoted wife, and the dying Christian".

The damages of £325 awarded to Edmund Leahy, with only a derisory six pence towards the enormous legal and other costs which must have been involved in the fifteen-day hearing, amounted at best to a pyrrhic victory, gained at great cost. The Leahys' careers in Ireland were already in ruins. Patrick and Edmund had lost their public offices early in 1846 and, with the collapse of their railway enterprises later that year, and the facts disclosed during the court case, they must have been completely discredited and disgraced in the eyes of their colleagues and private clients in Ireland. It is hardly surprising, therefore, that there is no record of any of them in the annals of Irish engineering in 1847 or the following years. There is no mention of them in the records of the Institution of Civil Engineers of Ireland from 1844 onwards and none of them was included in the 1849 list of the members and associates of the Institution.[4] The former family residence at Bruin Lodge, Cork, was acquired by the Great Southern & Western Railway Company in the early 1850s to facilitate the construction of the tunnel which was to bring the

Chapter IX— At the Cape of Good Hope, 1849-51

Dublin–Cork railway from its temporary terminus at Kilbarry into a new station at Penrose Quay;[5] the house served as a temporary terminus for the Cork & Youghal Railway in 1860-61 and was demolished to make way for the extension of that railway to a new terminus at Summerhill in 1861.[6]

Developing "our foreign possessions"

By early 1848, Edmund Leahy had taken up residence in London and was attempting to develop a new career for himself. Victorian London had already become a magnet for generations of middle-class Irish who had a prominent presence there in the arts, politics and the professions[7] but, rather than attempting to achieve fame and fortune in the metropolis itself, Edmund looked to the colonies to secure his future. He wrote initially in March 1848 from an office at 32 Charing Cross to Earl Grey, then Secretary of State for War and the Colonies, seeking to interest him in a scheme for "connecting our foreign possessions with home by means of Steam Communication".[8] The national advantages of such a scheme were obvious, according to Leahy, and he offered "his humble opinion of the ease and advantage" with which the Government could secure "a direct steam communication, at comparatively small cost, with the Cape of Good Hope Colonies and also the Mauritius Islands". At the time, the Peninsular and Oriental Steam Navigation Company (the P&O) was already operating paddle-steamers on the mail service between Britain and Alexandria, linking up by an overland route with the East India Company's steamers which provided a service from Suez through the Red Sea to Bombay. Leahy's proposal was that a new steamship service should be established linking Aden, at the entrance to the Red Sea, via Mauritius with the Cape, and with a branch service from Mauritius to New Holland (Australia), Van Diemen's Land and New Zealand.[9]

Although he seems to have had no experience whatever of the design, construction or operation of steamships, nor any connections with shipping or financial interests, Leahy had no hesitation in assuring Earl Grey that, *as a civil engineer*, he could speak with some confidence of the practicability of the scheme. He offered "to submit the project in minute detail, confident that I shall be able to establish clearly its great national and commercial importance", and offered references, if necessary, from many (unnamed) members of both Houses of Parliament and others. At the time, the security and reliability of routes to the colonies – and especially to India – were matters of great importance to British trade and imperial policy, and the development of some long-distance steamship services was already being subsidised, either directly or by mail contracts, by the Government in London and by some of the colonial governments. And so, in spite of his flimsy credentials, Leahy was invited by the Colonial Office to supply details of his project and of the cost.[10] When he did so, he received no more than a formal letter of acknowledgement expressing Earl Grey's thanks for the proposals; Leahy regarded this as an honour although nothing was done to implement his ambitious plans.[11]

In search of fame and fortune: the Leahy family of engineers, 1780-1888

The empire on which the sun never set did not reach its full development until the first quarter of the twentieth century partly because the Colonial Office was still cautious in the mid-nineteenth century about projects involving expansion and settlement of the African colonies and elsewhere. Ignoring this fact, Edmund Leahy offered his services in April 1848 for an expedition to the Cape of Good Hope "to prepare a general statistical survey and report on the colonies and to establish an extensive system of emigration to them from the United Kingdom under the direction of the Government".[12] He proposed that he should "proceed, with two parties of Assistants, to explore the country parallel with and along the Sea Shore" and for about 150 miles inland. "I propose", he went on, "to be allowed to construct about 30 Main Lines of Road, with suitable branches", at a cost of up £100,000 to be met from a rolling fund financed by a new land tax as well as import and export duties. Confident of the merits of his proposals, Leahy assured Earl Grey that the colony could provide surplus revenue of £2 million and more each year if his scheme were implemented, and added that "the lengthened experience I have had in constructing works such as above alluded to, gave me ample opportunities of witnessing the beneficial effects, of which indeed no one can doubt". But he warned that great care would be needed in selecting the best settlement sites and that there would be a need for skilled surveys for this purpose. His audacious proposal concluded with a passage which warrants quotation in full:

> I beg leave to propose that, say, one thousand emigrants (chiefly navvies, unemployed railway labourers) be sent out, say 500 on the 1st June, and 500 on the 1st July; and that they be immediately after arriving employed in constructing judiciously selected lines of road from the coast to the interior, in the settlements along Algoa Bay, Great Fish River, Port Natal, etc., etc; and I shall have no objection to undertaking the general direction of the work as Engineer, with of course, a proper staff of Assistants.

The official in the Colonial Office to whom this extraordinary proposition was referred must have had some difficulty in restraining himself when presenting his observations to the Secretary of State, but he confined himself to noting that "Lord Grey would not be disposed to delegate so vast an undertaking to an individual".[13] Some days later, Leahy was told, in language of the kind which is still the stock-in-trade of civil servants, that "there are no public funds from which assistance could be provided to enable you to enter on the extensive works you mention".[14]

Despite this rebuff, the Leahys decided to emigrate to the colony about whose potential for development Edmund had written in such glowing terms. The Cape had been occupied by Britain in 1795, after the Dutch East India Company had held it for 143 years as a victualling station for ships travelling from Holland to the Far East. It was returned to the Dutch in 1803 under the Treaty of Amiens, passed again into the hands of the British in

Chapter IX— At the Cape of Good Hope, 1849-51

1806 when war in Europe resumed, and formally became a British possession in 1815 as a result of the peace settlement in Europe. The importance of the Cape to Britain lay in its strategic position on the long sea route to India rather than as a base for territorial expansion of a kind which, at that time, the Government was anxious to avoid in Africa. British settlers were few in the early days but a Government sponsored emigration scheme brought over 4,000 men, women and children from Britain and Ireland to the colony between 1820 and 1822, including two groups totalling nearly 300 from Cork.[15] While subsequent schemes brought some additional Irish immigrants, nineteenth-century South Africa did not attract mass Irish migration at any stage.[16] The colony had a white population of some 76,500 by the early 1850s,[17] and was continuing to attract a steady flow of immigrants - 2,392 between 1842 and 1847 - some of them with passages paid for by the Government of the colony.[18] In 1848, the Cape was experiencing a period of sustained economic expansion, its frontiers were being stabilised, its towns were expanding and it appeared to offer the prospect of attractive new careers for men of talent and enterprise.

Arrival at the Cape

Edmund Leahy was accompanied to the Cape by his wife, Catherine King FitzGerald, whom he had married according to the rites of the Church of England in London on 8 July 1848,[19] and by his 68-year-old father and other family members. The party sailed from St Katherine's Docks, London on 8 January 1849 on the *Geelong,* which had been described in an advertisement in *The Times* on 28 November 1848 as a "fast-sailing clipper-built barque" of 500 tons, commanded by William Wyse, and with "most excellent accommodation for passengers".[20] The ship carried 133 tons of coal for the Cape Gas Light Company, sundry other cargo and a large volume of mail, with 22 passengers, including two children, in what was described as "good accommodation", and eight more in steerage. After 75 days at sea, the *Geelong* arrived in Table Bay in the late evening of Friday, 23 March 1849. The passenger list, as published in the Shipping Intelligence and Shipping Trade columns of the local newspapers,[21] included a Mrs Leahy, a Mr Leahy and "Misses (3) Leahy and Masters (3) Leahy", making it difficult to determine precisely which members of the extended family were in the group. It seems likely, however, that the three unmarried Leahy sisters (Helena, Susan and Anne) were in the party and, if "Mr Leahy" is taken to refer to Patrick, the three "masters" Leahy, may have been intended to cover Edmund, Denis and Matthew. Thus, only Patrick junior, by then President of St Patrick's College, Thurles, and the one married sister, Margaret (Ryan),[22] would have remained in Ireland in 1849.

In travelling to the Cape, Edmund Leahy's first objective seems to have been to obtain for himself the office of Civil Engineer to the Colony from which the long-serving Lt Col Charles Cornwallis Michell had been forced to resign in August 1848 after suffering a severe stroke.[23] With this in mind, Leahy had obtained a letter of introduction from the Colonial

Secretary to Major-General Sir Harry Smith, Governor of the Colony.[24] In addition, he brought with him a number of references, including one from Major-General Sir John Fox Burgoyne, former Chairman of the Commissioners of Public Works, and another from R L Shiel, MP for Dungarvan, who had represented County Tipperary from 1832 to 1841. However, when Leahy approached the Governor about the post, he found that it had already been filled by the appointment in December 1848 of Captain George Pilkington[25] and so he was forced to look to other possibilities to support himself and the family.

Railway surveys

Railway engineers and contractors were at work in many parts of the empire in the late 1840s, supported by English capital and technology. In 1849, the railway mania had yet to reach the Cape but just three weeks after Edmund Leahy's arrival there, the *Cape Town Mail* reported a significant new development:

> Mr Leahy, engineer - who arrived in the *Geelong* - has, it is understood, been surveying the country round about Hottentot's Holland, Stellenbosch, and the Paarl, with a view of laying down a railroad from Cape Town to these districts. Mr Leahy comes out, it is said, appointed by some English capitalists, to report upon the matter, and money will be forwarded to this colony to carry out the object if Mr. Leahy's report should be favourable.[26]

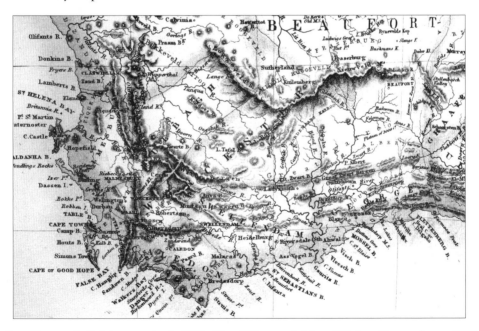

Cape Town and adjoining areas in the 1860s (*Royal Illustrated Atlas of Modern Geography*, ca. 1865).

Chapter IX— At the Cape of Good Hope, 1849-51

This report was accurate to the extent that it referred to Leahy's involvement in survey work but there was no foundation for the suggestion that he was backed by English capitalists: this was either journalistic speculation or, perhaps an interpretation put about by Leahy himself to attract local support for his project. Years later, he wrote that "I first proposed and surveyed the Railway from Cape Town to Wellington *at my own expense*".[27]

While Leahy was engaged in his surveys, a railway route from Port Elizabeth to Graham's Town was being surveyed by another engineer and there were reports that yet another scheme was being promoted to link Cape Town to the interior of the colony.[28] But Leahy seems to have succeeded in interesting the members of the Commercial Exchange at Cape Town in his particular scheme and they appointed a committee, of which he was a member, to consider promoting the project.[29] According to his own account of what transpired, the committee were unanimously of the view "that railway construction is especially suitable - indeed essential - to the development of the resources of the colony", having reached that conclusion "after laborious investigations" and careful surveys and reports from experienced engineers - presumably Leahy and his father.[30] In addition, the Committee established that there was enough traffic between Cape Town and the towns of Wynberg, Stellenbosch, Paarl, Wellington and Somerset-West to produce a dividend of 9% on the investment of £500,000 that would be needed to link these towns together, and with Cape Town and its port, by 70 miles of railway. It was hoped to raise the necessary funds in London, provided a 3.5% guarantee could be secured from the Imperial Exchequer and the Government of the colony.

As he had done a few years earlier in the case of his Irish railway schemes, Leahy rushed recklessly ahead with the promotion of this scheme and by August 1849 he had prepared a prospectus and submitted it to the Government of the colony. But he was sharply reminded by John Montagu, secretary to the Government, that while the Governor was prepared to give the matter every consideration, the prospectus "anticipates the steps which it may be found necessary to take and assumes that the Government have already gone into details".[31] Leahy was told that reference to an interest guarantee could not be made in the prospectus as this had not been officially sanctioned, and it was also pointed out to him that the Governor, who was named as Patron of the company, had not consented to act as such. And finally, the "flattering" proposal that he himself should be President of the company was declined by Montague: he had resolved years before not to be involved in, or connected with, any matter which might be brought before him officially "and from this rule, I never deviate". Ironically, it was his failure to set such standards for himself - and to abide by them - that had contributed to Leahy's loss of his public office in Ireland and was later to involve him in even greater controversy in Jamaica.

Leahy had financial difficulties of his own in 1849. He was sued by the Union Bank in August for payment of a debt of £40 sterling which he had incurred in May and, in default of any appearance by him, the Supreme Court gave judgment with costs for the Bank[32]. But in spite of this embarrassing setback, he seems to have become involved by the autumn in significant engineering activity at the Cape although it is difficult to establish the full facts. According to a document which he wrote in 1851, he had acquired "heavy steam-driven machinery consisting of wheel-work, cast-iron frames, etc" and was "connected with many works and improvements in the colony and was likely to derive considerable advantages from his residence there".[33] A central roads board had been set up in 1844 and the "vast advance" in road development which was being made in subsequent years was reported to have opened up substantial new opportunities for engineers and contractors.[34] Well-planned and engineered new roads were being built using local convict labour, existing roads were being extensively improved, new passes were being cut through the mountains in the interior of the colony and other obstacles to speedy travel were being eliminated.[35] Most of the credit for these works (which, according to the Governor, "would do honour to a great nation instead of a mere dependency of the British Crown") was due to John Montagu, the Colonial Secretary and Lt Colonel Michell, who had been Colonial Engineer from 1828 to 1848.[36] But, with his experience as an Irish county surveyor, Leahy would have been well fitted to play a role in this expanding programme - and may well have done so.

The convict agitation
The mails which had arrived with Leahy on the *Geelong* on 23 March 1849 brought news of plans for the early implementation of a scheme which had been developed by Earl Grey, the Colonial Secretary, to declare the Cape to be a penal colony to which convicts from Britain and political offenders from Ireland would be transported.[37] Plans of this kind had been under consideration for several years, the intention being that convicts would be engaged on infrastructural work, but when they were first publicised in 1848, petitions opposing them were signed by thousands of the inhabitants of the Cape and forwarded to London.[38] Once it became known in 1849 that the scheme was definitely to go ahead, there was an immediate response from the colonists, with hostile meetings and demonstrations being organised by the Anti-Convict Association and threats to boycott anyone who would assist in implementing the scheme or who would employ convict labour.[39]

When the *Neptune* anchored in Simon's Bay on 19 September 1849 carrying the first 282 convicts, the Governor refused to allow them to come ashore. But the agitation continued and intensified. With a view to securing the revocation of the decision to designate the Cape as a penal colony, local residents solemnly pledged themselves to deny supplies or services to the army, the navy and the civil departments and their officers, and to suspend all dealings of any kind with the Government.[40] The anti-convict feeling was "furious and universal" according to the Governor, and the whole colony was in a state of "intense anxiety and excitement", with the great mass of people trying to force him to send away the *Neptune*.[41] He went on to tell the Colonial Office that the universal combination

Chapter IX— At the Cape of Good Hope, 1849-51

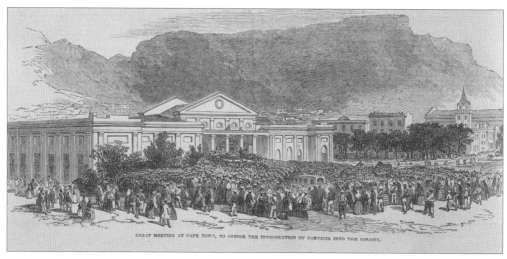

Great Meeting at Cape Town, to oppose the introduction of convicts into the colony, as depicted in The Illustrated London News *(ILN, Vol XV, No. 387, 25 August 1849)*

against the Government was a "most unprecedented proceeding"; that the difficulties it created were surmounted only by "very bold and independent action on the part of a few individuals who would not join in such extreme measures"; and that it was principally through the exertions of Captain Robert Stanford, "a country gentleman who is himself a large farmer", that he had succeeded in breaking the boycott and obtaining food and other essential supplies for the army, the navy, and the rest of the official establishment.⁴²

The 700-ton *Neptune* had left England in February but sailed first to Bermuda to pick up John Mitchel, the Young Irelander, who was to be transferred to the more healthy climate at the Cape because of his asthmatic condition; Mitchel had been detained on a hulk at Bermuda following his sentence in Dublin in May 1848 to 14 years' transportation. His *Jail Journal*, first published in book form in 1854, included several chapters on the agitation at the Cape in the course of which Mitchel expressed the hope that the colonists would "stand out to the last extremity" against what he regarded as "a brutal act of tyranny". He was well aware from contact with visitors to the *Neptune* and from reading the local newspapers of the progress of the colonists' campaign and noted at one stage that the Governor "would be already in sad extremity but for one or two desperately loyal individuals who are coming to his relief". Edmund Leahy was one of this small number of residents who supported the Governor but the fact that he is not named in Mitchel's account of the affair, as was Captain Stanford, suggests that any support he provided was not seen to be significant or did not attract public notice at the time.

Nevertheless, Leahy was bold enough to seek compensation from the Colonial Office for having done his duty, as he saw it, at the Cape. In a memorial of 2 June 1851, he deprecated the efforts which had been made "in a most unconstitutional manner" and by

In search of fame and fortune: the Leahy family of engineers, 1780-1888

a "dangerous combination of the colonists" to starve the troops and the navy and he believed it to have been his "simple act of duty, common to every British subject", to support the colonial government against the "unjustifiable, or more truly, rebellious, proceedings" which had been taken against it.[43] He asserted that it was only by the efforts of himself and Captain Stanford (in that order) that the near-rebellion had been frustrated and martial law averted, his own contribution, apparently, being to procure for the Governor a six months' supply of wheat and flour. This support was formally acknowledged by the Colonial Office[44] and by Governor Smith who described Leahy's conduct as "loyal and patriotic".[45]

On the other hand, the fact that Leahy supported the Governor seriously prejudiced his prospects of continuing to make a living at the Cape for, while the convict ship sailed for Van Diemen's Land on 19 February 1850 and "the triumph of the colonists was complete",[46] the boycott of those who had assisted the government was not lifted; on the contrary, it was resolved at a large public meeting that "the pledge remains forever in force and binding upon the consciences of us, and of our children, to the latest generations".[47] According to Leahy, he was "irretrievably injured" by the continuing boycott: his machinery and property was maliciously damaged, all business intercourse with him was broken off and he was "completely ruined" and "forced out of existence".[48] Patrick Leahy paid an even higher price for his son's loyalty to the Crown: according to Edmund, his father was "most cruelly persecuted by the opponents of the Government and my poor father received a personal injury from which he very soon died".[49] The older man, ten months after his arrival at the Cape, had been "waiting in expectation of some appointment from His Excellency, the Governor" and applied in January 1850 for the position of Civil Commissioner at Fort Beaufort which had just become vacant.[50] But no such appointment came his way; he died at the Cape on 5 October 1850[51] and was buried there in an unmarked grave. It was not until 1855, four years after he had left the Cape, that Edmund was sufficiently in funds to contemplate sending money to Dr Patrick

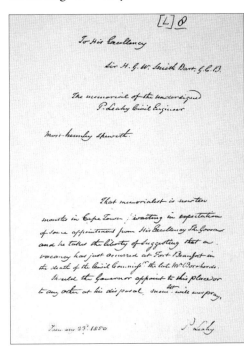

Patrick Leahy's memorial of 23 January 1850 to Governor Sir Harry Smith seeking the position of Civil Commissioner at Fort Beaufort (National Archives of South Africa, CO 4055 L 8).

Griffith, a Wexford–born Dominican who became the first resident Catholic bishop at the Cape, "for the purpose of having a tomb built over our poor father".[52] There is other evidence that the Leahys were in financial difficulties during their stay at the Cape. Edmund, who had borrowed money from his brother Patrick during this period,[53] told him some years later that "when we were in difficulties at the Cape, our poor father wrote to the Governor … who very kindly enclosed him a cheque for £12 without our poor father's even asking for it".[54] It seems that Edmund was unable to repay what he termed this "debt of honour" until April 1855 when he made arrangements with Patrick to have the sum repaid.[55]

The Kaffir War

Edmund and one of his brothers, probably Matthew, returned to London in February 1851,[56] travelling on one of the steamships which had just been introduced on the Cape Town to London route by the Union Castle line. However, he was obliged to leave the rest of the family behind, at least temporarily, "owing to his embarrassed and ruinous circumstances".[57] He took up residence in Peel's Hotel, Fleet Street, and soon began to bombard the Colonial Office with suggestions of one kind or another about public policy matters and with proposals for a variety of development schemes. These were designed, ostensibly, to improve conditions at the Cape but, no doubt, were also intended to provide professional engagements and an income for himself and perhaps his brothers.

Leahy's first set of proposals related to the eighth Kaffir War which had broken out in 1850, the latest in a long series of clashes between the Government of the colony and the Xhosa tribes on its eastern frontier. The cost of the war was greatly resented in England and Leahy clearly saw an opportunity to jump on the bandwagon by putting forward suggestions for a settlement. These ranged from military matters such as the use of light artillery rather than "ordinary troops", to the treatment of the native chiefs and the constitutional arrangements of the colony.[58] He warned the Colonial Office that "many parties at the Cape will endeavour to convert the calamity of the War into a means for amassing wealth" and offered to name large Dutch landowners who were profiteering due to the high price of foodstuffs in areas where crops had been lost because of "the present commotion".[59] Leahy's solution to this was that 20 cargoes of wheat and flour should be sent from London to the Cape and sold at prices well below the inflated levels being charged by the Dutchmen, but the only response from the Colonial Office was a short formal letter conveying Earl Grey's thanks for the communication.[60]

A coaling station at Natal

In advancing his ideas about the conduct and settlement of the Kaffir War, Leahy could have hoped to do no more than ingratiate himself with the Government in London. But his next proposals were of a different character. In a letter of 4 April 1851, he drew attention to the existence of large coalfields near Pietermaritzburg in Natal province and to the importance of having these worked effectively to facilitate steamship services to outposts of the Empire in India and Australia.[61] At the time, steamships bound for the East had to refuel several times on the journey from England and, as only low-grade coal was mined at the Cape for local consumption, colliers were still being sent from England to the various ports on the route, at heavy expense and with some risk of spontaneous combustion. All of this could be obviated, and time and money could be saved, according to Leahy, if local coals could be used to set up a coaling station at Natal.

Leahy told Earl Grey that if he "considered the foregoing suggestions material, I would be happy ... to report fully on their commercial practicability, with the view of carrying them into effect either directly by the Government or, with such encouragement as they would give, by private parties". Although the security and development of the sea routes to India was one of its major concerns, the reaction in the Colonial Office to Leahy's scheme was less than enthusiastic: "the existence of large coal deposits is news to me", was the comment of the official concerned[62] and the reply sent to Leahy pointed out that "the coal, supposing it to be as valuable as you represent it, can only be worked by private enterprise".[63] In the event, large-scale coal mining at the Cape had to await the diamond and gold rush of the 1870s.[64]

Railway development
Ignoring the clear indications from the Colonial Office that State investment in infrastructure or economic development was not consistent with Government policy, Leahy's next move was to attempt to revive the railway scheme he had developed in Cape Town and to obtain either a subsidy or an interest guarantee to enable the project to go ahead. In May 1851, Earl Grey agreed to meet Leahy and others to discuss his proposals for financing the scheme, but the outcome was that Leahy "learned distinctly" that Grey "was not prepared to undertake any such liability on the part of the Home Government".[65] However, Leahy did not give up the effort, suggesting that the Government of the colony would guarantee "a moderate percentage" on the investment required, and provide the land free, if the Colonial Secretary would sanction this.[66] Such an arrangement had already been approved when railways were first proposed in Ceylon, according to a Colonial Office official, but Earl Grey was not prepared to make a rushed decision: in a letter of 21 May, he told Leahy that he was "fully aware of the importance of railway development in almost every country" but he was unable "to judge the feasibility of such an undertaking at the Cape".[67] Besides, the matter was one for the Governor of the colony in the first instance, and interested parties should address themselves to him. Finally, Grey could not "hold out any prospect of Her Majesty's Government recommending to Parliament that any assistance should be granted from the British Treasury towards this railway".

This firm statement of policy by the Colonial Secretary did not end Leahy's hopes of

Chapter IX— At the Cape of Good Hope, 1849-51

direct involvement in railway development in southern Africa. Early in 1852, when there was a revival of interest in building a railway from Cape Town to Swellendam, a local newspaper suggested that Messrs Fox Henderson & Co, who were involved at that stage, had adopted Leahy's "calculations and elaborate surveys". The writer believed that Leahy's reports "showed very clearly that the work is not only practicable, but would also be profitable" and "if the engineer be not in error in his estimate of the cost of the work ... we imagine there will be but little difficulty in carrying out the design".[68] The newspaper's confidence in the project was based on the view that Leahy was "fully competent as he possessed high testimonials" and was "practically acquainted with the Iron Way since the first rails were laid down". With his brother Matthew and Sir Charles Fox, Leahy had a meeting with Earl Granville, the Foreign Secretary, at the end of December 1851, at which the railway proposals may have been discussed[69] and a scheme for building a railway at the Cape, using Leahy's plans, was put to the Colonial Government in the following year but not accepted.[70] Nevertheless, the first railway at the Cape, built some ten years later, seems to have been based to some extent on Leahy's surveys. The Colonial Government, with the support of the authorities in London, had begun planning for this in 1858[71] and agreed to provide the company involved with a 6% interest guarantee on a capital of £500,000.[72] Construction work on the 55-mile line began in March 1859 and it was completed for the Cape Town and Wellington Railway and Dock Company by E & J Pickering, a London firm of contractors, in September 1863.[73]

Road and other development

Having failed in his efforts to attract government support for railway development at the Cape, Leahy turned to suggesting other schemes "calculated to forward the improvement of the colony".[74] He believed that the resources of the area were "of a huge character" and that the colony could become "a source of considerable wealth and strength to Britain" if these resources were properly tapped. However, he argued, major improvements in internal communications were the key to both economic development and security – and if there had been even moderate progress in this respect "many of the embarrassing occurrences for which the colony has been and still is remarkable would not have occurred ...". Besides, if it were left to the colonists, matters would remain at a standstill because, unlike the early settlers, "the present races of inhabitants are becoming degenerate in every way". Late in 1851, he therefore revived his 1848 proposals for a major road improvement programme and suggested also that large reservoirs should be built in areas of seasonal rainfall. All of these works could easily be financed by a land tax, according to his submission, and responsibility for them should be entrusted to a separate new Board to which he would willingly offer his services to work out detailed implementation arrangements. At official level, the Colonial Office, although noting that Leahy seemed to be unaware of the development programmes already being pursued by the government of the colony, drew up a draft reply to Leahy which would have left the way open for him to submit more detailed proposals for consideration. But Earl Grey had clearly heard enough of Leahy's ideas. "Considering what a troublesome man Mr Leahy is inclined to be, a much shorter answer will be better", he instructed his officials; as a result, the reply to Leahy was curt and direct: "His Lordship cannot entertain any such proposals".[75]

Claims for compensation

Facing the prospect that his various development schemes would be rejected, Leahy adopted another course, demanding either to be rewarded directly for his support of the Governor during the anti-convict agitation of 1849 or to be compensated, directly or indirectly, for his consequential losses.[76] In June 1851, he sought appointment to the post of Chief Engineer of the Cape Colony (wrongly believing that it was about to be vacated) or "such other office as may be suitable", but he was told that there was no vacancy which could be offered to him. As an alternative, he then sought "pecuniary remuneration" for services, losses and injuries at the Cape, estimating that a payment of something over £20,000 would be appropriate.[77] He argued that the Government should not allow any private individual to be ruined for doing his duty as a British subject, and pointed to the fact that Robert Stanford had not only been knighted for his services at the Cape in 1849 but had received an award of £5,000 as compensation for his losses. But while the Colonial Office accepted that Leahy had aided the Government, it rejected his compensation claims out of hand in June 1851; officials simply did not accept that his losses were as great as he had claimed, or that whatever losses he may have suffered arose directly from his relatively minor role in the events of October 1849.[78]

Leahy continued to press his claims for some months, and representations were made on his behalf to Earl Grey by Mr Sergeant Murphy, MP for Cork city, who had been involved with him in promoting the Cork & Bandon Railway Bill in 1845. Finally, however, he seems to have decided that if there were to be any prospect of support for new ventures elsewhere, he should withdraw his claims against the Government, while still insisting that the part he played at the Cape was not unimportant.[79] He therefore told the Colonial Secretary in December 1851 that he greatly regretted "ever having troubled him on the subject" of compensation and, in reply, was complimented for having acted judiciously. But Earl Grey was careful to put it on record yet again that while Leahy's support of the Government at a time of great emergency was much to his credit, it could not be recognised as establishing any claims for compensation. Apart from another unsuccessful attempt in 1858 to obtain appointment as Colonial Engineer at the Cape, this exchange marked the end of Leahy's association with the colony. By then, he had developed new schemes which were to occupy him in Turkey for the next six years.

Chapter IX— At the Cape of Good Hope, 1849-51

REFERENCES

1. CDA 1856/12, letter of 21 December 1856 from Edmund Leahy to Patrick Leahy
2. CC, 4 November 1847; IRG Vol III No 54, 8 November 1847
3. CE, 30 July 1845; *Cork Southern Reporter,* 29 July 1845 ; CDA, Ordo of Archbishop Patrick Leahy, 1870 and 1872, entries for 28 July
4. List published by Samuel B. Oldham, Dublin, 1849
5. General Valuation of Rateable Property in Ireland, 1852
6. C Creedon, *Cork City Railway Stations 1849 - 1985,* Cork, Third edition, 1986, page 30 ; Stephen Johnson, *Lost Railways of Co. Cork,* Stenlake Publishing Ltd, 2005, page 25
7. R F Foster, "An Irish Power in London: making it in the Victorian metropolis", in Fintan Cullen and R F Foster, *"Conquering England": Ireland in Victorian London,* National Portrait Gallery, London, 2005, pages 12 - 14
8. CO 48/293, letter of 7 March, 1848 from Leahy to CO
9. The P&O began steamship services to Australia in 1856
10. CO 48/293, letter of 20 March 1848 from CO to Leahy
11. CO 48/322, memorial of 2 June 1851 from Edmund Leahy; letter of 15 April 1848 from the Colonial Office to Leahy, referred to in the memorial, is not included in the relevant volume of papers - CO 48/293
12. CO 48/293, letter of 26 April 1848 from Leahy to CO
13. CO 48/293, minute endorsed on letter of 26 April 1848 from Leahy
14. CO 48/293, letter of 4 May 1848 from CO to Leahy
15. E Bull, "Aided Irish Immigration to the Cape 1823-1900", in *South African-Irish Studies,* Volume 2, 1992, page 269; G B Dickason, *Irish Setlers to the Cape,* A A Balkema, Cape Town, 1973
16. Donal P McCracken, "The Odd Man Out: The South African Experience", in *The Irish Diaspora,* edited by Andy Bielenberg, Pearson Education Limited, 2000, page 252
17. *Return of the White and Coloured Population of the Colony of the Cape of Good Hope,* HC 1852 Vol XXXIII (124)
18. *Third Report from the House of Lords Select Committee on Colonisation from Ireland,* HC 1849 Vol XI (86)
19. Marriage Certificate issued by the General Register Office, RD St George Hanover Square, Vol 1, page 55
20. *The Times,* 24 November 1848; *Cape Town Mail,* 24 March 1849; *Cape of Good Hope and Port Natal Shipping and Mercantile Gazette,* 30 March 1849
21. ibid; no other Leahy arrivals at Cape Town were recorded in these newspapers between March and October 1849.
22. For details of Margaret Leahy's family, see Chapter XVII.
23. ODNB, biographical note by F G Richings
24. CO 48/322, memorial of 2 June 1851 from Edmund Leahy; General Sir Harry G W Smith (1787-1860) became Governor of the Cape Colony and British High Commissioner in South Africa in 1847. He had served in the Peninsular War, in the American War and in the Waterloo Campaign, and subsequently in Jamaica, the Cape and India. Dissatisfaction with his conduct of the 8th Kaffir War (1850-1853) led to his recall in March 1852.
25. CO 53/85, *Blue Book of Statistics,* 1848
26. *Cape Town Mail,* 14 April 1849
27. CO 48/392, letter of 21 August 1858 from Edmund Leahy to CO
28. *Cape Town Mail,* 12 May 1849
29. *Cape of Good Hope and Port Natal Shipping and Mercantile Gazette,* 9 January 1852
30. CO 48/322, memorial of 2 June 1851 from Edmund Leahy to CO
31. CO 48/322, letter of 29 August 1849 attached to memorial of 2 June 1851 from Edmund Leahy; National Archives of South Africa, CO 4047 nos. 57, 59 and 61
32. National Archives of South Africa, Record of Proceedings of Provisional Case, CSC 2/2/1/91, No 46
33. CO 48/322, memorial of 2 June 1851 from Edmund Leahy
34. *Copies or Extracts of Despatches relative to Convict Discipline and the Employment of Colonial Convicts in the formation and Improvement of Roads at the Cape,* HC 1850 (104) XXXVIII
35. *Cape of Good Hope and Port Natal Shipping and Mercantile Gazette,* 4 May 1849
36. *Illustrated Official Handbook of the Cape and South Africa,* London, 1893
37. *Cape of Good Hope and Port Natal Shipping and Mercantile Gazette,* 3 August 1849
38. *Correspondence with the Governors of the Cape of Good Hope and Ceylon, respecting the Transportation of Convicts to those Colonies, etc* HC 1849 Vol XLIII (217)
39. *Cape of Good Hope and Port Natal Shipping and Mercantile Gazette,* 29 June 1849

40. *Cape of Good Hope and Port Natal Shipping and Mercantile Gazette*, 19 September 1849, 12 October 1849
41. CO 48/299, Despatch No 193, 18 October 1849
42. ibid
43. CO 48/322, memorial of 2 June 1851 from Edmund Leahy; CO 48/392, letter of 21 August 1858 from Leahy
44. CO 48/392, minute of 30 August 1858
45. CO 48/322, memorial of 2 June 1851 and letter of 3 December 1851 from Edmund Leahy
46. *Cape of Good Hope and Port Natal Shipping and Mercantile Gazette*, 15 December 1850
47. ibid, 19 February 1850
48. CO 48/322, memorial of 2 June 1851 from Edmund Leahy
49. CO 48/392, letter of 21 August 1858 from Edmund Leahy
50. National Archives of South Africa, CO 4055, L 8, Memorial of P Leahy dated 23 January 1850
51. *Cape Town Mail*, 12 October 1850, death notice
52. CDA 1855/1, letter of 8 April 1855, Edmund Leahy to Patrick Leahy
53. CDA 1860/12, letter of 11 March 1860, Edmund Leahy to Patrick Leahy
54. CDA 1855/2, letter of 30 April 1855, Edmund Leahy to Patrick Leahy
55. ibid; CO 48/392, copy of letter of 26 May 1855 from Sir Harry Smith to Edmund Leahy
56. CO 48/322, memorial of 2 June 1851 from Edmund Leahy
57. ibid
58. CO 336, No 1, Cape of Good Hope Register 1850-1852, letter of 1 March 1851 from Edmund Leahy
59. CO 48/322, letter of 11 March 1851 from Edmund Leahy
60. CO 48/322, letter of 15 March 1851 from CO to Edmund Leahy
61. CO 48/322, letter of 4 April 1851 from Edmund Leahy
62. CO 48/322, note on letter of 4 April 1851 from Edmund Leahy
63. CO 48/322, letter of 11 April 1851 from CO to Edmund Leahy
64. *South Africa, Official Yearbook*, 1979
65. CO 48/322, letter of 12 May 1851 from Edmund Leahy; the Court Circular of 7 May 1851 (published in *The Times*) confirms that Matthew and Edmund Leahy had an interview the previous day with Earl Grey and that a Mr Harrison Watson had also attended.
66. CO 48/322, letter of 12 May 1851 from Edmund Leahy
67. CO 48/322, letter of 21 May 1851 from CO to Edmund Leahy
68. *Cape of Good Hope and Port Natal Shipping and Mercantile Gazette*, 9 January 1852
69. The Court Circular, as reproduced in *The Times*, 1 January 1852
70. FO 78/2163, printed pamphlet of 20 pages entitled *Leahy v. The Ottoman Government: A Statement of the Claim of Edmund Leahy, Esq., CE, upon the Ottoman Government, in the matter of the Belgrade and Constantinople Railway*, I R Taylor, London, (?) 1859
71. CO 48/391, letter of 10 August 1858 from the Railway Department, Board of Trade; CO 48/390, Despatch No 143, 24 July 1858
72. *Illustrated Official Handbook of the Cape and South Africa*, London, 1893
73. ibid; Fred A Talbot, *Cassels Railways of the World* (2 vols) London nd, pages 517 and 518
74. CO 48/322, letter of 17 November 1851 from Edmund Leahy to CO
75. CO 48/322, letter of 25 November 1851 from CO to Edmund Leahy
76. CO 48/322, memorial of 2 June 1851 from Edmund Leahy
77. CO 48/322, letter of 16 June 1851 from Edmund Leahy
78. CO 48/322, letter of 24 June 1851 from CO to Edmund Leahy
79. CO 48/322, letter of 3 December 1851 from Edmund Leahy

CHAPTER X

EDMUND AND MATTHEW IN TURKEY, 1852-58

"we are such good friends (particularly Reshid Pasha, the Grand Vizier to whom I may say I entirely owe our present independence) that I think they will not entrust those things to anyone from me".

Edmund Leahy, 1855[1]

Paving the way for the Orient Express

Railway building began in Belgium, France, Germany and Italy in the 1830s and soon afterwards in Switzerland and Austria. By 1851, it was possible to travel from the English Channel through to Vienna, capital of the Hapsburg Empire of Austria-Hungary and for centuries the hub of communication lines. This rail link, extending for some 700 miles, incorporated sections of track which had been planned and constructed on a piecemeal basis, largely by private companies. To the south and east of Vienna lay the vast Ottoman Empire of some 35 million people, with its capital at Constantinople guarding the gateway to the Orient. The European section of the empire included much of present-day Greece, Bulgaria, Romania, Moldova, Croatia, Serbia and Bosnia, while in Asia, the empire extended eastwards as far as Basra on the Persian Gulf, taking in Anatolia, Syria and most of Iraq. With one long extension eastward, Edmund Leahy proposed in 1851 to link Western Europe to Constantinople by railway, a project with little precedent in terms of scale and ambition even at the height of the railway mania.

Leahy's initial – and relatively modest – proposals for railway development in Turkey were formulated during the summer of 1851 while he was still engaged in extensive correspondence with the Colonial Office about his various schemes for developing the Cape colony. In August of that year, from his temporary base at Peel's Hotel, Fleet Street, London, he wrote to Lord Palmerston, the Foreign Secretary, submitting "on behalf of himself and other British subjects", a proposition to the Turkish Government to construct a railway from Constantinople to the town of Adrianople (now Edirne) situated about 150 miles north-west of the then Turkish capital.[2] As well as constructing this line, Leahy and his partners proposed to provide all of the rolling stock and "if practicable to connect

Chapter X— Edmund and Matthew in Turkey, 1852-58

London with Constantinople by a continuous line of Electric Telegraph". The whole enterprise was to be dependent on a guarantee from the Turkish Government covering the capital involved, and a free grant of the necessary land.

For strategic reasons, support for the Ottoman Empire was a fundamental part of Lord Palmerston's foreign policy, the major objective being to maintain the balance of power in the eastern Mediterranean and to protect the route to India and the far-eastern colonies from pressure or disruption by Russia. But despite this objective, it was unrealistic to expect much assistance for a Turkish railway project from a Government which was cautious, to say the least of it, about involvement in similar projects in Britain, Ireland and the colonies. The reply from the Foreign Office to Leahy's proposal was prompt and uncompromising: he was told that "Her Majesty's Government make it a rule not to take part in, nor to interfere in, private speculation of this kind".[3] But Leahy was not easily discouraged and so, instead of dropping his plans in the face of this refusal of Government support, he went on to develop a much more ambitious scheme which he presented to the Foreign Office in November 1851.[4] He claimed that in the previous three months he had prepared "general plans and sections of the entire country" from Belgrade to Constantinople, and from the Asiatic shore of the Bosporus through Aleppo to Baghdad and on to the eastern limits of the Ottoman Empire at Basra on the Persian Gulf. Bearing in mind the lack of reliable large scale maps, the difficult nature of the terrain, and the distance from Belgrade to Basra of over 1,800 miles, this must have seemed an incredible claim even to those who were familiar with the hasty surveys carried out at the time of the railway mania and the enormous output and workloads of Victorian railway engineers. Leahy was later to admit that he had done no more than prepare *in London* "an outline sketch plan of the probable course appearing to be practicable, according to the most reliable sources of information then obtainable, which, however were very meagre and scanty".[5]

Leahy was not the first to propose a major expansion of the railway system to improve communications with India. A British civil engineer, Sir Rowland Macdonald Stephenson (1808-95), who had been primarily responsible for promoting the East India Railway on which work began at Calcutta in 1850,[6] wrote to Lord Palmerston in January of that year proposing the construction of 2,800 miles of railway from the English Channel to the Persian Gulf, so as to create what he described as "a National Highway through Europe to the British possessions in Asia"; Stephenson was given letters of introduction to several governments along the proposed route to assist him in negotiations with them, but did not succeed in advancing his scheme.[7] Following so closely the proposals advanced by the more experienced Stephenson, Leahy's scheme was remarkably well received, perhaps because he had managed to come to some arrangement about the construction of the line with Sir Charles Fox, the eminent civil engineer and contractor, and with Thomas Brassey, the greatest railway contractor of the mid-Victorian era; according to his own bizarre account of events, he "introduced these gentlemen, as the intended contractors, to Lord Palmerston who was "satisfied with their respectability".[8]

In search of fame and fortune: the Leahy family of engineers, 1780-1888

According to Leahy, both the President of the Board of Control for India, Lord Broughton, and the East India Company promised support for the project and the Turkish Ambassador in London was also convinced that Leahy's inter-continental railway plans would "contribute mightily" to the prosperity of the Ottoman Empire.[9] In addition, Leahy claimed to have been told by the Ambassador that even if the Turkish Government did not take on the project, it would at least meet all the expenses incurred in preparing preliminary plans and surveys, as well as his fees as consulting engineer. Leahy then sought backing in England for the project, on the basis that he had been commissioned to travel to Constantinople "with the view of arranging and agreeing with the Turkish Government for the completion of this great work".[10] In a letter to Lord Palmerston, he stressed the economic and strategic advantages that would accrue: the line would open up enormous markets in Persia and central Asia for British manufacturers and, in addition, the proposed railhead at Basra (which could be reached by train from London in less than a week) was within a few days' steaming of India and Ceylon. Palmerston, who was always concerned to promote British interests abroad, may well have been attracted by these possibilities - according to Leahy "he highly approved" of the scheme[11] and granted him an interview at the Foreign Office to allow him to present and explain the plans.[12] As a result, Leahy was given a despatch for delivery to Sir Stratford Canning, the long-serving British Ambassador in Constantinople,[13] instructing him to afford his protection and good offices to Leahy, as far as he could properly do so;[14] this was a matter of some significance because Canning exercised an unusual degree of influence on the Sultan and his policies and had continually pressed for reform and modernisation measures, including improved communications.[15]

Before leaving for Constantinople, Leahy also sought to interest Earl Grey at the Colonial Office in his new project and offered, with Sir Charles Fox, to bring the railway plans to the Office for his inspection. In addition, he sought "any mark of his Lordship's approval with which he may honour me - as a professional man, some notice would be useful, besides which I should always feel proud of being able to show that, when it was needful, I was ready to aid the Government and was found among those who acted properly".[16] The reply (based on Grey's own manuscript notes) accepted that Leahy's conduct during the "great emergency" of 1849 at the Cape was much to his credit, and noted that his new undertaking was one of great importance. But that was as far as Earl Grey was prepared to go: the railway project had "no immediate connection with the business of the Colonial Office and would require more consideration than his Lordship has time to devote to it in order to enable him to form an opinion on the feasibility of the project. He will not, therefore, give you the trouble of bringing him the plans".[17]

Arrival in Constantinople

Edmund and Matthew Leahy arrived in Constantinople in February 1852, with some assistants.[18] At the time, the Ottoman Empire was already regarded as the sick man of

Chapter X— Edmund and Matthew in Turkey, 1852-58

Europe: it appeared to be moribund and crumbling and many observers did not expect it to survive for long in spite of the efforts of the Great Powers to keep it intact. There was an antiquated and hopelessly incompetent, chaotic, corrupt and disorganised administration and little progress was being made in the implementation of the Tanzimat – the reform movement initiated by Sultan Mahmut II in the 1820s. While Constantinople had been linked to Britain by steamship services for 20 years, leading to a steady influx of British visitors and businessmen, westernisation was still an objective rather than a reality in 1852. In such circumstances, it must have been difficult for newcomers like the Leahys, even with the letters of introduction they brought with them from London, to make the contacts necessary to advance their plans. However, with the assistance of Sir Stratford Canning and the Turkish Ambassador in London, they not only managed to obtain permission to go ahead with their railway surveys but also to procure a *firman* (effectively, a royal decree) from Sultan Mejid I who ruled from 1839 to 1861 directing provincial governors and other officials to give them all the assistance they might require in making the surveys. The full text was as follows:

> To all my Mufties, Governors, Vice-Governors, and all other of my great and provincial officers, who shall behold this, my illustrious name, we command you to give attention and to take care that our illustrious and Royal wishes shall be attended to. Now, know you all, that the eminent and learned English engineers, Edmund Leahy and Matthew Leahy, of distinguished birth and elevated parentage, have been recommended to my Royal and Imperial self by my illustrious and ever faithful ally the Ambassador of England, to whom all honour and glory be given to the end; and it is my Royal and Imperial wish that you all take care that those distinguished princes in their profession of engineering may receive every assistance and authority which they shall require, to enable them to make all examinations and measurements for planning the new mode of moving by steam through my empire, first from Constantinople to Adrianople, Philippopoli, Sofia, and Belgrade, and in all other places where they may wish to examine. Let all who behold this, my Royal and Imperial Command, give attention and take heed, and be fearful.[19]

With their assistants, the Leahys worked on the surveys of the Constantinople-Belgrade section of the proposed railway for most of 1852 and 1853. In July 1852, Lieutenant Colonel Neale, the British Consul at Varna on the Black Sea, reported to the Foreign Office that "two Englishmen, Messrs Edmund and Matthew Leahy, visited Varna on their return from surveying a route through Roumelia and Bulgaria for the purpose, as I am informed, of forming plans and estimates for a projected railway from Constantinople to the Austrian frontier and passing through those provinces".[20] Neale believed that the proposed railway could have important political and commercial results and was somewhat annoyed that the

Leahys would not disclose whether the project was being undertaken with the knowledge of the British or Turkish Governments - "the two gentlemen . . . observed a reserve respecting their plans at which I venture to express my regret". Nevertheless, Neale was able to report in some detail on the route proposed by the Leahys for a line from Belgrade to Constantinople (a distance of some 550 miles), with a branch about 160 miles long from Sofia to Orsova on the Danube, and he put forward his own suggestions for changes designed to take better account of local circumstances and traffic needs.

Turkey and adjoining countries, ca 1850, showing the route of the proposed Belgrade-Constantinople railway.

The Leahys were in Varna again in December 1852 when they told the consul that the first section of the line from Constantinople westward to Phillipopoli "could be considered as almost decided on by the Turkish Government" and that work on it might begin in the spring of 1853.[21] In addition, they claimed to have made accurate surveys for a railway from Ruschuk to Varna - a line which "presented no difficulties whatever". However, according to the consul's despatch to the Foreign Office, "they did not seem to entertain any immediate hopes that the line would be undertaken by the Turkish Government" and they dwelt at length on a cheaper option - a good stone road on which omnibuses might run on payment of a toll. The Leahys also told the consul that they had visited Bucharest and had found the Prince of Wallachia, who ruled that principality under the control of the Sultan, very willing and anxious to encourage a projected railway from Bucharest to Georgina, provided the line from Ruschuk to Varna was assisted by the Government in Constantinople. They had intended to travel onwards to Kustengee, on the Black Sea, to explore the practicability of reviving an earlier proposal to develop a canal from there to Cyerna-Voda on the Danube but, while bad weather forced them to return to Constantinople, it appears that a report and/or plans for a Danube canal were prepared by the Leahys a few years later.[22]

All of this would have amounted to a formidable programme of work, even for a large team of engineers and surveyors, and it must therefore be assumed that the Leahys' surveys and railway plans were of a very preliminary nature indeed. In any case, detailed survey work would hardly have been appropriate in advance of some financial commitment from the Turkish Government which, by early 1853, was already concerned with more pressing issues. Tension had built up in the Balkans and the series of events which was to lead to the Crimean War had already begun. In July 1853, the forces of Tsar Nicholas I crossed the river Pruth, marched into the principalities of Moldavia and Wallachia (now southern Romania), advanced to the Danube, and laid siege to Silistria, south of the river and located in today's Bulgaria. The Sultan declared war on Russia in October and soon afterwards his troops engaged the Russian armies in Wallachia. In this situation, there was little prospect that his Government would have been disposed to commit itself to major infrastructural development of the kind being promoted by the Leahys, some of it in the very provinces which had been overrun by Russia. Not surprisingly, therefore, when Charles Stokes, agent for Fox and Brassey, "offered favourable terms to the Ottoman Government for immediately making the part of the line from Constantinople to Philippopoli, and the remainder on the same terms, as soon as required by the Government, and in accordance with Mr Leahy's plans", the Government rejected the offer.[23]

The Crimean War
When Turkey and Russia went to war in 1853, the basic concern of the British Government was to maintain the integrity and independence of the Ottoman Empire so as to ensure the preservation of the overland route to India and to prevent Russia from gaining control of the eastern end of the Mediterranean and becoming a naval power in the area. Although Britain did not formally enter the war until March 1854, contingency plans were being made for some months before that. By a curious coincidence, two of the main figures involved in these early stages had connections with Edmund Leahy; both of them were later heavily criticised for their part in the conduct of a war which was characterised by indecisive leadership of the Allied armies and extraordinary incompetence and inefficiency in military operations.

The British Mediterranean fleet was moved to Besika Bay, close to the Dardanelles, in June 1853, and to Constantinople in October, under its commander Vice-Admiral Sir

James Whitley Dundas who, as liberal MP for Greenwich, had been a member of the House of Commons committee which examined and recommended Leahy's Cork & Bandon Railway Bill in 1845. In May 1854, the fleet landed British troops at Varna, under the command of Field-Marshal Lord Raglan, to prevent a possible Russian advance on Constantinople. No evidence has emerged to suggest that Leahy had any contact with Admiral Dundas during this period but he did make contact with Lieutenant-General Sir John Fox Burgoyne who, at the age of 72, joined Raglan's staff in Varna as General of Engineers and became his trusted adviser.

Having served as Chairman of the Commissioners of Public Works in Ireland from 1831 to 1845 and as first President of the Civil Engineers Society, Burgoyne was obviously familiar with Leahy and his Irish engineering. He had been serving as Inspector-General of Fortifications and head of the Royal Engineers when he was instructed to visit Constantinople early in 1854, to assist in the preparations for war and to advise on the fortifications and other works necessary for the defence of the city. It was on his recommendation that Varna was selected as a suitable base for the British and French armies and about 130,000 troops had moved to the area by July. But neither the British nor the French commanders had any reliable information about the locality and the maps available to them were hopelessly inadequate.[24] According to Edmund Leahy, he therefore gave all of his "plans and reports of Turkey to our Government, free of charge upon their solicitation" and he claimed that the plans had been used by Burgoyne and by Lord Raglan.[25]

Support for Leahy's claim is provided by a letter written by Burgoyne in 1858 which testified "to the extreme readiness" shown by Leahy at Constantinople "during the late War in the East to afford any information and to forward the public service in every way".[26] Burgoyne elaborated on this in a letter written in 1867, when he stated that "previous to the Crimean Campaign, and during the preparations of a warfare in Turkey" Leahy had placed at his disposal "some most interesting maps and surveys of the country that had been made and collected by him".[27] In 1867 also, General Lord Strathnairn who, as Colonel Hugh Henry Rose had served as chargé d'affaires at the embassy in Constantinople in 1852-53, wrote that Leahy did him very good service in his diplomatic post by giving him "true and very important information respecting political events which ended in the Crimean War".[28] There is also a letter dated 28 March 1854 from two Army staff surgeons acknowledging "the invaluable topographical information on European Turkey" which one of the Leahys had provided and which "is so calculated to benefit the Medical Department of the English Army at the present conjuncture".[29] Given that the allies had only rudimentary maps of the theatre of operations at the time, and the evidence that the Leahys had actually surveyed the area around Varna, it seems reasonable to accept that their survey work may well have assisted the British commanders in 1854 and, in particular, the troops of the ill-fated Light Brigade who were assigned to carry out reconnaissance of the Russian positions as far north as the Danube while the brigade was based at Varna.[30]

Chapter X— Edmund and Matthew in Turkey, 1852-58

In September 1854, after the Tsar had withdrawn his troops from the Danube on foot of a threat by Austria to join the war against him, an allied expeditionary force of some 50,000 men landed at Kalamita Bay in the Crimea, to begin an advance south-eastwards on Sevastopol, Russia's great arsenal and port, and the base of its Black Sea fleet. It was largely on Burgoyne's advice that this location, some 30 miles north of Sevastopol, was chosen as the landing site, and he was primarily responsible for the decision of the allied commanders, after the initial victory at the Alma on 20 September, not to attempt a direct assault on Sevastopol from the north, but to march around the city to Balaklava and attack Sevastopol from the south. This strategy was heavily criticised as the year-long siege dragged on, punctuated by the famous battles of Balaklava and Inkerman. As the army's chief engineer, Burgoyne had to carry a large share of the responsibility for the failure to provide adequate transport facilities and back-up supplies and services for the army in the Crimea, a failure which caused terrible hardship for the troops as the fierce Crimean winter of 1854-55 closed in. The "condition of the army before Sevastopol" and its suffering from cold and cholera contributed to the fall of Lord Aberdeen's government in January 1855 and led to the recall of Burgoyne to London a month later.

The inability of army engineers in the Crimea to plan and carry out the works necessary to allow provisions and materials of war to be moved efficiently from the port at Balaklava to the positions occupied by the army on the heights above Sevastopol led to the establishment of two civilian work forces, each of them led by Irish engineers. As chief engineer of the Civil Engineering Corps, James Beatty initiated the building of the Crimean Railway in January 1855, a project which was described some years later as "one of the greatest feats in railway making that has ever been known".[31] The 40-mile network from Balaklava to various points on the heights above Sevastopol was constructed in record time by engineers, contractors and several hundred navvies who, with their horses, engines and other equipment, were brought out from England in a fleet of large steamers. In the following August, the Army Works Corps of more than 700 men, with William Doyne as its superintendent, arrived in the Crimea and set about the construction of some 20 miles of all-weather road from Balaklava to the British and French positions. There is no mention of Leahy in Burgoyne's extensive published correspondence and journals relating to this period, nor is he referred to in the detailed reports sent to London by *The Times* correspondent, W H Russell;[32] neither is there any other evidence that Leahy visited the Crimea or had any direct involvement in these construction projects. However, as one who prided himself on his loyalty to the British Government, it is tempting to speculate that he could hardly have refrained from proffering advice on railway and road construction to Beatty and Doyne, as they passed through Constantinople; in April 1855, he was certainly in contact there with Thomas Brassey[33] who, with his partners, Samuel Morton Peto and Edward Betts, had been responsible for the organisation of the Civil Engineering Corps.

The siege of Sevastopol dragged on into the spring and summer of 1855, with regular bombardments and periodic assaults on the outer defences. The proceedings attracted widespread interest in Britain and Ireland where the newspapers, particularly *The Times,* were able, for the first time, to keep the public in touch with the latest details of a major war by reports telegraphed by special correspondents and sketches by war artists. Edmund Leahy's letters to his brother Patrick during this period provided first-hand reports and observations on some of the military operations. With the siege underway for six months, Edmund was speculating in a letter written at Easter 1855 about the beginning of a new allied bombardment of the town from a line of forts and entrenchments which had just been completed.[34] (A major bombardment did in fact begin on 9 April). In the same letter, he reported the arrival at Constantinople of "fresh troops every day on both tides, promising a bloody fight whenever the final struggle is made for Sevastopol". Some weeks later, he had "no news from Sevastopol - the siege drags itself slowly along without providing any suitable effect on the Russians".[35] Sevastopol finally fell on 8 September 1855, after a last massive artillery bombardment, but hostilities did not formally end until the Treaty of Paris was signed over six months later.

As the war dragged on, large numbers of sick and wounded British troops were evacuated to military hospitals at Scutari and other suburbs of Constantinople where, in the early stages, horrific conditions were experienced by injured and dying men. There is no evidence that the Leahys had any involvement in the work of refurbishing and improving these hospitals which went on during 1855. It was decided in the autumn of that year, after the fall of Sevastopol, that the English cavalry should leave the Crimea and transfer to winter quarters in the Bosporus, and the construction of suitable barracks and stables to accommodate up to 6,500 men and their horses became an urgent necessity. Brigadier-General Henry K Storks, who was in charge of the British headquarters at Scutari, instructed Captain E R Gordon of the Royal Engineers to lose no time in completing the works necessary to accommodate 1,500 of the cavalry there[36] and directed that the contract system should be used, as well as direct labour operations involving the Royal Engineers, to speed up the work. Edmund Leahy was awarded a contract to construct some of the new accommodation and, although work did not begin until the end of October, the correspondent of *The Times* was able to report on 27 December that barracks and stables had been constructed by Leahy in the form of "a parallelogram of 550 yards by 170, one side of which forms the barracks, and the three others the stables".[37] The report went on to say that the system of building adopted "is the one commonly used for the roofs of factories, combining strength with lightness", but it noted that lack of ventilation at the stables was already a problem and that adequate provision for drainage had not been made, leading to the possibility that the whole area would become an unpleasant sea of mud after a few days' rain. In other respects also, Leahy's performance seems to have created difficulties: according to a confidential report from Sir Henry Bulwer, the British ambassador, the contract for building the stables provided that Leahy "was to be paid by measurement" but he was found to have "charged double".[38]

Chapter X— Edmund and Matthew in Turkey, 1852-58

Railway Dispute

By 1855, whatever arrangements had existed between the Leahys and Sir Charles Fox had broken down. Edmund had been treated badly by Fox, according to a letter of his in April 1855 to his brother Patrick, and Fox's agent in Constantinople, Charles Stokes, "did me as much harm as he could".[39] But the letter went on to say that Leahy was still in contact with Thomas Brassey - "a better and more reasonable man for whom I have a great respect" - and expressed optimism about the prospects for his railway schemes, particularly because of his friendship with Mustafa Reshid Pasha, the Grand Vizier who was head of the central administration which controlled the Sultan's vast empire. Reshid Pasha was the greatest Turkish statesman of the period and had worked hard to promote westernisation and reform in Turkey but his plans to improve communications and to develop natural resources had to compete for funds with the extravagant and insatiable demands of the household of Sultan Abdul Mejid and his costly building projects, including the elaborate Dolmabahce Palace on the shores of the Bosporus; this extravagance impoverished the empire and forced the Government to take out a series of foreign loans at very high interest rates which it could ill afford to repay.[40] Nevertheless, Leahy was still confident in April 1855 that he would be engaged by the Government in connection with railway schemes and other construction works because "we are such good friends (particularly Reshid Pasha, the Grand Vizier to whom I may say I entirely owe our present independence) that I think they will not entrust those things to anyone from me; should they do so, I shall be obliged to go to London to negotiate those affairs on their behalf. In short, things look very satisfactory, Thank God".[41] However, a more realistic view of the likely outcome was held by the Constantinople correspondent of *The Times* who, in assessing the prospects for foreign investment in Turkey, warned against the risks inherent in advancing "magnificent schemes ... involving the outlay of millions" and the "extravagant ideas of sanguine Englishmen accustomed to Crystal Palaces and Britannia Bridges". As an example of what he had in mind, he cited the Leahys' proposed Constantinople–Belgrade railway, which, he wrote, would cost seven million pounds, run through a desert for hundreds of miles, probably never pay even a one per cent dividend, and ruin those foolish enough to finance it.[42]

In the event, Leahy's belief that his friendship with the Grand Vizier would be enough to ensure his engagement in the development of railways in the empire was not well-founded. When the Ottoman Government resumed their consideration of a possible rail link to Western Europe after the fall of Sevastopol, they were faced with a flood of proposals from British, French and Austrian investors and sensibly decided to initiate a public tendering process rather than accept any particular proposal. In October 1855, they published a detailed statement of the conditions under which they would be prepared to grant a concession for the construction and operation of the railway and set 1 April 1856 as the closing date for the receipt of tenders.[43] According to the local correspondent of *The Times*, W H Russell, the Leahys at that stage were the only civil engineers to have actually surveyed the whole line and while he believed that their report must be considered as a highly favourable one, calculated to attract foreign capitalists, he cautioned that "the attraction is likely to be neutralized by the unsatisfactory nature of the conditions imposed by the Turkish Government".[44] *The Times* went on to print extracts from the report which had been prepared by the Leahys describing the proposed line:

> The line of the railway is singularly direct between the two termini, and yet passes through the chief towns of European Turkey, including Adrianople, Philippopoli, Sophia, and Nish. It presents no extraordinary difficulties of construction excepting the passage of the Balkan, and even there we have been able to find a line adapted for locomotive power; but we must add that this portion requires the construction of some heavy bridging and numerous tunnels - none of the latter, however, exceed 2,000 yards in length. There is no gradient on the main line exceeding 1 in 59.3, and there is no curve of less than 400 yards radius.
>
> Previously to determining upon the best point for passing the summit level on the Balkan, we prepared a section of the entire of that range of mountains along its summits between Cape Eminé, at the Black Sea, and Mount Vitoab, south-west of Sophia. This section shows at one view the heights of all the passes of the Balkan, and, whilst it makes the greatest difficulty to be encountered in the formation of the railway quite apparent, it places the superiority of the line selected beyond any question of doubt.
>
> In the preparation of the surveys we have had to contend with unusual difficulties, owing to the great errors of the best published maps of Turkey. The German military map, which was supposed to be correct, is one mass of errors: in illustration of which we may mention, as regards the important towns of Dimotika and Tchirmen, that the former is placed to the east of the river Maritza, whereas it is actually on the west side; and the latter is placed to the north of the same river, while its true position is south of the Maritza, and 10 miles from where it is shown on the military map alluded to.
>
> We can state generally, after most careful examination, that this railway, throughout its entire length, can be made and worked with as much facility as any line in Europe of equal extent; and, looking at the great natural richness of the country which it traverses, as well as the existing population and traffic along its course, far exceeding those of any other part of European Turkey, we have no hesitation in saying that, if properly conducted, it promises to be very successful as a mercantile speculation.
>
> During our examination of the country previously to selecting the best course for the railway, we were much struck with the variety of geological formations that presented themselves, and with their probable mineral richness. A closer

Chapter X— Edmund and Matthew in Turkey, 1852-58

examination soon confirmed our previous suppositions; and we have the satisfaction of now forwarding specimens of silver, copper, lead, mercury, and iron, together with coal, salt, and saltpetre, which we prepared on the ground, and also specimens of the ore from which the metals were separated, &c.

While the tendering process was under way, Edmund Leahy offered his plans to the Porte for a sum of between £11,000 and £12,000. The plans were in three sections, the first showing the line from Constantinople to Adrianople, the second covering the section from there to Philippopoli and Sofia and the third showing the route from Sofia through Nish to Belgrade, some 700 miles in all. Each strip map, even if drawn to a small scale, must have been quite lengthy and it was necessary to lay them out on the floor of the Council Chamber while Leahy explained the various features - "the gradients of the line, the excavations and embankments, the bridges, viaducts, and tunnels ... and the works of art necessary for its completion".[45] But though the authorities expressed satisfaction with the plans, they refused to meet Leahy's financial demands. Subsequent negotiations in 1856 led to a verbal agreement under which the Government was to pay £1,000 immediately for the plans, and a further £6,000 as soon as they made any definite arrangements to construct the railway. According to Leahy, the terms of the agreement were subsequently set out in French by a Greek banker, George Zarify, a friend and business associate of Leahy's and who, notwithstanding the fact that he was also apparently a friend of the Sultan and confidante of many of those in government, has been described as "a rogue of doubtful reputation".[46] Fresh difficulties arose a few days later when it was suggested to the Government that the plans they were to buy from Leahy were merely copies and that the originals, which were in London, were the property of Messrs Fox Henderson & Co who had paid Leahy for his work in 1852-53. Although Leahy hotly refuted all of this and dismissed the allegations as mere intrigue, the Government took the matter seriously; they refused to make any payment to him, despite an intervention in his favour by the British Ambassador, and ultimately denied the existence of any binding agreement. Quite apart from this, reports emanating from Constantinople in 1856 suggested that the Government were simply not ready to make decisions on railway links between the empire and the rest of Europe.[47]

In 1857, Sir Austen Henry Layard (1817-94), who had been an attaché at the British Embassy at Constantinople from 1847 to 1852 before serving briefly as Under-Secretary of State for Foreign Affairs, was among those who were actively seeking a concession which would allow the Imperial Ottoman Railway Company to build a Vienna–Constantinople railway.[48] Layard had already established the Ottoman Bank to arrange funding for projects such as this and when Leahy heard of these developments, he wrote to remind Layard that he and his brother had "made this subject our study for the last five years".[49] He recalled that he had the pleasure on one occasion at Constantinople of showing Layard the preliminary plans and sections which he had sold to the Turkish Government the previous year, and he offered to assist in any way he could in progressing Layard's scheme. However, although the Turkish Government had granted a concession in 1856 to a group of British investors to build and operate a railway from Smyrna (now Izmir) to Aidin in Asiatic Turkey,[50] and some further concessions for lines along the western side of the Black Sea,[51] it was to take another 12 years for decisions to be made on the commencement of railway development between Constantinople and the west.

On his return to London in February 1858, and with no prospect that his railway plans would be adopted by any commercial concern, Leahy sought the intervention of the Foreign Secretary in an effort to force the Ottoman Government to pay him the £7,000 which he claimed was due to him for the plans. In a letter of 5 March, he blamed French intrigue and jealousy, and Turkish chicanery, for his predicament and requested the Foreign Secretary to direct that steps be taken to achieve a speedy and just settlement. However, the British chargé d'affaires at Constantinople to whom the letter was referred reported that Turkish ministers had asserted that Leahy had no claim on their government and a reply on these lines was sent to him in May.[52] In October 1859, having set out his claims in a twenty-page printed pamphlet which incorporated the opinions of lawyers who supported his case,[53] Leahy made a fresh appeal to the new Foreign Secretary, Lord John Russell, for intervention in his favour[54] and once again the matter was referred to the Embassy at Constantinople for a report.[55]

In March 1860, the new Ambassador, Sir Henry Bulwer, strongly refuted Leahy's claims in a lengthy official despatch to the Foreign Office and, in a separate personal note to the under-secretary, told him that he had heard "a very unsatisfactory account" of Leahy's "general proceedings" in Turkey, adding that he seemed to be not very scrupulous in his claims[56]. Sir Henry cast serious

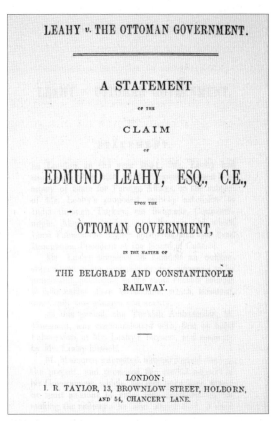

Title page of Edmund Leahy's twenty-page printed pamphlet setting out his claim on the Ottoman Government in respect of his plans for the Belgrade and Constantinople Railway.

Chapter X— Edmund and Matthew in Turkey, 1852-58

doubt on the evidential value of the "rough unsigned draft of an agreement, without signature or date", and in a tongue foreign to the Turkish officials, which had been produced by Leahy in support of his claim and he asked pointedly why Leahy had neither raised his claim, nor produced the document, until after the death of the Turkish official with whom the agreement was alleged to have been made. He went on to assert that, even if the document was authentic, it had never been approved by the Council of Ministers and sanctioned by the Sultan and was therefore unenforceable. "In all the circumstances", Sir Henry concluded, "it seems to me that Mr Leahy has no case; and if he is in any wise injured, he has himself only to blame – since people cannot come into foreign countries and carry on transactions in a manner which is not regular or binding in those countries, and then claim protection and support as if they had had custom and practice on their side". Bulwer sent a further despatch a month later, pointing out that Leahy never had any orders from the Porte to carry out railway surveys, that he had had received payments from the Fox Henderson agent in Constantinople for his work and that Fox Henderson insisted that the plans and surveys were therefore their property.[57]

Following some further correspondence with Leahy, and the submission of additional original documents to support his case, the legal advisers at the Foreign office took the view that he had a *prima facie* case and Bulwer was therefore instructed in September 1860 to arrange for the establishment of a mixed commission of English and Turkish officials to adjudicate on the issue. Given his own strong views on Leahy's claims, the ambassador seems to have assigned a low priority to this task and action was not taken in Constantinople until July 1861, after yet another letter had been sent by Leahy to the Foreign Secretary complaining about the delay and increasing his claim to £8,600, to allow for interest[58]. At first, Leahy protested that "with his long experience of Levanteen (sic) venality", he could confidently declare that a commission sitting in Constantinople would be "nothing short of a mockery of justice" and when, in a letter from Jamaica, he eventually agreed to the arrangement, he sought to have meetings of the commission held in London "being the most central for all parties".[59] Lord John Russell refused to ask the Turkish authorities to consent to this but enquiries made in London on his direction did lend some weight to Leahy's claims by establishing that a set of his railway plans had indeed been held in the Turkish embassy there for years. After further protests by Leahy about his inability to travel from Jamaica to attend the commission, the death of some of those who would have given evidence in his favour, and the need for the ambassador at Constantinople to represent and protect his interests[60], the proceedings eventually went ahead early in 1864, six years after he had first pressed the Foreign Office to take up his case. The matter reached its inevitable conclusion in the following June when Sir Henry Bulwer reported from Constantinople, one suspects with some satisfaction, that the commission had "entirely rejected" the claim.[61] Leahy was duly advised of this in July and his plans were returned to him, but he complained for some time afterwards that he had not received the sections "which showed the levels of the entire country between Belgrade and Constantinople, together with the gradients, and the excavations and embankments of the proposed line of railway".[62]

Apart from whatever payments were made to them by Fox Henderson in 1852-53 when the initial surveys were made, the Leahys' plans for railways in the Balkans brought them no reward and, when the time came for them to leave the area, they had not built a single mile of railway. The idea of linking Western Europe by rail to Constantinople, through Vienna and Belgrade, continued of course to engage the attention of railway promoters and engineers but it was not until 1868 that the Sultan awarded a concession to a Belgian Company to build the first section of the line. The concession was taken over in 1869 by a German-born financier, Baron Maurice von Hirsh, and when work began the following year, Edmund Leahy grasped the opportunity to seek to reopen his claims against the Ottoman Government. Writing to the Foreign Secretary in March 1870, he railed against the unfair and *ex parte* inquiry which had been held in 1864 and demanded that another full and fair examination of his case should be arranged, at which he would have an opportunity of giving direct evidence himself. The Foreign Office, however, replied promptly and brusquely, telling him that the Foreign Secretary "would not feel justified in again pressing your case upon the consideration of the Porte"[63] and that seems to have finally convinced Leahy to end what was, by then, a twelve-year-long campaign to recover the £7,000 which he claimed was due to him. A few years later, in June 1873, the first 350 miles of the railway from Constantinople were opened to traffic[64] but various uprisings in the Balkans, war with Russia and financial and other factors delayed completion of the route through Bulgaria until July 1888[65] when the way was finally open for the introduction of the *Orient Express* running between Paris and Istanbul. Edmund Leahy died in London three months earlier; had he lived, he would, no doubt, have claimed that it was his survey work in 1852-53 that had opened the way for the operation of what was to become Europe's most celebrated train.

The Pelion Mines

After their failure in 1853 to make progress with their various railway schemes, the Leahys turned to mining, and with some success. By early 1855, Edmund had been commissioned by the Porte to make a mineralogical survey of Turkey but the agreement was only a verbal one and there was a dispute about its terms before the Government agreed to pay 800,000 piastres (about £6,700 sterling) for the work. In 1856, Edmund and Matthew, together with George Zarify, were granted a 31-year lease of all mines and minerals in an area of over 20,000 square miles in the provinces of Thessaly and Epirus and in part of Macedonia - areas which now form part of modern Greece but which had continued to be part of the Ottoman Empire after Greece had gained independence in 1833. Under the agreement, the Leahys were to pay a sum of 70,000 piastres (some £600) annually to the Government, in lieu of all royalties and other dues.[66]

Chapter X— Edmund and Matthew in Turkey, 1852-58

Even before the lease was concluded, the Leahys had become heavily involved in the development of mines of the slopes of Mount Pelion, which rises to more than 5,000 feet above sea level near the town of Volos on the Aegean coast of Thessaly. In April 1855, Edmund expected that furnaces would be in full operation within a month, producing gold and silver, and yielding a good income; there were already 1,000 local men at work in the mines, besides some workmen who had been brought out from England.[67] It seems that Edmund was based for much of the time at the Hotel d'Angleterre in Constantinople, promoting his various projects, and he was listed in 1856 as one of the British subjects who enjoyed the protection of the British Embassy there.[68] Matthew, on the other hand, was more heavily engaged at the mines but Edmund was able to assure their brother Patrick, with whom he corresponded regularly, that Matt was well and in no danger from the "robber gangs" who operated in the countryside around the mines.[69] Matthew's letters confirmed that the country was "beset with robbers and blaguards (sic) and we never stir without eight or ten guards all around".[70] Edmund also held a poor opinion of the Greeks and believed that strong measures were necessary to deal with "the Banditti who infested the country"; he told Patrick, in a letter of 8 April 1855, that:

> The Greek is naturally a Bandit. Our Police have already shot 15 of them and hunt them night and day through the mountains like wild boar. In that last affair, 8 of the police were killed, they killing just an equal number of the robbers. Eventually, we will clear the country of these scoundrels but it is very disagreeable to be obliged to adopt such measures.
> These scoundrels of Bandits commonly bless themselves and make the sign of the Cross before coming to the fight; in short, such a combination of fanaticism, ignorance and villainy could not, I believe, be found elsewhere. The Turks are bad enough, but those rascally Greeks are about the lowest class of men of God's creation. Their cunning and quickness, which some people call natural talent, are not to be surpassed, but their vices surpass all other qualities.[71]

Quite apart from their problems with the bandits, the Leahy brothers did not have a particularly comfortable existence in Turkey. Edmund complained late in 1856 that "our affairs and connections here are far from being agreeable and I can tell you that I am not sleeping on a bed of roses".[72] Matthew, based at the Pelion Mines, had more reason to complain because the mines required "great care, supervision and constant fatigue on our part". He told Patrick at Christmas 1856 that "every morning we go away down the mountain and don't come back till night. The ground is too steep and rocky to ride and we return quite tired out. Sometimes we go off to see some of the mines and don't return for three or four days".[73] He was very lonely "for the want of newspapers" and was regretting the loss of "other periodicals of the day, transactions of the different societies, etc." which would have enabled him "to know what else may be going on in the scientific world". He missed his books on railways and various engineering subjects which were stored in London, and hoped that Patrick could arrange to have them sent out to him. And the Christmas season seems to have turned his thoughts to religious matters for he complained to Patrick about his failure to meet a request to send him "a couple of prayer books in English" - he had, apparently, asked for an ordinary prayer-book, a missal and a book of daily readings from the psalms.

Whatever about the personal hardship involved, the mines were obviously profitable for a time. By the end of 1856, they were yielding about £1,000 per month, according to Matthew, and "there is no knowing what the progress in this may be a few months hence".[74] Edmund was able to maintain a regular flow of funds to Patrick for family purposes, to pay off debts incurred between 1849 and 1851 at the Cape, and to clear his accounts with instrument-makers in London and in Cork.[75] When he learned from Patrick in 1855 that his health had failed, "just when prosperity begins to dawn on us", Edmund offered to arrange a relaxing holiday for him in Italy (where, he suggested, the change of climate would restore him to health) and to meet him in London or Paris on the way.[76] In September 1856, when travelling from Cashel to Thurles, Patrick was thrown "from a gig on the road with great violence" and took several months to recover from his injuries.[77] This gave Edmund another opportunity to press his brother to spend some time in Greece in the interests of his health, offering the prospect of "an excursion to the Vale of Tempé and all the classic quarters of Thessaly".[78] But Patrick declined these invitations, although he did visit England, France and Rome, and regularly took a holiday by the sea to help his bronchitis.[79]

Some months later, in May 1857, Edmund and a number of his friends undertook a week-long visit to the ruins of the ancient city of Sizicus (also known as Cyzicus) on a peninsula in the Sea of Marmara about 70 miles south-west of Constantinople. The city, which was destroyed by a series of earthquakes, the last one in 1063 AD, was believed to have been founded by Greek settlers in 756 BC but was incorporated in the Roman Empire's province of Asia in 133 BC and, some 450 years later, in the Byzantine Empire. According to his own account of the expedition, Leahy sailed from Constantinople to the port of Pandermo (now Bandirma) "in my steam-ship *Star*" and travelled onwards by means of "the best horses and saddles in the town" which were procured through the influence of the local Greek Orthodox Archbishop who accompanied the party. His memoranda of the visit describe some of the monuments which remained after the destruction caused by the earthquakes; these included the ruins of a large aqueduct over 100 feet high, massive broken columns (which may once have formed part of the temple of Emperor Hadrian, sometimes included among the wonders of the ancient world), massive fortress walls faced with blocks of black granite, and a number of large sarcophagi which were in remarkably good condition and still contained human skeletons. His notes

Chapter X— Edmund and Matthew in Turkey, 1852-58

drew attention also to some serious errors in the Admiralty charts of the seas around the peninsula, including the existence of uncharted submerged rocks on which many vessels had been lost. Leahy's memoranda (reproduced in full in Appendix II) were considered to be of sufficient interest by Sir Roderick Murchison, President of the Royal Geographical Society, and the most distinguished geologist and geographer of the day, to warrant publication in the Society's Proceedings in 1858.[80] The memoranda were also included in the bibliographical sources listed in the first major archaeological study of Cyzicus, published in 1910.[81]

"mines are not always to be reckoned upon .."

A despatch sent to London in May 1857 by the British Consul at Salonika and an accompanying letter from Vice-Consul Blunt at Volos forwarded extensive reports for the Foreign Office and the Board of Trade on the operations of the "English Mining Establishment on Mount Pelion". While the original reports do not seem to have survived in the Foreign Office files, it is clear from Blunt's covering letter that the Leahys' mining operations were then of considerable scale and significance. The reports are stated to have given details of "the privileges granted by his Majesty the Sultan to the proprietors" and of such matters as "the mechanisms and compartments of the Pelion Works; of the facilities of water supply, fuel and communication; of labour and capital; class of men employed; superintendence and protection; and finally, produce, results and larger features of operations".[82] Foreign Office papers also show that the Leahys were obliged to seek the assistance of the British authorities in coping with attacks on their mines. Another report from Vice-Consul Blunt in 1857 noted that Thessaly was "in a state of unaccustomed tranquillity" due to the measures adopted by the Ottoman and Greek Governments for the suppression of brigandage, but there were other dangers for the Leahys and their enterprise - "the works of the company are in such close proximity to the Greek islands that they and the persons and property engaged in them are constantly exposed to the danger of an attack from pirates, and this is a prospect which occasions a great deal of trouble and anxiety to the proprietors of the mines."[83] To assist the company, Blunt asked that British consulates in the region should collect information on the activities of pirates in the Aegean Sea and suggested that Britain and France should each maintain two small steam-propelled gunboats there, instead of occasional patrols by sailing ships. This suggestion was referred by the Foreign Secretary to the British Admiralty but the latter declined to act on it.[84]

It is difficult to understand why the Leahys abandoned their mining operations and were back in London by February 1858. There may well have been disputes with the Ottoman Government about mining rights and financial arrangements, just as there had been about railway schemes in 1856-57. It may be that the particular lodes which were initially exploited could no longer be worked successfully and that the Leahys were unable or unwilling to raise the necessary funds for the continued success and profitable operation of the enterprise. At the end of 1856, Edmund seems to have had some indication that difficulties lay ahead when he told Patrick that "I am far from sure that I am not near a good deal of trouble".[85] And while Matthew was pleased with the progress of the mines in December 1856, he cautiously noted that "mines are not always to be reckoned upon as one does not know for certain what the ground may give".[86] By the beginning of 1858, whatever the reason, the Leahys had ended their direct association with an enterprise which had promised so much and, apart from a brief visit by Edmund between September and November 1858, they withdrew entirely from Turkey. A printed report completed by William B Bray, a civil engineer, in September 1858, has all the appearance of a prospectus prepared in connection with a sale of the mines as a going concern. Bray noted that mining operations were well advanced at that stage and that extensive works, including a smelter, had been established; skilled labour was being provided mainly by Englishmen, with some Germans, Swiss and Poles, and ordinary labour was supplied by the local inhabitants who were "very efficient and economical".[87] In Bray's opinion, an outlay of about £15,000 was necessary for the efficient working of the mines but, subject to that, he advised that an annual profit of more than £100,000 could be achieved with careful and judicious management. It seems highly unlikely, however, in the light of what is known of their efforts to support themselves in subsequent years, that the Leahys succeeded in disposing of the mines on the basis of these optimistic profit projections.

REFERENCES

1. CDA 1855/2, letter of 30 April 1855 from Edmund Leahy to Patrick Leahy
2. FO 78/885, letter of 3 August 1851 from Edmund Leahy to Lord Palmerston
3. FO 78/885, letter of 11 August 1851 from FO to Edmund Leahy
4. FO 78/887, letter of 3 November 1851 from Edmund Leahy to Lord Palmerston
5. FO 78/2163, printed pamphlet of 20 pages entitled *Leahy v. The Ottoman Government: A Statement of the Claim of Edmund Leahy, Esq., CE, upon the Ottoman Government, in the matter of the Belgrade and Constantinople Railway*, I R Taylor, London, (?) 1859 (hereafter in this chapter referred to as "Statement of Claim"), page 3
6. Ian J Kerr, *Building the Railways of the Raj, 1850-1900*, Oxford India Paperbacks, Delhi, 1997, page 17
7. Sir Rowland Macdonald Stephenson, *Railways in Turkey: Remarks upon the practicability and advantage of railway communication in European and Asiatic Turkey*, John Weale, London, 1859
8. Statement of Claim, page 4
9. ibid, pages 3 -5, 19
10. CO 48/322, letter of 3 December 1851 from Edmund Leahy
11. ibid
12. FO 78/887, letter of 5 November 1851 from FO to Edmund Leahy
13. Canning (1786-1880) served in Constantinople in the 1820s and 1830s and from 1841 to the summer of 1852; he became Viscount Stratford de Redcliffe in 1852 and returned to Constantinople as ambassador between 1853 and 1858.
14. FO 78/887, letter of 5 November 1851; FO 78/851, despatch of 4 November 1851 to Sir Stratford Canning
15. Lord Kinross, *The Ottoman Empire*, The Folio Society, London, 2003, page 463 et seq
16. CO 48/322, letter of 3 December 1851 from Edmund Leahy to Earl Grey
17. CO 48/322, letter of 8 December 1851 from CO to Edmund Leahy
18. FO 611, Register of Names of Passport Applicants, shows that Edmund Leahy had obtained a British passport (No 10487) on 26 November 1851, but Matthew's name does not appear in the register; passports, however, were not mandatory at the time.

Chapter X— Edmund and Matthew in Turkey, 1852-58

19. Translation quoted in Statement of Claim, page 18
20. FO 78/903, Despatch No. 24, 17 July 1852
21. FO 78/903, Despatch No. 32, 27 December 1852
22. The catalogue of the library of the Royal Geographical Society, London, includes an entry for a document written by Edmund Leahy and entitled *Report of the Danube Canal, London, 1855* but, unfortunately, the document cannot now be found.
23. Statement of Claim, page 6
24. Clive Ponting, *The Crimean War*, Chatto & Windus, London, 2004, pages 79-80; Mick Gold, "The Doomed City", in Paul Kerr et al, *The Crimean War*, Boxtree, London, 1997, pages 39 - 40
25. CO 48/392, letter of 21 August 1857 from Edmund Leahy to the Colonial Secretary
26. C0 48/392, testimonial of 18 August 1858
27. CO 137/429, letter of 22 July 1867 from Lieutenant-General J F Burgoyne to Edmund Leahy, attached to letter of 21 August 1867 from Leahy to the Duke of Buckingham and Chandos
28. CO 137/429, letter of 16 July 1867 from General Lord Strathnairn to Edmund Leahy, attached to letter of 28 August 1867 from Leahy to the Duke of Buckingham and Chandos
29. CO 295/215, copy of letter signed by John Mitchell MD and William Linton MD, enclosed with letter of 31 January 1861 to CO from Denis Leahy
30. Piers Compton, *Cardigan of Balaclava*, London 1972, page 193
31. Arthur Helps, *The Life and Labours of Mr Brassey*, London, 1872; Brian Cooke, *The Grand Crimean Central Railway*, Cavalier House, Cheshire, 2nd edition, 1997; Philip Marsh, *Beatty's Railway*, New Cherwell Press, Oxford, 2000
32. George Wrottesley, *The Life and Correspondence of Field Marshal Sir John Fox Burgoyne* (two vols), London, 1873; W H Russell, *The War* (two vols), London 1855, 1856
33. CDA 1855/2, letter of 30 April 1855 from Edmund Leahy to Patrick Leahy
34. CDA 1855/1, letter of 8 April 1855 from Edmund Leahy to Patrick Leahy
35. CDA 1855/2, letter of 30 April 1855 from Edmund Leahy to Patrick Leahy
36. FO 195/452, Report dated 15 September 1855 from Brigadier-General H K Storks to Brigadier W R Mansfield, military adviser to the British ambassador at Constantinople; report dated 25 September 1855 from Storks to Viscount Stratford de Redcliffe
37. *The Times*, Report from our own Correspondent, Constantinople, 8 January 1856
38. FO 78/2163, personal and confidential letter dated 28 March 1860 from Sir Henry Bulwer to the Colonial Secretary
39. CDA 1855/2, letter of 30 April 1855, Edmund Leahy to Patrick Leahy
40. Joan Haslip, *The Sultan: The Life and Times of Abdul Hamid II*, Weidenfeld and Nicholson, London, 1958; John Freely, *Inside the Seraglio*, Penguin Books, London, 2000, page 259, 262-63
41. CDA 1855/2, letter of 30 April 1855 from Edmund Leahy to Patrick Leahy
42. *The Times*, 13 February 1855
43. *The Times*, 2 and 13 October 1855
44. *The Times*, 12 October 1855
45. Statement of Claim, pages 7-8
46. Joan Haslip, *The Sultan: The Life and Times of Abdul Hamid II*, Weidenfeld and Nicholson London, 1958; Zarify went on to become a friend of Sultan Abdul Hamid II who ruled from 1876 to 1909 and who, under Zarify's guidance, amassed a large personal fortune.
47. See, for example, reports in *The Engineer*, Vol II, 19 and 26 September 1856, pages 521 and 534
48. BL, Layard Papers, Vol CCV, Add. 39185
49. BL, Layard Papers, Vol LV, Add. 38985, f.135, letter of 16 February 1857 from Edmund Leahy to Sir A H Layard
50. The company formed to carry out this project - the Ottoman Railway Company – was chaired by Sir Macdonald Stephenson, managing director of the East India Railway Company.
51. Hyde Clark, "The Imperial Ottoman Smyrna and Aidin Railway", in *The Levant Quarterly Review*, No III, January 1861, Vol II, page 41
52. FO 78/2163, letter of 5 March 1858 from Edmund Leahy to the Foreign Secretary, the Earl of Malmesbury, and reply of 13 May 1858
53. FO 78/2163, *Leahy v. The Ottoman Government: A Statement of the Claim of Edmund Leahy, Esq., CE, upon the Ottoman Government, in the matter of the Belgrade and Constantinople Railway*, I R Taylor, London, (?) 1859
54. FO 78/2163, letter of 26 October 1859 from Edmund Leahy to Lord John Russell
55. FO 78/1426, Despatch No. 227, 1 December 1859

56. FO 78/2163, Despatch No 161 dated 27 March 1860 from Sir Henry Bulwer and personal letter from him dated 28 March 1860
57. FO 78/2163, Despatch No 201 dated 13 April 1860 from Sir Henry Bulwer
58. FO 78/2163, letter of 24 May 1861 from Edmund Leahy to Lord John Russell; Despatch No 488 dated 8 July 1861 from Sir Henry Bulwer
59. FO 78/2163, letters of 23 August 1861 and 8 September 1862 from Edmund Leahy to Lord John Russell and Under-Secretary A H Layard, respectively
60. FO 78/2163, letters of 22 April 1863 and 11 December 1863 from Edmund Leahy to the Foreign Office
61. FO 78/2163, Despatch No 169 dated 23 June 1864 from Sir Henry Bulwer
62. FO 78/2163, letter of 29 September 1864 from Edmund Leahy to Earl Russell
63. FO 78/2163, letter of 21 March 1870 from Edmund Leahy to Lord Clarendon and reply dated 28 March 1870
64. *The Times*, 7 July 1873
65. R J Crampton, *A Short History of Modern Bulgaria*, Cambridge, 1987
66. CDA, printed statement entitled *Mines of European Turkey,* (?) 1858
67. CDA 1855/2, Edmund Leahy to Patrick Leahy, 30 April 1855
68. FO 78/1242 1850, Register of Persons Enjoying Protection at Constantinople, 1856
69. CDA 1855/1, Edmund Leahy to Patrick Leahy, 8 April 1855
70. CDA 1856/12, Matthew Leahy to Patrick Leahy 21 December 1856
71. CDA 1855/1, Edmund Leahy to Patrick Leahy, 8 April 1855
72. CDA 1856/12, Edmund Leahy to Patrick Leahy, 21 December 1856
73. CDA 1856/12, Matthew Leahy to Patrick Leahy, 21 December 1856
74. ibid
75. CDA 1855/1 and 1855/2, Edmund Leahy to Patrick Leahy, 8 April 1855 and 30 April 1855
76. CDA 1855/2, Edmund Leahy to Patrick Leahy, 30 April 1855
77. Newman papers, Patrick Leahy to Newman, 12 September 1856, quoted in Christopher O' Dwyer, *The Life of Dr Leahy, 1806-1875*, MA thesis, Maynooth, 1970
78. CDA 1856/12, Edmund Leahy to Patrick Leahy, 21 December 1856
79. Christopher O' Dwyer, *The Life of Dr Leahy, 1806-1875*, MA thesis, Maynooth, 1970
80. Royal Geographical Society Journal Manuscripts Collection, JMS/9/123, *Memoranda of a visit to the Site of the Ruins of the Ancient City of Sizicus in Asiatic Turkey by E Leahy CE, 1857*; Proceedings of the Royal Geographical Society, London, vol 2, 1858, pages 376-378
81. The British archaeologist and ethnologist, Frederick William Hasluck, during his years as assistant director of the British School of Archaeology at Athens, published a series of papers about Cyzicus in various journals, culminating in a major study *Cyzicus: being some account of the history and antiquities of that city, and of the districts adjacent to it*, Cambridge University Press, 1910.
82. FO 78/1302, Despatch No 9 from the Consul at Salonika, 12 May 1857, to which is attached a letter of 4 May 1857 from Vice Consul Blunt at Volos
83. FO 78/1302, Despatch No 6 from the Consul at Salonika, 28 April 1857, forwarding a report dated 18 April 1857 from Vice Consul Blunt at Volos
84. FO 78/1302, Despatch No 5 from FO, 31 August 1857
85. CDA 1856/12, Edmund Leahy to Patrick Leahy, 21 December 1856
86. CDA 1856/12, Matthew Leahy to Patrick Leahy, 21 December 1856
87. CDA, printed report by William B Bray CE entitled *Pelion Mines and Works – Zagora, Thessaly, Turkey*

CHAPTER XI

FAMILY MATTERS, 1852-60

"I cannot in my conscience accuse myself of neglecting our poor sisters ... I cannot say that I am quite so blameless towards my unfortunate wife and it is hardly fair to throw those poor sisters between us whenever I show any disposition to do her some justice".

Edmund Leahy to Patrick Leahy, 1856[1]

A glittering clerical career

While his father and some of his brothers and sisters were attempting to build new lives for themselves in London, and later at the Cape of Good Hope and in Turkey, in the decade between 1847 and 1857, Patrick Leahy progressed rapidly in his career in the Catholic Church in Ireland. His appointment in 1847 as President of St Patrick's College, Thurles, where he had served for 17 years as a professor, strengthened the relationship he had developed with Archbishop Michael Slattery of Cashel and Emly whose friend and trusted confidant he had become. He acted as Slattery's theologian when the first National Synod of the Catholic Church for almost 700 years was held in the Thurles college in August–September 1850, leading to the adoption of a uniform and comprehensive code of ecclesiastical law for the country as a whole. His position as one of the three secretaries of the synod brought him into close contact with Paul (later Cardinal) Cullen who had returned from Rome to become Archbishop of Armagh and head of the Irish church earlier in the year. Leahy's profile at national level was maintained by his appointment as one of three secretaries of the Catholic University Committee which was set up at the synod, and by his service as one of the two secretaries of a second National Synod held in Dublin in May 1854. He played a leading part in the establishment in Dublin of the Catholic University[2] and was appointed Professor of Sacred Scripture at the University and Vice-Rector to John Henry (later Cardinal) Newman in May 1854.

By 1855, all of this must have led Leahy himself, and many others, to see him as the natural successor to the 72-year-old Archbishop Slattery who was then in poor health - indeed his appointment in that year as Dean of the archdiocese and Parish Priest of Cashel might well be seen as a move to enhance his limited pastoral experience with a view to the eventual succession. But matters were not proceeding as smoothly for Leahy in his personal and family life. In co-operation with his brother, Edmund, he had become the central figure in a web of financial arrangements designed to support on a continuing basis their three unmarried sisters who were in poor circumstances throughout the 1850s, and to provide occasional support also for their brothers, Denis and Matthew. By 1856, he seems to have become concerned that these arrangements and, in particular, the precarious financial position of his unmarried sisters and their perceived dependence on him, might affect his prospects of succeeding the ailing Dr Slattery. Moreover, he regularly found himself in conflict with Edmund about the extent of the provision which should be made for the sisters, and how the cost should be shared.

Archbishop Patrick Leahy 1806-75.

Caring for "the poor girls"
After the death of their father at Cape Town in October 1850, the three unmarried Leahy sisters, Helena (Ellen), Susan and Anne, then aged between 30 and 40, followed Edmund and Matthew back to London in 1851.[3] They remained there when the two brothers set out for Constantinople the following year and, throughout the 1850s, lived in various lodging

houses and apartments in the Chelsea, Brompton and Kensington areas.[4] Edmund's wife, Catherine, seems to have died shortly after giving birth to a son, Gerald, at sea in 1851, presumably on the journey back from the Cape[5]. He had married again by the mid-1850s, but his new wife, Juliet, who was much younger than he was,[6] also remained in London with his young son, living - whether as a guest or as a servant is not clear - at the home of Lady Adelaide Constance Lennox (wife of Lord Arthur Lennox, son of the Duke of Rutland) at 21 Ovington Square in the fashionable area of South Kensington.[7]

Edmund wrote in 1858 that his father had left the family "quite unprovided for ... and they are since a burden upon me as their only support".[8] While this was not entirely true, it does appear that, notwithstanding his own family commitments, Edmund was held primarily responsible by Patrick for the support of their sisters in the 1850s. He accepted readily enough that he should "do my duty to the poor girls", but he suggested to Patrick at least once that Denis, his older brother, should share the responsibility "for after all, he is as much a brother as I am".[9] In the mid-1850s, when his enterprises in Turkey were prospering, Edmund maintained a regular flow of funds to Patrick for family purposes. At Easter 1855, for example, he sent £100 and, three weeks later, a further £150.[10] Patrick was told to use these sums in whatever way he considered most useful for "the girls" and Juliet, but there were also more specific instructions: Juliet was to get a regular sum of £5 a month and, on one occasion, a special payment of £12 "to dispose of for her wants such as clothes for herself and Gerald and, if she wishes to give her mother anything for the expense she has been to her". Again, at Christmas 1856, Juliet and "the poor girls" were each to get additional payments of £20.[11]

There is no obvious explanation for Edmund's practice of conducting his financial transactions with his wife - and much of his other financial business - through the agency of his brother who, in effect, had been acting as Edmund's banker since 1850. While he was living at the Cape of Good Hope, Edmund had borrowed money from Patrick,[12] but a few years later the flow of funds was reversed: in April 1852, Patrick sent a draft for £250 to John Sadleir of the Tipperary Bank with instructions to return £150 to him and to lodge the balance of £100 to Edmund's credit at the Bank of England.[13] In April 1855, Patrick was given precise instructions about making various payments from funds which had been supplied by Edmund:

> These two sums or debts which I wish to pay through you if you have no objection to the trouble viz. £50 to Troughton & Simms, the mathematical instrument makers of Fleet Street, London; and £12 to Sir Harry Smith, the late governor of the Cape of Good Hope. His address is Lieutenant General Sir Harry G. Smith, Bart., G.C.B., Horse Guards, London but I write to him myself by this post as well as to Troughton & Simms enclosing orders on you for the above sums and when you hear of them please to honour my drafts.

Chapter XI — Family Matters, 1852-60

Sir Harry's is a debt of honour – when we were in difficulties at the Cape, our poor Father wrote to him and Sir Harry very kindly enclosed him a check for £12 without our poor Father's even asking for it. If it was not inconvenient to you to drop him a line asking (in) what bank you would lodge that sum to his credit, it would perhaps be the handsomest manner of acting. You need not write to the London people (T & S) as they will no doubt forward my order but which Sir Harry might not like to do. It is a pleasure to repay the latter.

The family finances led to some harsh exchanges between the brothers. In 1856, for example, when Patrick accused Edmund of favouring his wife, at the expense of his sisters, even though the amounts in question were not over-generous, he received a sharp response:

I think I have not been ungenerous to anyone having a family claim on whatever I have … If you would do me justice, I think you would be one of the first to admit that I have not been unmindful of a brother's regard for his poor sisters, but although you appear to entertain a different opinion, and although I have a great respect and regard for everything you tell me, yet my dear Pat, I cannot in my conscience accuse myself of neglecting our poor sisters. I am easy in my mind respecting such a charge. I cannot say that I am quite so blameless towards my unfortunate wife and it is hardly fair to throw those poor sisters between us whenever I show any disposition to do her some justice.[14]

Apart from purely financial matters, there were other issues in contention between the brothers. Edmund did not approve of the girls staying in London: "I fear they are not able to contend with the rogues of London and if they would occupy themselves in a respectable way in or near Dublin, I think it would be better for them".[15] But Patrick had other ideas; late in 1856, he pressed Edmund to bring at least one of the sisters to live in Turkey, apparently because he believed that their partial dependence on him was prejudicial to his position in Cashel, but Edmund replied that he could not see –

the connection clearly between your holding or resigning Cashel and my refusing to commit the folly of bringing one of the sisters to this country. They are all in England and you are in that wretched country, Ireland, and although your relative places might remain unaltered, you might well hold Cashel, if it was worth having, without bringing them, or any of them, with you there. I never asked you to take one of them … There is a difference … between generosity and folly and unless you would wish me to be worse than foolish, you certainly would not desire me to bring any of them here, at least if you understood (which you clearly do not) the difficulty of our position in this country.[16]

Matthew backed Edmund's view, asking Patrick "how could they come or live here - no one to speak to all day and sometimes for days"[17] - and this seems to have marked the end of that particular dispute.

When Archbishop Slattery died in February 1857, Patrick Leahy soon emerged as the choice of the great majority of the priests of the diocese, and the favourite of the powerful Archbishop Cullen, to succeed him. His appointment as archbishop was approved by the Pope in May 1857 and he was consecrated by Cullen in Thurles on 29 June, with Newman as preacher. As archbishop, he continued to make regular payments to his sisters who remained in London and, by agreement with Edmund, half of these payments were recouped from the accumulated funds which he had sent from Turkey. In some years, the archbishop meticulously recorded the payments in his Ordo: in 1859, for example, the entry "sent £5 to sisters" occurs each month and there were occasional payments of an additional £6 for rent. Similar payments were recorded in his Ordo for 1861.[18] The fact that he had to continue these payments caused some embarrassment for Patrick, one of whose critics, Rev James O' Carroll, surmised in his 1862-1864 diary that the payments were among the causes of the high expenses of his household[19] although the archbishop was not, of course, "the sole support" of the sisters, as O'Carroll's diary alleged.

In 1859, when Edmund said that he was "hard up", he was accused by Patrick of having attempted to meet the demands of a pressing creditor by drawing a bill on him and received what he described as an unreasonable, passionate and abusive letter from the archbishop.[20] This led to another dispute about money matters, with Edmund remarking sadly that "men do greatly change in time" and accusing his brother of applying for his own use funds which were to be held in Edmund's name in the Bank of Ireland. Patrick was reminded that "besides the larger sums", he had been sent "many smaller ones - some for £250, some for £100, some for £50, and so on, not one of each, but many". In addition, he had been given £300 or so to meet expenses which had arisen on his appointment as Vicar General of the diocese and Parish Priest of Cashel in September 1855. Allowing for payments made to Juliet, half of the agreed payments to his sisters, and sums paid to support Matthew and Denis in the late 1850s, Edmund calculated that Patrick still owed him about £650 in 1860. While he was prepared to accept that the archbishop's wants might surpass his income, taking account of the poor circumstances of his diocese, he suggested that Patrick should at least bear this debt in mind in arranging his financial affairs. However, the archbishop was himself in financial difficulties in 1860: his annual income was much less than had been received by his predecessor in the early 1830s, while the cost of living had doubled,[21] and he wrote at one stage of the possibility that he would be forced to sell his residence, dismiss his servants and go elsewhere.[22] It cannot have been easy, therefore, for him to meet Edmund's demands for repayment of the substantial sum which was in dispute between them.

Chapter XI — Family Matters, 1852-60

"Denis is not doing much good"

Denis Leahy, described as a land surveyor and civil engineer, had an address at East Main Street, Thurles, in 1846[23] and seems to have been living at Cabra, near Thurles, until the 1850s. He then moved to London and in May 1853, when he was 41 years old, married Elizabeth Mary Ann Longmore Anderson, the London-born 22-year-old daughter of a deceased naval officer, according to the rites of the Church of England.[24] From 1855 to 1857, he was listed under the name "Denny Leahy" in the commercial section of the *Post Office London Directory* as a civil engineer, with an address at 78, Stanley Street, Pimlico.[25] He was awarded an important contract worth £13,045 in October 1854 by the Metropolitan Commission of Sewers to construct the Ranelagh Sewer[26] to replace a 200-year-old watercourse which had become a vast open cesspit, the subject of frequent public complaint and newspaper editorials,[27] and the cause of cholera and other diseases. The new brick sewer, running from King's Road, Chelsea, and along by Sloane Square and the Royal Hospital to the Thames, just west of Chelsea Bridge, was to be 3,725 feet long, with a diameter of more than 9 feet for most of its length. The contract, which was one of the largest awarded by the Commission in 1854, included the construction of 72 linear feet of a 19 feet high river wall, with an outlet 9 feet in diameter and fitted with cast iron flap valves.[28] In June 1857, a report from J W Bazalgette, engineer to the Metropolitan Board of Works (which had succeeded the Commission of Sewers in 1855) noted that the sewer had been completed and contained no hint that it was not functioning satisfactorily.[29] Nevertheless, for reasons which are not recorded in *Hansard*, Viscount Torrington called in the House of Lords in July 1858 for returns to be made by Bazalgette giving full engineering details of the sewer before and after the alterations made under what he called *Leahy's Contract*.[30] These were duly provided in July[31] and the matter seems to have attracted no further notice in the Lords or in the press.

Edmund Leahy remained in touch with Denis during the 1850s and, in a letter from Turkey in April 1855, told Patrick that he feared "Denis is not doing much good", adding that "if the railway etc. succeeds, I can do something for him".[32] Five years later, Denis Leahy's practice cannot have been a very successful or remunerative one for he was again the subject of correspondence between Edmund and Patrick who was continuing to make payments to him from funds provided by Edmund.[33] Apart from underwriting these payments, Edmund seems to have given up any hope of helping his brother to find his feet: he had told Patrick in 1856 that while he would have been glad to do some good for Denis, he feared that "any effort I could make for him would not be very successful".[34]

At one stage in the 1850s, Denis seems to have intended to join his two brothers in Turkey and a reference which he produced in 1861 states that "in Turkey, he was engaged with his brothers in surveying lines of railway for the Turkish Government, and in working the lead and silver mines of Thessaly".[35] This reference, however, cannot be relied on: there is no other evidence that Denis worked in Turkey and, if he did so, it could have been no earlier than December 1856, when Matthew told Patrick, in one of his letters from Turkey,[36] that Denis had "no notion of coming here" and "even if he came, I am afraid he would not do much good for himself or for us. However, he is a fool in his old days. If he came here, he certainly could do little more than act as a clerk, at least for some time, as he knows nothing of the works going on here and less of any foreign language". This latter begs the question of whether Edmund and Matthew spoke either French, the language in which much of the official business of the Ottoman Government was conducted, or Greek, which would presumably have been the normal medium of communication between management and the local workers at the Pelion Mines.

"Matt's wants seem to require some aid"
After his return to London from Turkey in 1858, Matthew Leahy got into financial difficulties and joined the list of family members who had to be supported by Edmund and by Patrick. In November 1859, for example, Edmund, who was then in Jamaica, told Patrick that "Matt's wants seem to require some aid" and he directed the archbishop "to let him have what his wants require, debiting me with the amount".[37] The Archbishop then sent £50 to Matthew, carefully noting in his Ordo that this sum was "advanced by order of my brother Edmund".[38] About the same time, he sent Matthew a further £5, apparently from his own funds.[39] Following some further payments by Patrick from the funds which he held for Edmund, the latter wrote angrily from Jamaica in March 1860 that "Matt's extravagance is unwarrantable" and suggested that the archbishop should "endeavour to get him a first class professional place like inspector of railways".[40] In the meantime, Edmund was willing to go on supporting his brother; although he told Patrick that "hard up would express my state better than anything else," he allowed him to continue to give Matt "a reasonable amount for support" from the funds which he held on Edmund's behalf.

As one of the leaders of the Irish Catholic hierarchy, Archbishop Patrick Leahy was in regular direct contact with the Chief Secretary for Ireland about a variety of issues, and travelled to London from time to time for meetings with other senior members of the cabinet. He may well have used the opportunity presented by such a meeting to make representations on the lines advocated by Edmund or, indeed, he could have taken up his brother's case in writing with a member of the Government, just as he did some years later when Edmund himself was in difficulties in Jamaica.[41] At the time, many public appointments were filled without advertisement or competition, and when a vacancy arose, it was open to the political head of the Department of State concerned to appoint the individual of his choice. There is no evidence that Archbishop Leahy used whatever influence he had in London to advance Matthew's prospects of securing a public appointment but, given the apparently effortless manner in which he secured an appointment in Trinidad in 1860, only a few months after it had been suggested to the archbishop that he should endeavour to "get him a first class professional place", the possibility cannot be ruled out.

Chapter XI — Family Matters, 1852-60

REFERENCES

1. CDA 1856/12, letter of 21 December 1856 from Edmund Leahy to Patrick Leahy
2. At one stage, Leahy pointed out the advantages of locating the university in the spacious building which was already available in Thurles but Newman disagreed, commenting that the building was located "on a forlorn waste, without a tree, in a forlorn country, and a squalid town" (quoted in Christy O' Dwyer, "The Beleaguered Fortress: St Patrick's College, Thurles, 1837-1988", in *Thurles: The Cathedral* Town, Geography Publications, Dublin, 1989, page 240).
3. The other sister – Margaret – had married in the 1840s; for details see Chapter XVII.
4. *Post Office London Directory*, 1855-1858; at the time of the 1861 Census of England, Ellen and Anne Leahy, described as fund holders, were lodgers at 33 Westmoreland Place, London (RD Kensington RG09/2/83/54).
5. The Census return for 1871 gives Gerald's age as twenty and, in the "where born" column, the entry is "Atlantic Ocean"; a search of the indexes to the Births and Deaths registered in England in 1850-53 has not turned up any entries for Gerald or Catherine Leahy.
6. No record of the marriage of Edmund and Juliet can be traced in the index to marriages for the years 1851-55 maintained by the General Register Office, London but the 1871 Census of England (Parish of Camberwell, Borough of Lambeth, RG 10 732) lists Juliet as Edmund's wife, her place of birth being given as London, and her age as 34, which can hardly have been correct if they had married by 1852.
7. CDA 1855/2, letter of 30 April 1855 from Edmund Leahy to Patrick Leahy; Lord Arthur Lennox (1806-1864) was a Lt Colonel in the army, sat in Parliament from 1831 to 1848 and was Clerk of the Ordnance in 1845-1846; in 1863, Edmund Leahy referred to him as one of his "distinguished friends".
8. C0 48/392, letter of 21 August 1858 from Edmund Leahy to CO
9. CDA 1856/12, letter of 21 December 1856
10. CDA 1855/1, letter of 8 April 1855 from Edmund Leahy to Patrick Leahy; CDA 1855/2, letter of 30 April 1855 from Edmund Leahy to Patrick Leahy
11. CDA 1856/12, letter of 21 December 1856 from Edmund Leahy to Patrick Leahy
12. CDA 1860/12, letter of 11 March 1860 from Edmund Leahy to Patrick Leahy
13. CDA 1852/1, letter of 16 April 1852 from John Sadleir, Great Denmark Street, Dublin, acknowledging receipt of an earlier letter from Patrick Leahy; Sadleir's Tipperary Bank crashed in February 1856 after Sadleir's suicide in London, resulting in substantial losses for many hundreds of its depositors, including a number of Tipperary clergy – see James O'Shea, *Prince of Swindlers: John Sadleir MP 1813–1856*, Geography Publications, Dublin, 1999.
14. CDA 1856/12, letter of 21 December 1856 from Edmund Leahy to Patrick Leahy
15. CDA 1855/2, letter of 30 April 1855 from Edmund Leahy to Patrick Leahy
16. CDA 1856/12, letter of 21 December 1856 from Edmund Leahy to Patrick Leahy
17. CDA 1856/12, letter of 21 December 1856 from Matthew Leahy to Patrick Leahy
18. CDA, Ordo of Dr Leahy, 1859, 1861
19. CDA, O'Carroll Diary, entry following 31 January 1864
20. CDA 1859/35 and 1860/12, letters of 26 November 1859 and 11 March 1860 from Edmund Leahy to Patrick Leahy
21. CDA, O'Carroll Diary, 1 December 1863
22. Christopher O' Dwyer, *The Life of Dr Leahy, 1806-1875*, MA thesis, Maynooth, 1970, pages 639-642
23. *Slater's National Commercial Directory of Ireland*, 1846
24. Marriage certificate issued by the General Register Office, London, RD Marylebone, Vol 1a, page 870
25. *Post Office London Directory*, 1855, 1856, 1857
26. *Metropolitan Commission of Sewers - Accounts, in Abstract, of Receipts and Expenditure during the Year 1854 etc, Schedule of Contracts for the performance of Special Works*, HC 1854-55 Vol LIII (175)
27. See, for example, *The Times*, 7 September 1854
28. London Metropolitan Archives, MCS/219/138, contract dated 20 October 1854 with plan and three sections of the Ranelagh Sewer
29. *Report of the Metropolitan Board of Works for the year to 30 June 1857, report of 10 June 1857 by J W Bazalgette*, HC 1857 (Sess 2) Vol XLI (234)
30. Journal of the House of Lords, 21 July 1858
31. *Returns by the Engineer of the Metropolitan Board of Works ... regarding the Ranelagh Sewer, House of Lords Sessional Papers 1857-58* Vol XVIII (221) (265)
32. CDA 1855/12, letter of 30 April 1855 from Edmund Leahy to Patrick Leahy
33. CDA 1860/12, letter of 11 March 1860 from Edmund Leahy to Patrick Leahy; Ordo of Dr Leahy for 1861

34. CDA 1856/12, letter of 21 December 1860 from Edmund Leahy to Patrick Leahy
35. CO 295/215, reference dated 12 February 1861 from John Scott Russell
36. CDA 1856/12, letter of 21 December 1856 from Matthew Leahy to Patrick Leahy; testimonials provided by Denis in 1861 included a copy of a letter of 28 March 1854 from Army Staff Surgeons in Constantinople expressing thanks for assistance provided in Turkey but the copy omits the Christian name of the Leahy to whom it was addressed, leaving open the possibility that the letter was being used by Denis in error or to deceive.
37. CDA 1859/35, letter of 26 November 1859 from Edmund Leahy to Patrick Leahy
38. Ordo of Dr Leahy, 1859, entry for 31 December
39. Ordo of Dr Leahy, 1859, entry for 19 December
40. CDA 1860/12, letter of 11 March 1860 from Edmund Leahy to Patrick Leahy
41. *House of Lords Sessional Papers 1864* Vol XIII (254), letter of 17 August 1863 to the Duke of Newcastle

CHAPTER XII

MATTHEW AND DENIS IN TRINIDAD, 1860-63

"then might it be said with truth that the colonial service had cost this one family very dearly indeed".

Archbishop Patrick Leahy, August, 1863[1]

A vacancy in Trinidad

Following a dispute with the Governor of the colony, Lewis W Samuel was suspended from his post as Superintendent of Public Works in Trinidad in September 1859 and returned to London later that year.[2] Claiming that his health had been severely tested by the tropical climate, he obtained sanction from the Colonial Office for an extended stay in London but, before the deadline set for his return, he resigned his office on 5 April 1860.[3] The resignation was immediately accepted by the Colonial Office where the official concerned noted that there would probably be "a suitable candidate" on the list of applicants maintained by the Duke of Newcastle, then Secretary of State for the Colonies.[4] The Colonial Office files do not disclose whether Matthew Leahy's name was on this list but there is evidence that the Earl of Carlisle, who was then Lord Lieutenant of Ireland, made representations to the Duke in his favour - as Matthew delicately put it, the Earl "takes an interest in my behalf".[5]

On learning of Samuel's resignation, Lieutenant-Governor Walker of Trinidad was concerned that steps should be taken quickly to appoint a new Superintendent of Public Works; a lieutenant of the Royal Engineers who had filled the post temporarily had resigned and left for England, and Walker told the Colonial Office that there was "no independent professional person in the island who is qualified for the office and of whose services I can avail myself".[6] He added that in his view "it would be difficult to induce any civil engineer of competent skill to come out here on the insufficient pay and allowances assigned to the office". But Matthew Leahy was not deterred from seeking the post either by the level of pay or by the notoriously difficult climate. By 1860, he had been living at various addresses in London for two years, with no obvious source of income apart from the contributions he received from Edmund and Patrick. He had been in touch with the

Chapter XII — Matthew and Denis in Trinidad, 1860-63

Colonial Office in March about the possibility of obtaining an appointment in the colonial service[7] and, on 16 April, "understanding that a vacancy exists or is about to" in the post in Trinidad, he applied in writing for it.[8] However, he wanted to keep open the possibility of obtaining a better post, asking that, should the Duke of Newcastle "favour me now with this appointment, yet when a more lucrative office of the kind should be at Your Grace's disposal, my qualifications and the services which I have rendered to the Government may be remembered".[9]

In support of his application, and to satisfy the Duke of his professional competence and of his qualifications, Leahy submitted a number of testimonials on 31 May. The three which have survived[10] all came from prominent figures in the engineering profession of the day:

> General Sir John Fox Burgoyne, then serving at the War Office as Inspector General of Fortifications, was very happy to provide a testimonial in favour of Leahy: "his examination for a County Surveyor in Ireland, and his occasional services there, prove a considerable degree of qualification for the management of public works; and frequent communication with him and some indirect knowledge of his proceedings since, lead me to have much confidence in his professional competence". (The fact that the Office of Public Works reported in 1845 that Leahy had been examined for a county surveyor post "and found deficient"[11] raises questions about the value of this and, indeed, other testimonials provided for the Leahys by Burgoyne in the 1860s).

> Charles Vignoles, then a Fellow of the Royal Society and nearing the end of a distinguished engineering career, had pleasure in "strongly testifying to the talent and character" of the applicant. He was peculiarly fitted "from his every experience as a civil engineer, for an appointment to manage any public work of importance - and his habits and business methods and practical engagements abroad fit him for any Government appointment as connected with public operations".

> John Scott Russell, Vice-President of the Institution of Civil Engineers, the marine engineer and shipbuilder who had contracted in 1853 to build Brunel's *Great Eastern*, expressed the conviction that "your experience as a civil engineer particularly qualifies you to superintend the Public Works of the Colonies".

On 6 June, the Duke of Newcastle, in a manuscript note on Leahy's file, indicated that "I will appoint this gentleman - I have made careful enquiries as to his qualifications".[12] A formal offer of appointment was issued on 19 June stating that the salary for the post was to be £400 a year, with £150 for travelling expenses and a free passage to the colony.[13] Leahy immediately accepted the post and, having already applied for and obtained a passport,[14] indicated his intention to travel to Trinidad as soon as possible. He sought a free passage also for his wife[15] and a contract passage for his sister, Susan, "should either or both come with me",[16] but when told that a free passage was available only for himself, and faced with a bill of nearly £63 for two contract passages, he notified the Colonial Office that he would be accompanied only by his sister and that they wished to travel via Jamaica, presumably to visit Edmund, who had taken up a position there in 1859.[17] On 23 July, Leahy was in Liverpool "to see some family members" but found that it was "impossible to go by Jamaica due to unforeseen difficulties".[18] Finally a passage to Trinidad at public expense was ordered for him on the Royal Mail Steam Packet Company's paddle-steamer which was due to leave Southampton on 2 August and a contract passage was ordered for Susan.[19]

"Mr Leahy arrived here yesterday"

A despatch dated 22 August 1860 from Lieutenant-Governor Walker to the Colonial Office reported that "Mr. Leahy arrived here yesterday"[20] and on the same day, the *Trinidad Royal Gazette* published notice of Leahy's formal appointment as Superintendent of Public Works.[21] Leahy was then 42 years old and for the first time in his career had a

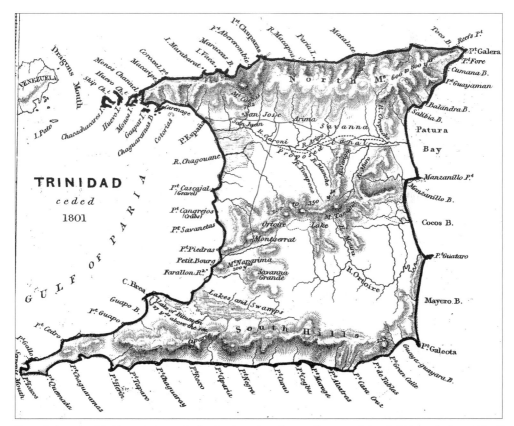

Trinidad in 1844 (from **Maps of the Society for the Diffusion of Useful Knowledge, Vol. II,** London, 1844).

secure, if not particularly attractive or well-paid, position in the public service. He was responsible for all public works on the island and, as the Governor noted, "so far as expenditure is concerned, the Department of Works is the most important branch of the Government".[22] But this must be taken in context for despite all the trappings of empire, the colony which Leahy was to serve was a small one: the island is roughly 50 miles from north to south and 30 miles broad and its total land area of less than 2,000 square miles is only the size of an average Irish county and much less than the area of County Cork where Leahy's original public works experience had been gained. In 1860, no more than 4.5% of the land of Trinidad had been reclaimed, 6.7% of it was swamp, and the balance was mostly impenetrable forest and jungle.[23] With its tropical climate, the island presented a particularly inhospitable environment for Europeans, even if it was not regarded as "the white man's grave" (as were parts of Africa); medical services were poor, sanitation was primitive and there was a high death rate from tropical and other diseases, including yellow fever, malaria, typhoid, cholera, dysentery and tuberculosis. In 1854 alone, 7,483 persons, or roughly 10% of the population, died of cholera.[24] By 1861, the total population had increased to 84,438 (including 5,341 of European origin) but overall numbers were being maintained only by the introduction of thousands of indentured Indian and Chinese coolies to work on the cocoa and sugar-cane plantations in place of the liberated slaves.[25]

After it was ceded to Britain by Spain in 1802, Trinidad was administered by a Governor appointed by the Crown, acting in conjunction with an appointed legislative council and subject to detailed supervision by the Colonial Office in London. Under the general control of the Governor, the Superintendent of Public Works was responsible for the design, construction and maintenance of roads and bridges; waterworks and sewerage schemes; military barracks, gaols and police stations; hospitals; government offices and other public buildings; and piers, harbours and navigational works. When Leahy took up the post in August 1860, it had been held temporarily by three different engineers in the year since his predecessor's suspension, and work on a number of major projects had naturally been delayed. During the following four months, the pace of activity was stepped up, with tenders being invited for a variety of projects, such as improvement works at the New Public Buildings, including water closets and new apartments; improvement works at the Lunatic Asylum in Port of Spain, the island's capital; the erection of Bonding Stores in Port of Spain; and repairs to the police stations at Arima and Arouca.[26] It is likely, however, that most of the engineering and architectural work required to bring these projects to tender stage had been done by Leahy's predecessors or by the long-serving staff of the Office of Public Works (an overseer and a chief clerk-draftsman) before his arrival in the island.

A "melancholy duty"

On 17 December 1860, the Lieutenant-Governor sent the following despatch to the Duke of Newcastle:

> It is my melancholy duty to report to Your Grace the death of Mr Matthew Leahy, the Superintendent of Public Works.
>
> This sad occurrence took place on the evening of Friday the 14th inst., within less than four months after his arrival in the colony.[27]

The cause of Matthew's death is not recorded, but since he was only 42 years of age, it seems likely that he succumbed to one of the tropical diseases for which contemporary medicine could provide no cure. After his death, his sister Susan remained on the island for some months and had to be supported by her brother, Patrick, before returning to London.[28] The fate of the wife – if such existed – whom Matthew may have left behind in London is not known.

Matthew's death does not mark the end of references to him in the files of the Colonial Office. His London creditors had apparently pursued him to Trinidad and, after his death, sought to have their claims met by the Government. A London purveyor of meat submitted an unpaid account for £27 12s. for meat supplied to Matthew and suggested - but without success - that his loss should be recouped from public funds.[29] John Weale, who had a well-known bookshop at High Holborn in London, appeared to have a stronger case: he claimed that Leahy had purchased some engineering books from him, at a cost of £16 9s., and felt entitled to recoupment because "these books were for the public service, selected with care and ceremony by Mr Leahy just previous to his departure, for immediate use as Government Engineer".[30] But there was no record of the books in the Office of the Superintendent of Public Works in Trinidad and so the Colonial Office declined to recoup Weale. However, they felt that the bookseller should not suffer the loss and, as they had done with the butcher, sent him the address of Leahy's brother, Denis, in the expectation that he might clear off the debt.[31]

"I beg ... to ask most respectfully to be allowed to succeed him"

In reporting the death of Matthew Leahy, Lieutenant-Governor Walker asked that an early appointment should be made from London.[32] He feared that "this untimely death of Mr Leahy will again expose us to much public inconvenience" and repeated the view that the salary attached to the post might not be sufficient to secure the services of a competent professional man. The Colonial Office lost no time in setting the selection machinery in motion but did not accept that the salary should be increased: instead, the official dealing with the matter observed that "the sufficiency or insufficiency of the salary will appear when endeavours are made to fill up the appointment".[33] Once more, the file was referred to the private office of the Secretary of State, "in case they might have a candidate on the Duke of Newcastle's list" and, again, a Leahy was selected by the Duke to fill the vacancy.

While neither Edmund nor Matthew had a very high regard for their brother's skill or competence as a civil engineer, Denis himself had other ideas - and was able to produce copious references attesting to his ability and qualifications. And so, a fortnight after news of his brother's untimely death had reached London, Denis applied for the vacant post in Trinidad in a letter written on black-rimmed notepaper:

> Having received the melancholy news of my dear brother's death ... I beg to offer my services and to ask most respectfully to be allowed to succeed him in that office. I can submit the same or equal testimonials as to my engineering character and capability. I hope and trust, therefore, that this application will receive the favourable consideration of His Grace, the Duke of Newcastle, and that His Grace will suspend the appointment in order that I may obtain from my friends the necessary letters and recommendations to be submitted to His Grace.[34]

By to-day's standards, this job application was remarkably lacking in detail, with no effort made to provide a curriculum vitae or particulars of experience or achievements. The testimonials submitted subsequently by Leahy - although many of them came from well-known engineers - did little to make up for this deficiency:[35]

> General Sir John Fox Burgoyne obviously had very little knowledge of Denis Leahy's engineering but was prepared to provide him with a reference "from my knowledge of the professional merits of your father and brother and my acquaintance with yourself". He went on: "I know you to belong to a good stock, that your father and brother always spoke of you as an able co-operator with them and I have heard of works in which you have been engaged, and when your proceedings were spoken of with respect".

> Richard B Grantham, a prominent English consulting engineer who had served in the 1830s as a county surveyor in Ireland, testified that he had known Leahy since he entered the profession, although he had not been in communication with him "of late years". He could do no more than declare that Leahy "has been engaged in conducting public works in the capacity of Engineer, and I believe him to be efficient".

> John H Taunton, another well known English engineer who had worked on railway surveys in Ireland in the 1840s, was slightly more forthcoming, stating that Denis, from his knowledge and experience of public works, would be well qualified to succeed his brother.

> John Hawkshaw, yet another London-based consulting engineer, stated that he and Leahy were in the same office in early life (they had both worked with Alexander Nimmo as young men) and "from what I have known of you, I should think you would be able to fill the office you seek quite satisfactorily".

HK Connell, Locomotive Superintendent of the London & North Western Railway, was also of opinion that Leahy was well qualified for the post in Trinidad.

John Scott Russell certified that Leahy was "a civil engineer and surveyor of great experience, having been from boyhood engaged with his father, with the late Alexander Nimmo, with his two brothers, and on his own account, in the survey and execution of various public works of importance, including railways, harbours, public buildings, roads, bridges, mines, etc., not only in England and Ireland but also on the Continent. In Turkey, he was engaged with his brothers in surveying lines of railway for the Turkish Government, and in working the lead and silver mines of Thessaly".

In addition to these testimonials from prominent English engineers, Leahy was able to offer references from a number of county surveyors in Ireland. The Wexford surveyor, James Barry Farrell, had met Leahy in 1845: "he was then, with his father and brothers, extensively engaged in projecting Irish railway lines". Horace Townsend, surveyor for Queen's County (Laois) and previously for Tipperary North Riding, recalled that Leahy had been a contractor for road works in North Tipperary about 1843 and had completed his contracts satisfactorily. The surveyor for East Limerick, Thomas Kearney, bore testimony "to the high qualifications of your father, yourself and your two brothers as Civil Engineers and Architects from the year 1834 to the time of your leaving Ireland". And William Henry Deane, at the time county surveyor in Tyrone, and formerly in South Tipperary, thought that Leahy, whom he had known since 1834, was "a civil engineer, surveyor and inspector of works of first class order" and he attributed his own advance to the professional knowledge imparted to him by Leahy.

Denis succeeds his brother

Denis Leahy was living with his wife, two sons and a 20-year-old servant girl at 2 Westbourne Terrace, Islington, London, when the census of England was taken on 7 April 1861.[36] Fortunately for him, his qualifications and suitability for the appointment in Trinidad were not seriously called into question as there was, apparently, no other contender for the post. On 15 June, the Duke of Newcastle decided that Leahy should be appointed[37] and this was formally announced in the *London Gazette* of 2 July. Leahy accepted the post on terms and conditions similar to those which had applied to his brother (a salary of £400 a year and £150 for travelling expenses) although it was clear that an increased salary would have been offered if this were necessary to fill the post. When he was instructed to proceed as soon as possible to Trinidad and told that a free passage would be available on the steam-packet which left Southampton twice each month,[38] he told the Colonial Office that he wished Elizabeth, his 30-year-old wife, a son aged five months, and a female servant to accompany him, "provided each can obtain a passage at the expense of the Colony or the Government"; the eldest boy,

Chapter XII — Matthew and Denis in Trinidad, 1860-63

Albert William Denis, then aged six, was to remain in London.[39] When told that a free passage was available only for himself, he dropped the idea of bringing out a servant and asked that passages for himself and his wife and younger son, William Henry, should be arranged on the Royal Mail Steam Packet which was due to leave on 17 August.[40] But there was to be another hitch in his arrangements. The public auction of his furniture occupied more time than he had anticipated (it was not completed until 14 August) and he did not expect to receive the proceeds for some time; as a result, he was forced to ask the Colonial Office to transfer the family's passages to the steamer which was to leave on 2 September.[41] The official involved agreed to this but directed that "Mr. Leahy ought to be told in pretty decisive terms that he should have told us this before" and that any loss to the Government of Trinidad would have to be borne by him.[42]

Leahy arrived in Trinidad in September 1861 and formal notice of his appointment as Superintendent of Public Works appeared in the *Trinidad Royal Gazette* towards the end of that month.[43] At the time, the island's economy was in a particularly depressed state. The preference given to West Indian sugar, on which the economy was heavily dependent, had been removed by Britain in 1854, exposing the planters to free trade, full competition and bankruptcies.[44] By 1862, because of the very poor sugar crops in 1860 and 1861, "every interest in the island had languished", according to a despatch from Governor Robert Keate, and there was pressure on the colonial finances and on individuals.[45] The estimates of expenditure for 1862 were therefore severely pruned and virtually all new works had been eliminated, apart from a new Port of Spain sewerage scheme to which the government was committed; in addition, the total provision for the maintenance of barracks, gaols, government buildings, waterworks, police stations, hospitals, navigation works and lighthouses was cut to only £2,270, with a further £2,250 for the maintenance of roads, wharves, jetties and bridges.[46] These restricted budgets would have allowed little scope for even the most skilled and competent of engineers to make any significant impact on the infrastructure of an island where large areas had still to be surveyed and mapped.

One of Leahy's first tasks was to complete the work of improving and enlarging the lunatic asylum at Port of Spain for which tenders had been invited by his brother in October 1860.[47] This assignment took on special urgency in November 1861 when the Governor learned that Prince Alfred, the 18-year-old second son of Queen Victoria, was to visit the colony during the following January.[48] The intention was to complete the new wing (designed for 14 inmates) and fit it out temporarily to accommodate the Prince and his suite. At a later stage, Leahy was directed to prepare plans for adapting the basement of the asylum to provide extra cells; he reported that this might easily be done at an extra cost of some £382 and submitted detailed estimates for other modifications costing a further £2,183.[49] In both cases, the estimates were signed in a clumsy and faltering hand, with none of the flourishes which characterised the signatures of so many contemporary engineers.

Because of the prevailing budgetary restrictions, the only major engineering work with which Leahy was directly involved in Trinidad was the construction of a sewerage system to serve Port of Spain where there was a "most pressing necessity for sanitary improvements" according to the General Board of Health.[50] An ordinance for establishing this system had been passed in October 1858 but the project was retarded by the changes in the Office of Superintendent of Public Works and little had been done until 1862 beyond getting the pipes out from England; the estimated total cost of the system was £15,000 and £9,804 was allocated to meet expenditure in 1862.[51] Leahy spent the first few months after his arrival in Trinidad "taking his levels and preparing his sections and detailed working drawings" for the sewerage scheme.[52] But the Office of Public Works had no proper instruments for the task and, because of "the great nicety of measurement and of levelling" that was involved, the Governor had to requisition new instruments from London to meet Leahy's needs.[53]

In the event, the dry season of 1862 (January to May) was virtually at an end when work began and, almost immediately, the project became embroiled in major controversy, culminating in a petition signed by a large number of the householders and merchants of Port of Spain seeking to have the project aborted. This was supported by a resolution of the legislative council. It was claimed that "the underground sewers already laid down to carry off the impurities of the Hospital and Gaol have proved to be a failure", that there would regularly be blockages in the new pipes, that the whole system would be fraught with dangers to the health of the town, and that it would be impolitic and cruel to proceed with it.[54] The petitioners also argued that it was unwise to start the work just as the wet season was due to begin, and that the scheme would cost far more than the £15,000 which had been budgeted for in 1858. Finally they objected strenuously "to the idea of trenching the public thoroughfares of the town and cutting through private properties".[55]

Governor Keate strongly defended the project. He acknowledged that because of "gross misuse of the water closets at the hospital and gaol by their ignorant inmates, much difficulty had been experienced in keeping clear the service pipes" but this was only "a little defect of detail" and would be overcome by the use of "lidded" pipes which would allow obstructions to be cleared.[56] He called for a report from the Office of Public Works on the effectiveness of the existing systems and was advised by Leahy that "judging from what I have seen, and the quantity of matter which I have observed discharging at the Jetty at high water mark, I should pronounce the sewers from the Hospital and Royal Gaol to answer the purposes for which they were originally intended very well indeed".[57] Additional stoneware drainage pipes were requisitioned in July from the Agents General for the Crown Colonies in London[58] and the main sewer along one of the principal streets of the town was being laid in August.[59] However, it was clear by mid-September that the project was again falling behind schedule and that expenditure in 1862 would amount to less than one-third of the budget of nearly £10,000 which had been set aside for it.[60] Thereafter, the project seems to have become a total failure, judging by the comment of an English visitor

Chapter XII — Matthew and Denis in Trinidad, 1860-63

25 years later that "of sanitary arrangements there seemed to be none" even though the town then had a population of up to 40,000.[61]

A despatch sent to the Colonial Office in September 1862 reported that "the sewerage works and the new wing to the Lunatic Asylum are the only public works of any importance in progress, or likely to be so, in 1862 except an extra water supply to St James' Barracks".[62] This latter was a relatively small project for which a supply of three-inch iron pipes had been requisitioned by Leahy from London in August.[63] Planning was going ahead, however, for other projects; for example, a site for a police station at the village of La Brea, near the famous asphalt lake, was acquired and plans for new works at the Colonial Hospital in Port of Spain were completed in August. These included a large crudely-drawn layout plan at a scale of one inch to 50 feet, and plans and elevations at a scale of one inch to 10 feet for a two-storey building, 55 feet by 24 feet, to provide Matron's quarters, a carriage shed and storage space. The plans and the cost estimates which came to £2,164 were signed *D. Leahy, Supt. Pub. Works* in the same faltering hand used in his earlier work.[65] Other works proposed at the hospital comprised the installation

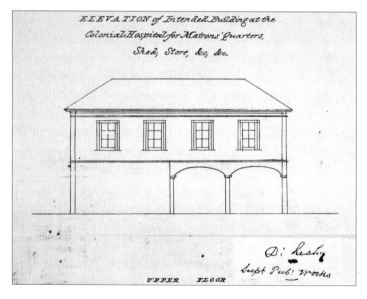

A drawing entitled "Elevation of Intended Building at the Colonial Hospital for Matron's Quarters, Shed, Store etc etc ", signed by Denis Leahy (CO 295/218).

of water closets, improvements to the sewerage system, an ulcer shed and a new internal road network; these were to cost an additional £2,425.[66]

Although moves were afoot to develop both "tram roads" and railways in the island during his tenure of the post of Superintendent of Public Works, there is no evidence that Leahy had any direct involvement in promoting such development in Trinidad. In the mid-1850s, the Legislative Council had adopted *An Ordinance for encouraging the formation of Tram Roads*[67] and proposals for a tram road, with a number of branches, were current in 1857.[68] Tenders were invited in 1859 for the transport of tramway materials and navvies to the island,[69] construction of the tramroad was reported to have made some progress by the end of 1862,[70] and it seems possible that Leahy had some supervisory responsibility for the work.[71] However, before this or any other projects could be brought to fruition, Denis followed his brother Matthew to an early grave; Governor Keate reported with regret to the Colonial Office that he had died on 21 December 1862.[72] Leahy was little more than 50 years of age and, according to his widow, Elizabeth, his death occurred "after a short illness brought on by the severity of the climate".[73] The diary of Rev James O'Carroll records that about 30 priests attended a Requiem Office for Denis in his brother's cathedral in Thurles on 22 January 1863, although he "was married to a Protestant"; the diarist noted also that many of those in attendance had remarked that "Thurles was not the happiest selection to hold an Office as, while residing there and at Cabra, his was not the most edifying life".[74]

A widow's economy with the truth

Following her return to London in 1863, Elizabeth Leahy claimed that she and her two children were left "totally unprovided for" and as there was then no formal system of pensions for the widows of those who died in the colonial service, she wrote to the Duke of Newcastle on 16 June 1863 appealing for assistance:

> I therefore humbly solicit Your Grace to use your interest to procure me a gratuity from the Government to enable me to procure some means of living and also to obtain for my eldest boy, aged eight years, a presentation to the Blue Coat School, where he can receive an education suitable to his position and which I shall never be able to afford him. I beg to refer you to Sir John Burgoyne, who had known my husband many years, for a confirmation of the truth of my statement.[75]

The Colonial Office sent the widow's appeal to the Treasury to consider the possibility of awarding her a grant from the Royal Bounty Fund, observing sympathetically that "if a donation could be afforded from the Fund it would apparently be well bestowed".[76] Mrs Leahy received a non-committal interim reply - anything more, in the view of the cautious senior official dealing with the case in the Colonial Office, "might raise expectations which might be disappointed".[77] In the event, the Treasury officials, who were concerned as always with precedent, rejected the application, notwithstanding an admission that the circumstances were painful; they ruled that "if, as in the case of Mrs Leahy, the mere circumstances of a public official having died from illness brought on by the severity of the climate, and leaving a widow and two children unprovided for, was to be established as a

Chapter XII — Matthew and Denis in Trinidad, 1860-63

ground for making a grant out of Royal Bounty, it is obvious that the amount available for such purposes must be largely increased, in order to meet the number of such cases that constantly arise".[78]

In light of this correspondence, one might wonder how the 32-year-old widow and her two sons managed to survive after 1863. It appears, however, that Elizabeth had been economical with the truth when she wrote to the Colonial Office on 16 June for she had married John Henry Bannister, a London surgeon 20 years older then herself, on 4 June and, with her children, went to live at his home at 436 Oxford Street.[79] The fact that some of her late husband's letters to the Colonial Office in 1861[80] came from this address suggests that the couple were already known to one another before Elizabeth's return from Trinidad. Sadly, however, Elizabeth died in October 1864[81] and while Bannister married again in February 1868[82], his stepsons, Albert and William Leahy were still living with him and his young wife and son at Oxford Street in April 1871.[83] Ten-year-old William Henry (Harry) was listed as a scholar in that year's Census return and sixteen-year-old Albert was described as a draper's assistant, a designation which is consistent with the fact that Archbishop Leahy's Ordos for 1869 and 1871[84] note his address as 232-234 Regent Street, where the firm of Dickins & Jones, Linen Drapers and Hosiers, was located.[85]

While the fate of William Henry Leahy has not been established,[86] his elder brother went on to have a distinguished medical career. By 1878, having studied in London, Paris and Strasburg, Albert had become a member of the Royal College of Surgeons, like his stepfather, and he went on to gain his Fellowship in 1881 and an MD in 1893.[87] After some years in a variety of posts in Charing Cross and other London hospitals, and with a number of publications in medical journals to his credit, he gained first place in open competition for an appointment in the Indian Medical Service and his appointment as a surgeon in the Bengal Army was announced in the *London Gazette* in March 1883.[88] He had reached the rank of Lieutenant Colonel before his retirement in June 1903, possibly on medical grounds,[89] following which he returned to live in London where he married Ulrica May Fitzpatrick in 1908.[90] He died at his home at Iddesleigh Mansions, Victoria Street, Westminster after a long illness on 17 July 1917,[91] and was survived by his wife who died in London in August 1928.[92]

REFERENCES

1. *House of Lords Sessional Papers 1864* Vol XIII (254), letter of 17 August 1863 to the Duke of Newcastle, Secretary of State for the Colonies
2. *Trinidad Royal Gazette*, Vol 25, No 39, 21 September 1859; CO 295/211, letter of 8 February 1860
3. CO 295/211, letter of 5 April 1860 from L W Samuel
4. CO 295/211, note on letter of 5 April 1860 from L W Samuel
5. CO 295/211, letter of 31 May 1860; the seventh Earl of Carlisle, George William Howard, 1802-1864, served as Chief Secretary for Ireland 1835-1841 and as Lord Lieutenant 1859-1864.
6. CO 295/209, Despatch No 81, 6 June 1860

In search of fame and fortune: the Leahy family of engineers, 1780-1888

7. CO 382/13, letter of 1 March 1860 from Matthew Leahy
8. CO 295/211, letters of 1 and 16 April, 1860
9. CO 295/211, letter of 16 April 1860 from Matthew Leahy
10. CO 295/211, testimonials enclosed with letter of 31 May 1860
11. NAI, CSORP 1845/W/1288, letter of 29 January 1845
12. CO 295/211, note on letter of 31 May 1860 from Matthew Leahy
13. CO 295/211, letter of 19 June 1860 from CO to Matthew Leahy
14. FO 611, Register of Names of Passport Applicants, shows that Matthew obtained a passport (No 23448) on 8 June 1860; there are no entries in the register for a wife or sister.
15. If, when and whom Matthew may have married has not been established; his name does not occur in the index to marriages registered in England between early 1858 (when he returned to London from Turkey) and December 1860.
16. CO 295/211, letters of 21 June and 17 July 1860
17. CO 295/211, letter of 19 July 1860
18. CO 295/211, letter of 23 July 1860
19. CO 295/211, letter of 25 July from CO to Matthew Leahy
20. CO 295/209, Despatch No 119 of 22 August 1860
21. *Trinidad Royal Gazette*, Vol 26, No 34, 22 August 1860
22. CO 295/210, Despatch No 175, 17 December 1860
23. CO 295/210, Despatch No 180, 22 December 1860
24. CO 295/214, Despatch No 179, 5 December 1861
25. CO 295/214, Despatch No 179, 5 December 1861; CO 295/218, Despatch No 134, 15 July 1862
26. *Trinidad Royal Gazette*, Vol 26, Nos 35-50
27. CO 295/210, Despatch No 175, 17 December 1860; a death notice in the *Cork Examiner* on 23 January 1861 gave 15 December as the date of Matthew's death and that date was noted in Patrick's Ordos for 1864 and 1872 as the anniversary of the death.
28. Ordo of Dr Leahy for 1861, 8 February, records a payment of £20 to Susan in Trinidad.
29. CO 295/215, letter of 7 February 1861 from William Slater
30. CO 295/215, letter of 12 April 1861 from John Weale; it was Weale who published Edmund Leahy's *Practical Treatise on Making and Repairing Roads* in 1844.
31. CO 295/213, Despatch No 85, 6 June 1861; CO 295/215, letter of 14 February 1861
32. CO 295/210, Despatch No 175, 17 December 1860
33. CO 295/210, CO note on Despatch No 175, 17 December 1860
34. CO 295/215, letter of 30 January 1861 from Denis Leahy
35. CO 295/215, testimonials submitted with letters of 12 and 15 February 1861 from Denis Leahy
36. Census of 1861, RD Islington, RG09/135/15/29
37. CO 295/215
38. CO 295/215, letter of 22 June 1861 from CO to Denis Leahy
39. CO 295/215, letters of 5 and 22 July from Denis Leahy; Albert William Denis Leahy was born on 20 July 1855 at 78 Stanley Street, Pimlico, London, Birth Certificate issued by the General Register Office, London, RD St George Hanover Square, Vol 1a, page 196.
40. CO 295/215, letter of 29 July 1861 from Denis Leahy; William Henry Leahy was born on 14 February 1861 at 2 Westbourne Terrace, Barnsbury Park, London, Birth Certificate issued by the General Register Office, London, RD Islington, Vol 1b, page 243.
41. CO 295/215, letter of 15 August 1861 from Denis Leahy
42. CO 295/215, minute of 16 August and letter of 20 August 1861 from CO to Denis Leahy
43. *Trinidad Royal Gazette*, Vol 27, No 38, 23 September 1861
44. Philip Sherlock, *West Indian Nations - A New History*, London, 1973
45. CO 295/218, Despatch No 112, 4 June 1862
46. CO 295/213, Despatch No 120, 6 August 1861
47. *Trinidad Royal Gazette*, Vol 26, No 43, 23 October 1860
48. CO 295/214, Despatch No 176, 23 November 1861
49. CO 295/219, Despatch No 195, 17 November 1862
50. CO 295/218, Despatch No 150, 4 August 1862
51. ibid

Chapter XII — Matthew and Denis in Trinidad, 1860-63

52. ibid
53. CO 295/216, Despatch No 41, 21 February 1862
54. CO 295/218, Despatch No 150, 4 August 1862
55. CO 295/218, petition of 27 June 1862
56. CO 295/218, Despatch No 150, 4 August 1862
57. CO 295/218, report of 24 July 1862, included in Despatch No 150, 4 August 1862
58. CO 295/218, Despatch No 136, 17 July 1862
59. CO 295/218, Despatch No 150, 4 August 1862
60. CO 295/219, Despatch No 167, 17 September 1862
61. J A Froude, *The English in the West Indies*, Longmans, Green and Co., London, 1888, page 64
62. CO 295/219, Despatch No 167, 17 September 1862
63. CO 295/219, Despatch No 159, 23 August 1862
64. CO 295/219, Despatch No 183, 21 October 1862
65. CO 295/218
66. CO 295/218, Despatch No 154, 18 June 1862
67. CO 300 (68) 1857
68. *Trinidad Royal Gazette*, Vol 23, No 4, 23 January 1857
69. *Trinidad Royal Gazette*, Vol 25, No 49, 30 November 1859
70. CO 295/220, CO official's minute on letter of 4 December 1862 from the Trinidad Railway Company
71. Conventional railway development in Trinidad came somewhat later (although preliminary surveys for a railway network on the island were made as early as 1846) and both Matthew and Denis Leahy were in their graves before construction of Trinidad's first railway began.
72. CO 295/220, Despatch No 215, 23 December 1862
73. CO 295/225, letter of 16 June 1863 from Elizabeth Leahy of 23 Percy Street, Bedford Square, London to the Duke of Newcastle
74. CDA, O'Carroll diary, 23 January 1863
75. CO 295/225, letter of 16 June 1863 from Elizabeth Leahy to the Duke of Newcastle
76. CO 295/225, CO official's minute of 30 July 1863
77. CO 295/225, letter of 29 July 1863 and minute endorsed on it
78. CO 295/225, minute of 19 August 1863 from the Treasury to the CO
79. Marriage Certificate issued by the General Register Office, London, RD Pancras, Vol 1b, page 216; John Henry Bannister was a native of Hampshire who had gained membership of the Royal College of Surgeons (London) and the Licentiate of the Society of Apothecaries (London) in 1837; he had been living at Oxford Street for at least ten years before his marriage to Elizabeth. Bannister died in London on 8 May 1887 (Medical Directory for various years).
80. For example, CO 295/215, letter of 15 August 1861
81. Death Certificate issued by the General Register Office, London, RD Strand, Vol 1b, page 375
82. Marriage Certificate issued by the General Register Office, London, RD London Middlesex, Vol 1c, page 111
83. Census of Population 1871, civil parish of St Anne, Borough of Westminster, RG 10 144
84. CDA, Ordo of Archbishop Leahy for 1869 and 1871
85. Dickins & Jones, one of the West End's leading department stores for more than 150 years, closed in January 2006.
86. The death of a William Leahy, aged 14, was registered in London in January 1875, following an accident at Rotherhithe some months earlier, but it is not certain that this was William Henry (Death Certificate issued by the General Register Office, London, RD St Olave, Southwark, vol 1d, page 144).
87. *Medical Directory*, 1893, 1901; *The Medical Who's Who*, 1914
88. *London Gazette*, 23 March 1883
89. *London Gazette*, 21 July 1903; *The Quarterly Indian Army List for January 1, 1912*, Superintendent of Government Printing, Calcutta, 1912
90. Index to Register of Marriages, September quarter 1908, RD Westminster, page 1195
91. Death Certificate issued by the General Register Office, London, RD St George Hanover Square, Vol 1a, page 422; *The Times*, 19 July 1917
92. Death Certificate issued by the General Register Office, London, RD Hendon, Vol 3a, page 316; *The Times*, 24 August 1928

CHAPTER XIII

EDMUND LEAHY IN JAMAICA, 1858-63

"I never was a subordinate in my life".

Edmund Leahy, 1858[1]

Cables under the oceans

Soon after his return to London in February 1858 after his six years in Turkey, Edmund Leahy resumed where he had left off in 1852, bombarding the Colonial Office with requests for aid for schemes of one kind or another which cannot have amounted to more than vague notions of his own, conceived without any serious thought or analysis and with no evidence of financial backing or other support. In May, just as the telegraph cable required for a second attempt to connect Valencia, County Kerry, to Newfoundland was about to be loaded onto cable-laying ships in Plymouth, Leahy wrote to Lord Stanley who had become Colonial Secretary a few months earlier, proposing the establishment of telegraphic communication from England "by sea line" to Gibraltar, Ascension Island, Cape Town, Melbourne, Sydney and Van Diemen's Land (Tasmania); all of this, he assured Stanley, would be feasible and practicable, as well as being "remunerative in a mercantile sense" and eminently useful from a national point of view. The cost of the scheme was put by Leahy at about £6 million, but he believed that this vast sum would be readily subscribed "provided a fair amount of Government aid can be obtained in the form of an annual payment for the use of the lines".[2]

In reviewing this submission, Colonial Office officials must have been aware that the technical and commercial feasibility of long-distance cables under the oceans of the world had yet to be established. The longest submarine cable then in existence extended for only about 300 miles in the relatively shallow waters of the Black Sea between Varna and Balaklava, and the first attempt at laying a cable from Valencia to Newfoundland, a distance of about 1,900 miles, had failed in the previous year. However, instead of dismissing out of hand a scheme for laying up to 15,000 miles of cable which had been proposed by a man who had no experience of electrical engineering and no obvious back-up, the reply from Lord Carnavon, Under-Secretary for the Colonies, simply advised

Chapter XIII — Edmund Leahy in Jamaica, 1858-63

Leahy that a scheme of such an extensive nature should be submitted to the Treasury[3] which had, in fact, agreed in 1857 to pay an annual sum of £14,000 to the Atlantic Telegraph Company should it succeed in bringing a trans-atlantic cable into operation[4].

A steamship service on a "strange route"
Leahy's next proposal, submitted to the Colonial Office in August 1858, involved an equally unrealistic scheme - the establishment of a steamship service to Vancouver via the North-east Passage ie, going by way of Norway, the Barents Sea and the Bering Strait to the west coast of America, a voyage which no ship had yet succeeded in completing.[5] This proposal may have been influenced by the fact that the more difficult North-west Passage from the Atlantic to the Pacific, around northern Canada, was in the news at the time, with F L McClintock leading a mission to establish what had become of the ill-fated expedition of 1845 from which Sir John Franklin and more than 100 others were never to return.[6] While admitting that the seas north of Siberia were closed by pack ice during the winter months, Leahy suggested optimistically that the travel time for the 8,000 mile journey during the other months could be less than 30 days and this, he contended, would be of immense value to the colony of Vancouver Island and the newly established colony of British Columbia. To test the practicability of his proposition, Leahy told the Colonial Secretary that he was "ready at once to start a steamer on the line, as a pioneer, provided you approve of my idea and that the Government will agree to pay the expenses in case of success". However, notwithstanding the fact that thousands of people were flocking to British Columbia following the discovery of gold in the Fraser valley in 1857, the Colonial Office official who dealt initially with the letter was not impressed: "this is indeed a strange route", he suggested to the Permanent Under-Secretary, Herman Merivale, adding that "the projector must go alone, I think - it can scarcely be imagined that any person would prefer this way of reaching the other side of North America". Merivale and the Colonial Secretary agreed and directed that the project should be "civilly declined".[7]

Leahy, clearly, had no understanding of the difficulties and practicalities of the voyage from the Atlantic to the Pacific "by rounding the northern termination of the old world" as it was described at the time. In fact, it was to take another 21 years for the first west to east navigation of the North-east Passage to be completed by the Swedish steamship, *Vega*, and even then the voyage was described as "one which will always rank as one of the greatest geographical feats" of the century;[8] the route did not come into regular use by shipping until the 1930s, when a fleet of large Russian icebreakers became available to keep it open.

Seeking a position at the Cape
Leahy's efforts to find some gainful employment for himself took a new direction when news of the death of Captain Pilkington, Colonial Engineer at the Cape Colony, reached

him in August 1858. The post was one of the best-paid engineering positions in the colonial service, carrying a salary of £1,000 a year and an allowance for travelling expenses, and Pilkington had been appointed to it in December 1848, a few months before Leahy had arrived in the colony in the hope of securing it. Having lived at the Cape for two years and having "studied the capabilities of the colony very closely" during that time, Leahy considered himself eminently suitable to succeed Pilkington and his application for the vacant post was with the Colonial Secretary even before the Governor's despatch notifying the vacancy was received in the Colonial Office on 3 September.[9] In advancing his case, he set out his engineering career in some detail; he made rather dubious claims to have been "engineer conjointly with Mr Robert Stephenson MP and also with Mr Vignoles" on several projects, he described his work as county surveyor in Ireland in grossly exaggerated terms and referred to his extensive railway survey work in Turkey. He also submitted a testimonial from Lieutenant-General Sir John Fox Burgoyne who stated that, although he held "a very favourable opinion" of Leahy's qualities as a professional man, "it is not in my power to give such specific testimonials to your qualifications as might be most valuable to you".[10] While offering to submit further testimonials, if these were needed, Leahy's claims on the office were heavily dependent on other considerations; in particular, he argued that his support for the Government during the convict agitation in 1849 at the Cape would justify his selection and he was supported in this by representations from Captain Mervyn Archdall, MP for Fermanagh since 1834.[11]

The Earl of Carnarvon, Under-Secretary for the Colonies, when advised by his officials that Leahy had indeed aided the Cape Government in 1849, suggested to the Colonial Secretary, Sir Edward Bulwer-Lytton, that, "supposing all professional qualifications are up to the mark, *ceteris paribus*, any person who has suffered on behalf of the late Government has a claim to be preferred".[12] However, instead of making an appointment on a patronage basis, the Colonial Secretary followed the advice of the Governor[13] and invited the Council of the Institution of Civil Engineers to select "the most eligible candidate" from the five (including Leahy) who had applied for the post, or to nominate another suitable engineer.[14] A special meeting of the Council was convened on 20 September to review the applications[15] and the five members present decided unanimously to recommend John Scott Tucker (1814-82) who, by coincidence had been in South Africa in 1851, just after Leahy's departure, to report on a proposed railway at Cape Town;[16] Tucker was duly appointed to the post a few days later.[17]

An offer of an appointment in Jamaica
Lord Carnarvon's view that Leahy was entitled to preferential treatment clearly carried weight because, without even waiting for the Institution of Civil Engineers to complete its assessment of the applications for the post at the Cape, it was decided to offer him an appointment in Jamaica for which he had not applied and for which there were numerous

Chapter XIII — Edmund Leahy in Jamaica, 1858-63

other well-qualified applicants. Leahy accepted this appointment on 18 September 1858[18] and the Governor of Jamaica was formally notified later that month in a despatch which made it clear that Leahy was to fill one of three new road engineer positions, one for each of the island's counties - Cornwall, Middlesex and Surrey. The Governor had been pressing for these appointments to be made since early 1858 and had transmitted the many applications he had received to the Colonial Office so that a selection could be made by the Secretary of State.[19] Among these was an application from P J Klassen, a native of Vienna, who had worked as Assistant Engineer under the Leahys on the Cork & Bandon Railway in 1845-46 and whose application was supported by a testimonial provided by the Leahys in 1846.[20] Another applicant was John H Brett of Dublin, son of Henry Brett, a former colleague of Leahy's in the ranks of the Irish county surveyors and who himself was later to become a county surveyor.[21]

The letter of offer sent to Leahy on 15 September 1858 referred to a position as "Civil Engineer in Jamaica to superintend the Main Roads" and advised him that before accepting it he should call to the Colonial Office to peruse the Act under which the post had been created.[22] Had he acted on this advice, he would have realised that the post on offer to him was not what the Governor described as "the highest and most lucrative appointment to which a salary of £1,000 is attached"[23] but one of the other two road engineer posts which carried salaries of £600 a year. The second post of this kind was accepted by Francis Dawson, a 22-year-old engineer who had studied at King's College, London, served in the Crimea with the Army Works Corps in 1855-56 and worked on the Midland Railway in England.[24] The third and most senior post was to be held jointly with the post of Colonial Engineer and Architect, giving a total salary of £1,000 a year; this was offered to Robert Barlow Gardiner, a 40-year-old native of Jamaica who had trained as an engineer in England and had worked there and in Italy, South America and Portugal, mainly on railway development.[25]

Leahy was in Constantinople in October and on his return to London in November had "some few arrangements" to make[26] with the result that his departure for Jamaica was postponed until 17 December when passages at the contract rate had been arranged by the Admiralty for himself and his wife and son on the Royal Mail Steam Packet Company's steamship which was scheduled to begin its 21-day voyage from Southampton that day.[27] However, within a few days of his arrival on the island in early January 1859, Leahy "felt compelled to tender his resignation" in a letter addressed to the Colonial Secretary.[28] He explained that the Governor had made it clear to him that his position was to be that of roads engineer for the county of Surrey and that his "professional position should be officially subordinate to another Engineer, not yet arrived, who is to be Chief Engineer". And he went on to proclaim: "I never was a subordinate in my life – the evidences of my qualifications which I forwarded with my application ... as well as the letter of appointment ... which you honoured me with, left the conviction that I was not to be a Subordinate Engineer, of secondary position, such as that now mentioned by the

Edmund Leahy's letter of 10 January 1859 to the Colonial Secretary protesting about the inferior position to which he had been appointed in Jamaica (CO 137/347).

Governor". But notwithstanding these protestations, and because he had incurred "very serious expenses on the faith of the appointment", Leahy seems to have opted to function as one of the road engineers, at least for a time.

Four months later, fate took a hand when Gardiner, who had arrived in Jamaica in March to take up the position of Colonial Engineer and Architect,[29] resigned from that office and from the position of Chief Roads Engineer with responsibility for roads in the county of Middlesex, the largest of the three counties and containing almost half of the total area of the island; Gardiner was apparently "a man of intemperate habits",[30] and died at Spanish Town on 14 June.[31] This left the way open for the Governor to seek approval from the Colonial Office for the appointment of Leahy to both these offices, with John Parry, a native of Jamaica, replacing him in the Surrey road engineer post.[32] These arrangements were approved in July by the Duke of Newcastle, who had just become Colonial Secretary, giving Leahy the appointment he desired, with a combined salary of £1,000 a year and an allowance of £1 a day when travelling.[33]

Chapter XIII — Edmund Leahy in Jamaica, 1858-63

A turbulent and depressed colony

With an area of 4,244 square miles, Jamaica measures 146 miles from east to west and between 22 and 51 miles from north to south. The island is mountainous and heavily forested in the interior, with low coastal plains and swamps, scattered hills and plateaux and numerous fast-flowing rivers. Although situated in the tropics, the climate varies considerably from the plains to the plateaux and the mountains. The island became a British colony in 1655, when it was seized from Spain, and by 1861 it had a population of

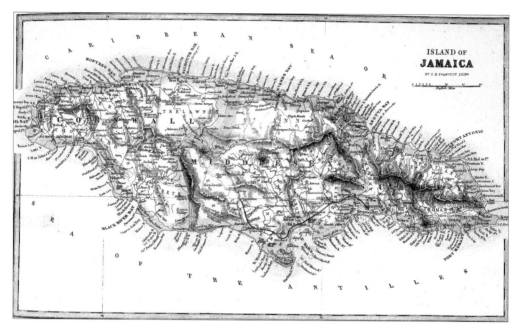

*Jamaica in the 1860s (**Royal Illustrated Atlas of Modern Geography**, ca. 1865).*

over 441,000 people, many of them emancipated slaves or their descendants; the white population numbered less than 14,000, of whom 471 were natives of Ireland. With a form of government which was almost unique in the British colonies, Jamaica was described by one of its Governors "as the most turbulent and unmanageable of all her Majesty's dominions".[34] While the Governor, as the representative of the Crown, was head of the Government of the colony and agent of the Imperial Government in London, the colonists were proud of the fact the island had, for some 200 years, a House of Assembly with 47 members, even though these were elected by a tiny minority of the island's population – less than 2,000 people who held or occupied property, paid direct taxes, or had salaries or investments, above certain thresholds; historians have described this Assembly as "a byword for inefficiency and corruption"[35] and "incapable of, and an impediment to good government".[36] In addition to the Assembly, there was a Privy Council of 15 members, a nominated Legislative Council of up to 17 members drawn from the landed interests on the island, and a salaried Executive Committee, comprising up to three members of the House and one from the Legislative Council, all selected by the Governor, to advise him on the administration of the affairs of the colony. The island also had a well-developed system of local administration, with a vestry or parish board in each of its 22 parishes.

When Leahy arrived there in 1859, Jamaica was beginning to recover from a long period of depressed economic conditions. The decline in the island's importance as a military and naval station at the end of the Napoleonic wars, the complete abolition of slavery in August 1838 following the enactment of legislation at Westminster in 1833, and the loss in the early 1850s of preferences for sugar and other goods sold on the English market, all had adverse economic effects. Sugar production fell sharply, the old plantation economy which had been heavily dependent on slavery had broken down, sugar estates were abandoned or went out of cultivation, land values fell dramatically, and basic infrastructure was run down through lack of investment. Epidemics of cholera and smallpox in the early 1850s had killed up to 60,000 people, poverty was rampant, violent crime was common, and living conditions, especially in the towns, were deplorable. On top of all of this, the political situation was a difficult one;[37] many members of the Assembly were politically and personally hostile to Governor Charles Henry Darling[38] who had taken up office in July 1857, and there was continuing serious disagreement between him and the Assembly on constitutional as well as administrative matters.

Road improvement works
Leahy's position of Chief Roads Engineer and the other two road engineer posts were created by a Main Roads Act under which responsibility for the most important roads on the island (including turnpike roads) was transferred from the parishes to a central body of Main Road Commissioners.[39] Expenditure on roads and bridges in 1858, the year before Leahy and his colleagues took up duty, was limited to minor repairs and was described by the Governor as "inconsiderable". In 1859, however, spending on the reconstruction and improvement of main roads and bridges was increased dramatically to a total of £56,000, financed from borrowing; as a result, the Governor was able to report that "a great improvement in the means of communication has undoubtedly taken place, and is universally admitted".[40] By the end of 1860, after a further £59,442 had been spent, the Governor reported that the main roads were now generally "in excellent order" and expenditure in the following years was therefore cut back to £15,800 in 1861, about £30,000 in 1862 and £6,000 in 1863.[41]

It is not possible to determine Leahy's personal contribution to this improvement programme but, as always, he was not slow to advertise and overstate what had been achieved. By 1863, he was asserting that the main roads in Jamaica were "upon the whole and with little if any exception, as good as any roads in Europe", even though they had been "almost impassable" when he came to the island; "it was a sort of mockery to call them roads", he wrote, "for they were absolutely nothing but wheel tracks in the earth, for the greater portion".[42] In reality, however, there seems to have been widespread dissatisfaction with Leahy's performance as Chief Roads Engineer. There were frequent complaints that the roads in Middlesex County for which he was directly responsible were in poor condition and that substantial funds had been wasted on them. It was also alleged that he was seldom to be seen at locations where road works were going on, that he failed to give proper instructions to road overseers and that months elapsed between his visits to some parts of his county.

Bridges

Although a cast-iron bridge over 80 feet long, between cut stone abutments, was constructed over the Rio Cobre river at the eastern end of Spanish Town in 1801 (and is now the oldest surviving bridge of its kind in the western hemisphere), Edmund Leahy's efforts 60 years later to provide a number of similar bridges in Jamaica were miserable failures. There were many rapid and dangerous rivers needing to be bridged in the island, among them the Dry River (known also as the Rio Minho) which runs southwards to the sea about 15 miles west of Spanish Town. This was a formidable river which could rise as much as 37 feet above its bed in a few hours after heavy rainfall and, when in flood, it cut off all east-west communication along the south of the island. In September 1859, Leahy told the Main Road Board that the erection of a bridge over this river would be perfectly practicable and, two months later, the Board set aside £6,000 for the project and directed him to commence the work as soon as practicable.

On several occasions in the following 18 months, Leahy was called on to provide plans, specifications and estimates for the bridge, but he failed to do so. Finally, in July 1861, he laid before the Board letters from John Scott Russell (then operating in London as a consulting engineer and ship builder) relating to the fabrication of an iron bridge, and he was authorised to have the work put in hands immediately. From Russell's letters, it is clear that Leahy had not sent him any plans or specifications, but had left it entirely to him to propose a design for an iron bridge; for the span of 120 feet that was envisaged, Russell suggested either lattice girders or plate girders, connected by cross ties and by overhead arched ribs, and assured Leahy that he had completed a large number of bridges of this kind for the East Bengal Railway. In the course of the following two years, land was acquired for the approaches to the bridge, and work began on the construction of approach embankments and on foundations for the masonry piers. By early 1863, however, when the ironwork for the centre span had already been shipped from England to Jamaica, serious doubts had arisen about the manner in which the project was being carried on. In April, the Governor asked to see the plans, specifications and estimates, but it emerged that there were none. At the same time, it became clear that bricks which were being manufactured on site for the construction of the piers and abutments were of poor quality and that the limited amount of work done to date had absorbed almost the entire budget for the project. All of this forced the Main Road Board to suspend the works. An iron bridge with a central span of 150 feet and two others of 75 feet each was eventually completed at May Pen in 1874, ten years after Leahy had left the island.[43]

Efforts to construct another large bridge over the Great River, which ran to the sea near Montego Bay on the north of the island, provided another example of Leahy's incompetence and lack of expertise as a bridge engineer. Dawson, the road engineer for Cornwall, had provided a plan, specification and estimate for an iron girder bridge in October 1861 and soon afterwards, Leahy, as chief engineer, wrote to John Scott Russell ordering "iron girders etc" for the bridge, giving the span as 112 feet between abutments; as before, no specifications, plans, sections or other directions were given to Russell except that the cost per ton of iron was to be the same as had been quoted for the Dry River Bridge. It transpired that the bed plates ordered by Leahy did not fit the abutments and the ironwork for the bridge, when delivered at Montego Bay from London, was found to be far too heavy for the abutments, weighing 56 tons as against the 12 tons which had been allowed for in Dawson's original plans.

Public buildings
In his capacity as Colonial Engineer and Architect, Leahy boasted in 1863 that he had constructed "a good Lunatic Asylum out of a mass of useless brick walls upon which over £36,000 had been previously wasted." The foundations of this building had been laid as far back as 1843, but work was delayed by lack of funding.[44] Confirmation that alterations to the incomplete structure were "taken seriously in hand "in 1859 and 1860 is available in the Governor's annual reports to the Colonial Office and it became possible for male lunatics to be accommodated in the new building in November 1860. Work continued in the following three years and the female section was completed in 1862, bringing the total cost of the works carried out by Leahy to about £8,000.[45] While it seems to have been accepted on the island that Leahy's alterations had improved the building and its ventilation, the fact was that the asylum had been built on the basis of plans prepared in England and under the superintendence of a clerk of works sent from England. Completion of the asylum seems to have been the only substantial building project carried out by Leahy as Colonial Architect, although he was responsible for additions and routine maintenance works at the general and female penitentiaries, the public hospital, the public buildings in Spanish Town and some local courthouses.

Leahy was called on in September 1862 to prepare plans and estimates for a new complex of public offices on a waterfront site at Kingston where a large number of commercial buildings had been destroyed by a major fire in the previous April. He submitted an estimate for £17,500 and a plan carrying his own signature, and the Executive Committee arranged for this to be displayed in the library of the House of Assembly so that the members might comment on it. But some time later, when the coloured ink faded in the strong sunlight, the plan was seen to be a print of the building provided at Kensington for London's Great International Exhibition of 1862,[46] with the domes and some other features obliterated by the coloured ink. When challenged on what one of the members of the Assembly described as a childish deception, Leahy's facile explanation was that he had a "peculiar right" to use that print because it was very like one of his own drawings of one of the facades of St Sophia, the great church built in Constantinople by the Emperor Justinian over 1,300 years earlier!

Business conducted in a "loose and careless manner"
Within 18 months of Leahy's appointment to the separate offices of Chief Roads Engineer and Colonial Engineer and Architect, Governor Darling reported with regret that the combination of the two offices in one incumbent had been a mistake: "I am sorry to say that experience has shown that the combined duties far exceed the power of one person, however highly qualified, efficiently to perform".[47] Two year later, Governor Eyre was complaining about "the loose and careless manner in which Mr Leahy has conducted the business of his department generally". The Assembly's dissatisfaction with his management and supervision of the roads programme led to the enactment in 1862 of legislation, which was greatly resented by Leahy, giving the parish boards greater supervisory powers in relation to the programme. In addition, the Assembly voted to appoint a committee to investigate the expenditure and general proceedings of the Main Road Board, and the serious irregularities that were alleged to prevail there. Leahy's carelessness and incompetence in the field of bridge design had been exposed by his conduct in relation to the Dry River and Great River bridges, and whatever degree of public confidence he retained must have been eroded by his childish effort to deceive the Assembly in relation to the design of the single most important building project which was proposed during his term of office.

All of this suggests that Leahy's tenure of his official positions in Jamaica must already have been under threat by 1863 when major controversy developed arising from his personal participation in a tramway construction project. This involved direct conflicts of interest between his public duty and his private interests – conflicts of a kind which had led to his downfall in West Cork almost 20 years earlier and which led in Jamaica, as in West Cork, to his departure from the scene leaving behind a costly infrastructural shambles. Leahy's activities in Jamaica also contributed to an almost complete breakdown in the working relationship between the Governor and the House of Assembly to which he himself was elected in 1863, and were even alleged to have played a part in the events leading up to the Morant Bay rebellion of 1865. The tramway scandal, as it came to be known, is the subject of the next chapter.

In search of fame and fortune: the Leahy family of engineers, 1780-1888

REFERENCES

1. CO 137/347, letter of 10 January 1859 from Edmund Leahy to CO
2. CO 323/253, letter of 4 May 1858 from Edmund Leahy to Lord Stanley
3. CO 323/253, letter of 8 May 1858 to Edmund Leahy from Lord Carnavon
4. John Steele Gordon, *A Thread Across the Ocean*, Simon & Schuster, London, 2002, page 66
5. CO 6/28, letter of 16 August 1858 from Edmund Leahy to Sir Edward Bulwer Lytton, Colonial Secretary
6. David Murphy, *The Arctic Fox, Francis Leopold McClintock, Discoverer of the fate of Franklin*, The Collins Press, Cork, 2004, pages 111 - 142
7. CO 6/28, minutes endorsed on letter of 16 August 1858 from Edmund Leahy to Sir Edward Bulwer Lytton, Colonial Secretary
8. Captain A H Markham RN, "The Arctic Campaign of 1879 in the Barents Sea", in *Proceedings of the Royal Geographical Society*, New Series, Vol II, 1880
9. CO 48/392, letter of 21 August 1858 from Edmund Leahy to the Colonial Secretary
10. CO 48/392, testimonials dated 18 August and 2 September 1858, attached to letters of 21 August 1858 and 2 September 1858 from Edmund Leahy to the Colonial Secretary
11. C0 48/392, letter of 26 August 1858 attached to letter of 29 August 1858 from Edmund Leahy to the Colonial Secretary
12. CO 48/392, memorandum of 31 August 1858
13. CO 48/390, Despatch No 143, 24 July 1858 from Governor Sir George Grey to the Colonial Secretary
14. CO 48/392, letter of 13 September 1858
15. Archives of the Institution of Civil Engineers, London, Minutes of the Council of the Institution
16. Memoir of John Scott Tucker, *PRICE*, Vol 71, 1882-83, page 418
17. CO 48/390, Despatch of 24 September 1858 from CO to the Governor of Cape Colony
18. CO 137/342, letter of 18 September 1858 from Edmund Leahy to CO
19. CO 137/336, letter of 26 March 1858 from Governor Darling to the Private Secretary to the Colonial Secretary
20. CO 137/336, letter of 12 February 1858 and testimonials sent by P J Klassen to Governor Darling
21. CO 137/336, letter of 15 February 1858 and testimonials submitted by John H Brett to Governor Darling
22. CO 137/342 and CO 138/72, letter of 15 September 1858 from CO to Edmund Leahy
23. CO 137/336, letter of 26 March 1858 from Governor Darling to the Private Secretary to the Colonial Secretary
24. CO 137/336, letter of 26 February 1858 from Francis Dawson to Governor Darling; Dawson died in Jamaica in 1871 (Memoir in *PRICE*, Vol LIX, 1879-80, Part I).
25. CO 138/71, Despatch No 32 from CO to Governor Darling; CO 138/72, letter of 15 September 1858 from CO to Robert Barlow Gardiner; Memoir in *PRICE*, Vol XIX, 1859-60
26. CO 137/342, letters of 18 October and 9 November 1858 from Edmund Leahy to CO
27. CO 138/342, letter of 26 November 1858 from Lord Carnavon to Edmund Leahy
28. CO 137/347, letter of 10 January 1859 from Edmund Leahy to CO
29. CO 137/344, Despatch of 4 March 1859 from Governor Darling to CO
30. CO 137/378, Despatch No 17 dated 20 January 1864 from Governor Eyre to the Duke of Newcastle
31. Memoir of Robert Barlow Gardiner, *PRICE*, Vol XIX, 1859-60
32. CO 137/345, Despatch No 70 of 23 May 1859 from Governor Darling to CO
33. CO 138/71, Despatch No 7 of 4 July 1859 from the Duke of Newcastle to Governor Darling
34. *Papers relating to the Disturbances in Jamaica, Despatch of 9 December 1865 from Governor Eyre to the Colonial Office*, HC 1866 Vol LI (3594)
35. W L Mathieson, *The Sugar Colonies and Governor Eyre 1849-66*, London, 1936
36. Lord Olivier, *Jamaica: The Blessed Island*, Faber & Faber, London, 1936
37. W J Gardner, *A History of Jamaica from its discovery by Christopher Columbus to the present time*, Elliott Stock, London, 1873
38. Sir Charles Henry Darling (1809-70) became Lt Governor of the Cape Colony on 20 January 1852 in the absence on military duties of the Governor, Sir George Cathcart. He took up duty as Captain-General and Governor-in-Chief of Jamaica on 24 July 1857, and in September 1863 became Governor of Victoria from which, following a period of conflict with the Legislative Council, he was recalled in April 1866.
39. CO 138/71, Despatch No 32 from CO to Governor Darling

Chapter XIII — Edmund Leahy in Jamaica, 1858-63

40. *Reports for the Year 1859 .... on the State of Her Majesty's Colonial Possessions, Report from Governor C H Darling dated 26 December 1860*, HC 1861 Vol XL (2841)
41. Reports from Governor C H Darling in *Reports for the Years 1860, 1861, 1862 and 1863 .... on the State of Her Majesty's Colonial Possessions*, HC 1862 Vol XXXVI (2955); HC 1863 Vol XXXIX (3165); HC 1864 Vol XL (3304); HC 1865 Vol XXXVII (3423)
42. CO 137/373, letter of 29 June 1863 from Edmund Leahy to the Duke of Newcastle
43. Joseph C Ford, *Handbook of Jamaica, 1910*, Government Printing Office, Jamaica, 1910; according to the *Jamaica Gleaner* of 15 April 1871, the iron bridge originally designed by Leahy and shipped from England in 1863 was to be "erected across a stream in the locality".
44. ibid
45. Reports from Governor C H Darling in *Reports for the Years 1860, 1861, 1862 and 1863 .... on the State of Her Majesty's Colonial Possessions*, HC 1862 Vol XXXVI (2955); HC 1863 Vol XXXIX (3165); HC 1864 Vol XL (3304); HC 1865 Vol XXXVII (3423); CO 142/76, *Blue Book of Statistics for the Colony of Jamaica, 1862*
46. The Great International Exhibition Building, designed by Captain Francis Fowke, was built on a 23 acre site at South Kensington, London; it was 1152 feet long, with two crystal domes, each 260 feet high and cost £300,000 to construct.
47. *Reports for the Year 1859 .... on the State of Her Majesty's Colonial Possessions, Report from Governor C H Darling dated 26 December 1860*, HC 1861 Vol XL 2841)

CHAPTER XIV

THE JAMAICA TRAMWAY SCANDAL, 1862-65

"I perceive from the correspondence that Mr Leahy is one of the two chief shareholders in a tramroad, receiving from the Government various advantages ... Such a position appears to me at first sight to be clearly incompatible with that of a Government Engineer, and I shall be glad to receive an explanation ... "
The Duke of Newcastle, Secretary of State for the Colonies, May 1863[1]

The background
In November 1845, the Jamaica Railway - the first railway in the Empire outside the United Kingdom itself - was opened by a private company at a cost of £222,500 which was financed by capital raised in England. The line ran for 13 miles from Kingston, the island's commercial capital and main port, to the administrative capital at Spanish Town, and was constructed and operated on conventional lines, using the standard British gauge of 4 feet 8½ inches. It owed its development to William Smith of Manchester and his brother, David, who had bought up a number of estates in Jamaica and was a resident there[2]. While it was obvious that the colony could not sustain a substantial railway network costing up to £20,000 per mile to build, the enthusiasm which followed the opening of the railway generated numerous proposals in the 1840s for less expensive light railway or tramway schemes in various parts of the island with a view to reducing the difficulties and cost of transporting sugar from the plantations to the ports. All of these failed, however, for lack of capital and government support.[3] So too did the Jamaica South Coast Railway Company with which the Smiths were associated; this was incorporated in London in 1859 but its plan to construct a line from Spanish Town to Old Harbour never got off the ground.[4]

In Britain and Ireland, there was a revival of interest in the late 1850s and 1860s in the idea of constructing tramways on public roads.[5] Tramways Acts were enacted in 1860 and 1861 establishing procedures under which tramway construction could be authorised in Ireland, somewhat similar legislation applying to Scotland and England was enacted at the same time, and a number of street tramway systems were in operation in English cities by the early 1860s[6]. In line with these developments, a Tramway Act was enacted in 1858 in Jamaica allowing advances to be made from public funds for the construction of tramways but, because of a clause requiring the repayments to be charged on the lands through which a tramway was to pass, the Act remained a dead letter.

In November 1859, when the Main Road Board was discussing the improvement of a road near Port Maria, Edmund Leahy recommended that, as an alternative, the Board should authorise him, as Chief Roads Engineer, to construct a publicly-owned tramway which, in his opinion, would cost very little more, and they agreed to do so. More than 15 years earlier, Leahy had canvassed the idea of building railways or tramways on each side of existing roads: in *A Practical Treatise on Making and Repairing Roads*[7], he had suggested that on such railways, "small wagons, for heavy burdens, linked together, might be drawn, and carriages for swift conveyance might be received in cradles suited to railways". In this way, he claimed, coaches and chaises could travel for any convenient distance on public railways, using either horses or steam engines as motive power, and turn off onto a common road. However, when given the opportunity to put these ideas into practice in 1859, Leahy failed to do so. In the following year, in his private capacity, he prepared plans and estimates for a number of plantation owners who wished to construct a tramway but that project fell through for lack of support. He then put together a scheme which was designed to bring substantial financial rewards for himself.[8]

The genesis of the tramway fraud
Arguing that tramroads would "afford immense convenience, as compared with common roads," and were often found to be "little inferior to railways", Leahy set out in 1861 to promote the idea of building tramways on existing roads in Jamaica, claiming that tramways being built in India and France on separate alignments were costing £3,000 per mile and that the island economy could not support outlay of this kind. In July, he privately engaged a local surveyor to make a survey for a tramway along the roads from Spanish Town via Old Harbour to Porus, a small township about 40 miles away, and to prepare the necessary plans and sections. In that year also, David Smith, manager of the Jamaica Railway, advised the Governor that, as he was unable to raise the capital himself, it would be necessary for the Government to advance funds at 6% if the line were ever to be extended. In the following year, there was severe depression on the island: the American Civil War had badly affected trade, low prices were being obtained for exported produce and large fires had devastated parts of the commercial centre of Kingston. As part of the response, it was decided to make another effort to promote the construction of tramways on the island and an Act was passed under which up to three-quarters of the cost could be advanced by the Government where tramway projects were carried out by private companies.

Once the Act was passed, Leahy invited Smith to become involved in his scheme for a tramway along the centre of the road to Porus but Smith initially refused because, as he put it, "Leahy was treading on my ground". But Leahy persisted, telling him that his surveys had shown that the levels were very favourable, that the Governor was most anxious to have the scheme carried out and would agree to the use of the main road and allow Leahy to join in the undertaking. Smith and Leahy then devised a scheme which would allow them to capitalise on the legal provisions under which three-quarters of the *estimated* cost of a tramway could be advanced on its being shown that one-quarter of the cost had been incurred by the promoters: by preparing and submitting inflated estimates of the cost of construction, and by attempting to obtain three-quarters of that estimated cost by way of government loan, they planned to achieve a situation in which a tramway could be completed without any capital investment by themselves. Remarkably, they believed that they had the support of some of the members of the Main Road Board for this deceitful scheme under which they were, in effect, to be treated as contractors for the construction of a tramway, of which they would become the sole owners at no cost to themselves - and even to make a profit on the construction work, if costs could be kept below three-quarters of the inflated estimates. These arrangements were not, of course, recorded in the minutes of the Board (to do so, in Leahy's view "would only be giving a handle to the opponents of the tramway") and it was tacitly agreed that they would not be publicised.

In March 1862, Leahy obtained sanction from Governor Darling to his "becoming interested in the execution of the contemplated tramroads". He had sought this sanction on the basis of a false assertion that some of the promoters considered his participation to be essential for the success of the project and that he was anxious to assist them "both professionally and pecuniarily"; one year later, however, he admitted that it was he who had recommended the tramway in the first instance: "indeed I might say I originated it". Darling's letter of sanction was dated 24 March 1862, the day before he returned to England on a year's leave of absence and handed over responsibility for the government of the colony on a temporary basis to Lieutenant Governor Edward Eyre.[9] It seems likely that Darling was pressurised by Leahy into issuing this letter without having had an opportunity of considering the detail or the implications of what was proposed; he was later to insist that his sanction had not been intended to cover Leahy's involvement as an investor or as promoter of the scheme and that, had he continued as Governor "I should not have permitted the Colonial Engineer to occupy the position he was allowed to occupy ... [and] the opinions of all of the civil engineers in the world would not have induced me to concur in the plan adopted by which the tramway appears to have been laid down in the centre of the main road".

Public launch of the scheme
The *Prospectus of the Jamaica Tramroad* was published in March listing Governor Darling as Patron, David Smith as Manager, and Leahy as Consulting Engineer in Chief. The

Chapter XIV — The Jamaica Tramway Scandal, 1862-65

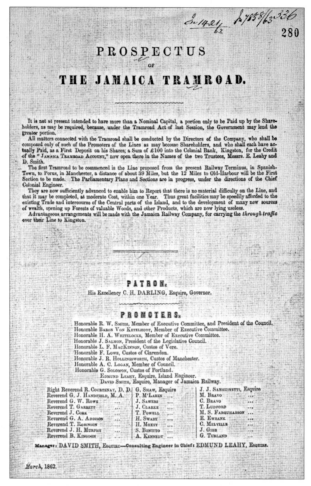

Prospectus of the Jamaica Tramroad, as published in March 1862, listing Edmund Leahy as one of the promoters and as Consulting Engineer in Chief (from CO 137/373).

tramway company was to have only a nominal capital and prospective shareholders were invited to pay an initial deposit of £100. The plans and sections for the 40-mile line were said to be in an advanced state, and the prospectus promised that the line would be completed at moderate cost within a year, opening up new sources of wealth and bringing substantial benefits for trade. In the event, although Leahy wrote to some 40 of "the country gentlemen" and others who might be interested, none of them were willing to become shareholders, with the result that a company was never formally established and Smith and himself were left to carry on as the sole promoters of the enterprise. On 19 June they submitted a petition seeking the approval of the Main Road Board (of which Leahy was an *ex-offico* member) for the construction of the tramway, stating that they were desirous of making it "at their own expense". It was to be built on a three-foot gauge and horses, initially, were to provide the motive power for the carriage of both passenger cars and wagons for freight. Smith's share of the cost was to be provided by a friend of his in England while Leahy put it about that he was "a gentleman of private means" at least sufficient to bear his share of the cost; at a later stage, when Smith was having second thoughts about the project, Leahy offered to buy him out for £1,000.

In support of the petition, Leahy submitted plans which were little more than an enlargement of the existing road maps and gave no details of levels, gradients, cuttings, embankments, or bridges. His estimates showed the cost of the entire undertaking to be £72,000, broken down under 14 headings covering materials and works, and separate estimates were supplied for each of the 10-mile sections into which the project was divided. Remarkably, however, the costs shown under each of the 14 headings were identical for each of the four sections: excavation and filling, for example, was estimated to cost exactly £1,041 13s 4d on each section even though the topography varied substantially from one section to another. In relation to receipts and operating costs, Leahy's submission was equally deficient. His estimates of freight volumes and passenger numbers were purely guesstimates and the fares proposed were three times as great as the railway rates. In all, he suggested that the tramway would bring in almost £14,000 a year, and after deducting capital repayment costs and working expenses (including the keep of 30 horses), he projected an annual profit of no less than £3,870 for Smith and himself.

Opposition develops

Notwithstanding the fictitious estimates and deficient plans, the Main Road Board approved the scheme on 9 July, and directed that work should commence on or before the following 1 August and be completed not more than one year later. To make this possible, an order was illegally made on the same day waiving the statutory provisions under which some months were to be allowed for the consideration of objections to the proposal. At that stage, however, a formidable opponent came forward to challenge the scheme. This was George Price (1812-90), descendant of a Welsh army captain who had been involved in the capture of Jamaica from the Spanish in 1655 and had settled there, building up substantial estates in the colony. Generations of the Price family held seats in the House of Assembly and Sir Rose Price, George's father, was knighted in 1815. George himself was proprietor of Worthy Park, a large estate in the interior of the island, Custos (or head of the magistracy) of St Catherine's Parish in which Spanish Town was situated, and a member of the Legislative Council. Although he was in favour of the concept of a tramway, he suspected from the beginning that the whole scheme was designed to allow the promoters to make a profit on the construction work itself. He questioned the cost estimates, asked how the figures for each of the 10-mile sections could be exactly the same, criticised the lack of detailed plans and specifications, and strongly challenged the proposal that the tramway should be laid down in the centre of the most important road in the island, only about 13 feet wide on average and bearing no comparison with the road from Paris to Versailles which had been cited by Leahy as an example of a road on which a similar tramway was being worked satisfactorily.

Instead of seeking independent engineering advice at that stage, the Main Road Board simply asked Leahy for a report, in which, predictably, he dismissed the objections as groundless and declared that "all experience of like work establishes the great superiority of the middle of the roadway". But Price did succeed in forcing the Board to allow time for the submission of objections and followed up by presenting a petition on behalf of the inhabitants of his parish objecting to the construction of the tramway along the centre of a relatively narrow road, bounded by deep, wet and dangerous ditches; this, they argued, would be a nuisance, a dangerous irritant to the general public and seriously inconvenient and dangerous for road traffic, especially at night. The objectors pointed out, quite reasonably, that routine road repairs and resurfacing would be likely to interfere with use of the tramway by bringing the road surface above the level of the rails and, conversely, that the rails would either sink, or be left standing well above the road surface, if that was eroded by rain and floods.

Price made repeated efforts in the latter part of 1862 to persuade Governor Eyre that he should become involved directly in evaluating the objections to the tramway but he refused to do so. Instead, he took the unwise step of endorsing the project in his speech at the opening session of the House of Assembly in November, even though the formal approval process had yet to be completed; noting the economic difficulties caused by the fall in the value of sugar production and losses due to drought, he told the House that he regarded the tramway as "one of the most useful and important enterprises" that had been initiated in Jamaica because it would open up regular and cheap communication with some of the most fertile and flourishing of the country districts and, in that way, help to revive the economy. Some months later, just before final approval was given to the project, Eyre again took the line that he would not "feel justified in interfering to stop the progress of a work which he believes will be of great benefit to the country". This left the way open for the Main Road Board, after cursory consideration of the objections, to make a detailed final order in February 1863 allowing construction of the tramway to go ahead, with rails laid down in the centre of the public roads; the only significant conditions were that the upper surface of the rails was to be no more than one inch above or below the surface of the road, and that roadways, nine feet wide, were to be provided on each side.

Political developments

While the tramway project was under consideration, the political situation in the island was deteriorating. Throughout 1862, the Town Party in the Assembly refused to cooperate with the Executive Committee drawn from the Country Party; obstructionist tactics were adopted to prevent any business being transacted and a vote of no confidence was passed, declaring that the Committee were incompetent and unfit to run the affairs of the colony. By January 1863, Governor Eyre felt that he had no option but to dissolve the Assembly and order an election. Taking advantage of established practice in Jamaica which permitted public officers to hold seats in the Assembly, Leahy decided to stand for election against the outgoing members (who were allies of George Price) in the parish of St Dorothy, with Smith as his running mate. There were allegations that the decision of the two men to go forward was effectively a condition of the granting to them of the tramway concession, in return for which they were to support the outgoing Executive Committee by their votes in the new Assembly. While Smith denied this, he made no secret of the fact that, if he had not expected to make a substantial profit out of the tramway, he could not have afforded to meet his election expenses of £1,000 which, presumably, would have included funds to engage in the bribery for which elections in Jamaica were notorious.

With political antagonism now added to the suspicion and personal animosity which had developed between Leahy and Price, the latter complained bitterly to the Governor that Leahy was devoting his time to canvassing when he should be engaged in supervising the public works but, while Eyre agreed that the Colonial Engineer should not be a member of the Assembly, he felt unable to prevent this as his predecessor had approved the practice. The Governor was advised in February by the local justices of the peace that a very considerable strengthening of the police force would be necessary in St Dorothy, where Leahy was standing, in order to maintain order and preserve the peace. They reported that "very great excitement is prevalent at this moment among the people" in regard to the election and that it would be "very severely contested from party feeling and local influences". The threat to law and order was obviously taken seriously as the Inspector of Police was immediately directed to reinforce the St Dorothy's force by three sergeants and 24 privates and also to send 60 rural constables there. The election went ahead on 10 March after what Price described as "an unusually large amount of money" had been expended; 100 of the 125 voters on the register voted, and Leahy and Smith gained a comfortable victory over the outgoing members. Overall, the 47 members of the new Assembly were returned by only 1,457 voters, while nearly 440,000 other residents of the island had no voice or influence in the election, prompting Eyre, when reporting the outcome to the Colonial Office, to add the wry comment: "and yet Jamaica is said to possess Representative Institutions".

From the Governor's point of view, and from the standpoint of effective government, the election made matters worse; the new Assembly which met on 24 March was just as bitterly divided as before and the outgoing Executive Committee were soon forced to resign. Eyre then had no option but to appoint a new Committee comprising leading members of the former opposition party and men who were personally antagonistic towards himself. For Leahy, notwithstanding his own success and that of his partner in

Chapter XIV — The Jamaica Tramway Scandal, 1862-65

A RETURN showing the Result of the last Elections of Members of Assembly for the different Parishes in Jamaica.

Parish.	Date of Election.	Names of the Candidates.	Whether elected or not.	No. of Votes for each Candidate.	Total No. of Voters who voted at the Election.	No. of Voters on the Register.	Remarks.
St. Catherine	9th Feb. 1863	William Thomas March	Elected	60	60	106	Resigned, A. H. Lewis elected 1863.
		Francis Robertson Lynch	Do.	60			
		Charles Hamilton Jackson	Do.	60			
St. John	12th Feb. 1863	Isaac Levy	Elected	21	23	26	
		Peter Alexander Espeut	Do.	14			
		Charles Anthony Price	Not elected	9			
St. Dorothy	10th March 1863	Edmund Leahy	Elected	72	100	125	
		David Smith	Do.	65			
		Andrew Henry Lewis	Not elected	42			
		Isaac Lawton	Do.	20			
St. Thomas ye Vale	23d Feb. 1863	Edward Kinkead	Elected	39	41	35	
		Hiam Barrow	Do.	29			
		Isaac Lawton	Not elected	11			
Clarendon	5th March 1863	Jacob Jacob Sanguinetti	Elected	27	31	42	
		Moses Bravo	Do.	27			
Vere	16th Feb. 1863	Charles McLarty Morales	Elected	21	21	35	
		James Harvey	Do.	22			
Manchester	20th Feb. 1863	John Reed Hollingsworth	Elected	21	21	48	
		Michael Muirhead	Do.	20			
St. Mary	17th Feb. 1863	Alexander Joseph Lindo	Elected	29	29	37	Called to the Council. Jas. M. Ferguson elected 30th Oct. 1863.
		George Geddes	Do.	29			

Return showing the result of the elections to the House of Assembly in Jamaica in February and March 1863, including the election of Edmund Leahy to one of the seats for the parish of St Dorothy (House of Lords Sessional Papers 1864 Vol XIII (254)).

securing membership of the Assembly, the outcome of the election spelled disaster. George Price, the most outspoken critic of his tramway scheme, and his political allies, were now effectively in control of the government of the colony. In addition, because of petitions presented by the defeated candidates in St Dorothy alleging that Leahy and Smith had been guilty of bribery and other corrupt practices, the Assembly refused to allow Leahy to take his seat until overruled six months later by the Governor and the law officers in London.[10]

Complaints about the construction work

The election was still in progress in March 1863 when construction work on the tramway began and it quickly became the subject of controversy and complaints; as one observer put it, "violent feeling had exhibited itself in the locality" because of the extent of the interference with the road, the failure to provide a roadway nine feet wide on each side of the rails, and the fact that the rails were being laid down in what was described as "a serpentine manner", in order to minimise the gradients. It became clear also, as work progressed, that the permanent way designed by Leahy was to take a novel form: the rails, weighing about 25 lbs per yard, had an unusual cross-section and were being fastened with spikes to longitudinal sleepers, consisting of short unshaped lengths of tree trunks, about six inches in diameter, with the bark still on them, and laid end to end directly on the clay subsurface of the road. Daniel McDowell, the Jamaica Railway Company's engineer, who might be expected to have some understanding of these matters, expressed serious reservations about the manner in which the work was proceeding and about the suitability of the rails and sleepers: "it is utterly impossible", he wrote, "that rails nailed to a piece of

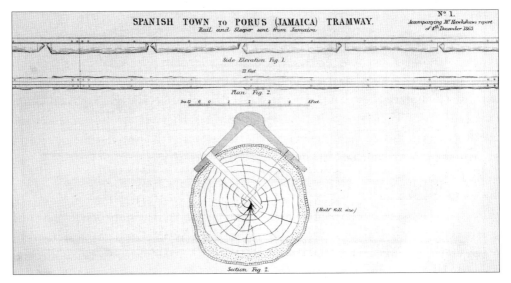

Drawings prepared to accompany the report of John Hawkshaw, consulting engineer, in December 1863 on the Spanish Town to Porus Tramway as designed and partly constructed by Edmund Leahy (House of Lords Sessional Papers 1864 Vol XIII (254)).

rough round timber would remain in their proper places while loaded waggons were continually running backwards and forwards over them, without cross-ties of some description being applied to keep the proper gauge". However, when Smith took up these points with Leahy, the latter was indignant that anyone should question his professional skill "especially in such a simple matter as laying down a tramway". But Smith reported the difference of opinion to a member of the Executive Committee, telling him that if the tramway were to be condemned it should be done at once, in which case he would be prepared to pay his share of the cost and thus suffer the penalty "for connecting myself with an incompetent partner".

Smith's unease led to a request to Francis Dawson, the road engineer for Cornwall and effectively a subordinate of Leahy's, to prepare a report on the progress of the works and he did so on 27 May, expressing general satisfaction with what was being done. He

Chapter XIV — The Jamaica Tramway Scandal, 1862-65

reported that the rails were almost exactly similar to an early form of the rail developed in the 1840s by W H Barlow,[11] with whom Dawson himself had worked in England; this type of rail had been designed on the principle that a heavy rail (90 lbs to the yard) with a very wide base could be laid directly on ballast, making sleepers unnecessary. Dawson, however, seems to have overlooked the facts that the Barlow rail had not been found to be successful and was never widely used, and that Leahy's rails were much lighter, averaging from 25 lbs to 27 lbs per yard. While he admitted that he had never seen round sleepers, with the bark still on them, being used on a railway, he believed that the rails and sleepers, if properly laid down and ballasted, would provide a sound, safe and durable tramway; he went even further, suggesting that the wooden sleepers could be dispensed with altogether, except at the rail joints where they were needed as a substitute for iron fish-plates. In light of this report, and being satisfied that £4,500 had by then been spent on the tramway, the Executive Committee allowed an advance of this amount to be paid to Leahy and Smith at the end of May. Up to that point, the operation was still viewed by those in power as a *bona fide* one but, before long, the whole scheme began to fall apart and the fraudulent nature of the project came into the open.

The fraud exposed

Leahy's ambiguous position in relation to the tramway, and his determination to maximise his earnings from the project in the short term, was exposed when he demanded payment of fees of £1,975 from public funds for having prepared the plans, specifications and estimates, even though he and Smith, rather than the Government, were to be the owners of the tramway. He claimed that fees at this level were appropriate for a professional person such as himself and would be paid in England for the preparation of railway plans. When asked to spell out exactly what services he had personally performed, and those which were performed by a deputy or assistant, Leahy replied that the engineering services were *done by myself*. This led the Governor to respond that "it is difficult to conceive how you could personally have performed service of so much difficulty, and which must have occupied much time, without neglecting your proper duties as Road Engineer and Colonial Architect;" he wondered also how could sanction be justified for "so great an anomaly as allowing a paid officer of the Government to claim for private work an amount far exceeding his official salary – and this, too, is only one of the private services which the engineer has executed". Contradicting what he had said earlier, Leahy then submitted that he had "accepted the place of consulting engineer" and that, as such, he had not personally made the surveys and plans but had employed others to do so at the usual remuneration.[12] But the Governor rejected Leahy's claim on the basis that, if he had a case at all, his fees should be met by the tramway company and that the fees claimed were in any case grossly excessive.

In search of fame and fortune: the Leahy family of engineers, 1780-1888

Claiming that £13,500 had already been spent on materials and earthworks by the middle of May, and that about four miles of the tramway had been laid down, Leahy and Smith pressed for further advances from public funds. Their demands led to the appointment of a Tramway Commissioner who reported that it was impossible to place any reliance on the original estimates as a basis for public funding; materials, for example, had cost much less in all cases than had been estimated and the quantities required in the case of some materials had been over-estimated by a factor of four. Leahy then complained that the Executive Committee had placed himself and Smith "in a regular fix" by attempting, as he put it, "to repudiate the distinct understanding" that three-quarters of the *estimated* cost would be paid to them on completion of the tramway, irrespective of the actual cost. At the same time, questions began to be raised in the newspapers, particularly in the *Jamaica Tribune* (a disreputable paper according to Leahy) about the generous financial arrangements which had been made with the company; a response from Smith, published in the *Jamaica Guardian* of 30 May, instead of allaying public concern, added fuel to the flames: he clearly stated that the true cost of constructing the tramway was expected to be £1,350 per mile and that the estimate of £1,800 per mile was submitted so as to throw the total cost of construction on public funds – "the country required a tramway, and they were willing to pay for it". In effect, he admitted that he and Leahy never intended to spend any money of their own on the project and had submitted a fictitious set of estimates in order to "make the island find the necessary funds" because "no man in his senses would undertake the construction of such a work unless with the prospect, and indeed certainty, of being very well paid for it".

When asked by the Governor to say whether he agreed with Smith, Leahy adopted a different approach, while still insisting that the understanding had been that £54,000 would be advanced from public funds, regardless of the actual cost of construction. He revealed for the first time that his estimate of £1,800 per mile contained hidden allowances of as much as 25% for "the cost of competent professional supervision", as well as for profit "to those who undertake the responsibility of execution" and for interest on capital invested up to the time of the opening of the line. He conceded that no specific amounts had indicated for any of these items and that their inclusion (by inflating the figures for other items) had not previously been disclosed - because nobody had raised the question - but he argued that his action was consistent with the normal practice followed by engineers and accountants. This may have had some validity in so far as the question of interest was concerned, but there was obviously no basis for including contractor's profit, where the same persons were acting as contractors and promoters of the project and were to become its owners.

Chapter XIV — The Jamaica Tramway Scandal, 1862-65

A situation "clearly incompatible with that of a Government Engineer"

Leahy made a fatal mistake at this stage by demanding that the Governor should take action to protect his good name and professional reputation against the publication by George Price of what he described as slanderous statements about him. When Eyre refused, stating that the issue was one which could only be settled in a court of law, Leahy insisted that the papers should be referred to the Colonial Office, thus disclosing to them for the first time that he had been allowed to have a personal interest in the tramway project. The response from the Duke of Newcastle left little room for doubt as to his views on this arrangement; he told Eyre that the situation in which Leahy was one of "the two chief shareholders in a tramroad receiving from the Government various advantages … appears to me, at first sight, to be clearly incompatible with that of a Government Engineer", and he demanded an explanation as to how this situation had been allowed to arise.

The Governor was spurred into action by this implicit rebuke. He was well aware at that stage that mistakes had been made by permitting Leahy to be involved in the tramway project and, as one biographer has put it, he had been "foolish and pig-headed not to have looked into the estimates before he did".[13] His first response to the Colonial Office was a defensive one, pointing out that the preliminary arrangements in relation to the tramway had been made while Governor Darling was in office, that it was he who had sanctioned Leahy's involvement, and that as he himself had been appointed only on a temporary basis, it would not have been proper for him to upset the arrangement. But he went on to tell the Colonial Secretary that experience had shown that it would have been better if Leahy had never been allowed to become involved, that serious difficulties had already arisen, that he feared that still greater embarrassments were in prospect, and that the whole operation might be "characterised as an attempt to commit a gigantic fraud upon the public".

Eyre moved quickly after that, telling Leahy that his connection with the tramway was detrimental to the public interest, was creating dissatisfaction in the public mind and was causing the main roads in his area to be neglected. Leahy rejected a suggestion that he should step aside from his official positions until his connection with the tramway ceased, insisting that "by assisting this tramway" he was doing great good to the island and no more than his official position demanded. At the same time, realising that action was likely to be taken against him, and on the basis, presumably, that attack is the best form of defence, Leahy addressed a letter to the Duke of Newcastle seeking his removal to "a similar office … in any other part of Her Majesty's dominions". He claimed that there was an evident disposition to downgrade his office, that the Executive Committee were hostile to him, that he had been subjected to "countless annoyances" in the carrying out of the tramway scheme, and that under an amendment of the Main Road Act, he would be "subservient to a number of petty parish boards composed of the poorest people in the Island" and forced to accept "degrading duties after five years laborious and difficult service and risking health and life". All of this was presented on the entirely false premise that he had been "*induced* to come out to Jamaica" by the Home Government!

Conduct "incompatible with your continuance in office"

On 30 June the Governor advised Leahy that he and the Executive Committee "find it impossible any longer to accord to you their confidence, or to regard your conduct … as otherwise than highly reprehensible, unbecoming an officer of the Crown, and incompatible with your continuance in office". Leahy was given a statement outlining seven charges against him but, as his replies were not considered satisfactory, the Privy Council of the Colony was summoned to meet on 14 July to deal with his case. The Council considered the seven charges seriatim, inviting Leahy to reply to each charge and to answer questions. In summary, the charges were:

- That he had prepared misleading and exaggerated estimates of the cost of the tramway containing hidden allowances for profit and interest, amounting to as much as 25% of the cost;
- That he had attempted to recover £1,975 in professional fees for engineering services which were said to have been performed by himself in person, whereas the services had largely been provided by others;
- That he had falsely claimed that the surveys and plans for which the fees were claimed had been performed with the sanction of the Main Road Board;
- That he had falsely claimed that the surveys and plans had been prepared on the authorisation of the Main Road Board and Governor Darling;
- That he had neglected his duties as Colonial Engineer;
- That he had published in the *Jamaica Guardian* a statement which insulted and questioned the competence of the members of the Main Road Board;
- That he had prepared misleading plans and estimates for the proposed public offices at Kingston.

When all of the charges had been dealt with, the Privy Council unanimously resolved, after a six-hour hearing, that charges 1 to 4 and 6 had been substantiated and that they warranted Leahy's suspension from his offices as Colonial Engineer and Architect and Chief Roads Engineer. The Governor sent the entire file to the Colonial Office seeking approval to Leahy's dismissal on the basis that he lacked rectitude and veracity and could not be entrusted with responsible duties. Leahy sought leave of absence to travel immediately to England so that he could put his case in person to the Duke of Newcastle, adding that, as his health had been seriously affected by the charges, the change of climate would be good for him. So confident was he of being granted leave that he sold off his horses, carriages and other effects but leave was refused, because there were issues to be resolved relating to the

Chapter XIV — The Jamaica Tramway Scandal, 1862-65

accounts for various public works and because he had not handed up maps, plans and other documentation relating to works which were going on under his supervision (it subsequently emerged that there were no such documents available in many cases).

Leahy set out his defence in a long and detailed memorial which, together with the comments of the Executive Committee and the Governor, was sent to the Duke at the end of July. Having attempted to refute each of charges of which he had been convicted, he complained that he had been persecuted by the new Executive Council and that the general state of society on the island was not at all favourable to persons in public office, leaving them open to groundless charges based on "a mix of political and personal dislikes". In an effort to influence the outcome of his case, he sent a copy of his memorial to his brother, Patrick, and asked him to forward it to the Duke of Newcastle. In doing so, the archbishop was careful to state that it would not become him to offer any remarks on the merits of the case while adding that Edmund's own statement was "so full and so clear as to leave nothing wanting to the completeness of his defence". He went on:

> If I might venture to make any observation in reference to so delicate a matter, it would be one suggested by the reading of this paper. It is, that should Mr Edmund Leahy's removal, compassed by (I will not say unworthy means, but by) the exercise of unmerited severity – should it follow in a few months after the death of his two brothers, who filling the same situation in the Island of Trinidad, died prematurely within the space of two short years, the victims at once of the climate and of professional duty – then might it be said with truth that the colonial service had cost this one family very dearly indeed.

Neither Leahy's own memorial nor the plea of his brother were sufficient to stave off the inevitable verdict; the archbishop was told that Edmund had had a fair and full opportunity of defending his conduct and that "if the Duke has not been able to arrive at any other conclusion than that which was unanimously adopted by the Lieutenant-Governor and Privy Council of Jamaica, it has not been from any want of a desire to show consideration for a family whose connexion with the public service of the Colonies has been attended with such unfortunate results". In a despatch to the Governor on 8 September, the decision that Leahy should be removed from office was endorsed by the Colonial Office and he was formally dismissed on 6 October.

Leahy's duties as Colonial Engineer and Architect had already been assigned on a temporary basis to John Parry, county engineer for Surrey, and another temporary appointment was made to fill Leahy's position as county engineer for Middlesex. In the following January, Eyre asked the Colonial Office to select a permanent replacement, as many important and expensive public works had been in abeyance since Leahy's dismissal. He pointed out that under a local Act, the Colonial Engineer and Architect was required "to produce evidence of his having been admitted and allowed to practice as a civil engineer and architect by some Society or Institution established for that purpose in the United Kingdom or the Continent of Europe", raising questions as to why Leahy, who had no such qualification, had ever been appointed to the post. In addition, and arising from experience with Leahy's predecessor, the Colonial Office was urged to make such "inquiries into the habits and character of any person who may be selected as will preclude the possibility of the appointee being given to habits to the indulgence of which a tropical climate is perhaps an incentive and which, I am sorry to say, are but too common amongst the public servants in this Colony".[14]

Review of the tramway plans

Work on the tramway ceased in mid-July 1863, with about eight miles of it partially laid and expenditure of more than £10,000 incurred on labour and materials. Ballasting had been carried out on only one mile of the route, the road had not been levelled and none of it had been widened to the extent required. No company had actually been formed and registered, no shares had been taken and no capital subscribed; instead, Leahy and Smith had secured all of the materials on credit which they were able to pay off in part on receipt of the £4,500 government loan. Over and above this, Leahy claimed that capital of over £7,000 was "buried in the tramway, which capital is formed entirely of my private means with those of Mr D Smith" and, in an effort to extricate himself from this financial mess, he offered to surrender their entire interest in the tramway to the Government on equitable terms which, for him, meant full recoupment of the entire outlay together with some additional payment "in consideration of our trouble".[15] Shortly after that, Smith advised the Governor that Leahy had assigned his entire interest in the tramway to him - but he had still not done so three months later; in addition noting that alarm had been expressed about the condition of the road and that many people objected to the fact that the tramway was to occupy the centre of the road, Smith offered to take up all of the rails and sleepers and relay them at the side of the road, if the costs were met by the Government.[16]

Because of what he described as the considerable doubts which had arisen about "the sufficiency or stability of the tramroad", Governor Eyre asked the Colonial Secretary to have a report on the project prepared by the Railway Department of the Board of Trade and to respond to a list of 18 specific queries about the rails, sleepers and wheels, specimens of which were sent to London.[17] The Railway Department in turn engaged John Hawkshaw, President of the Institution of Civil Engineers, to provide a professional opinion on the matter[18] and he did so in a report which reached Jamaica early in January 1864. Hawkshaw advised that the tramway should not occupy the centre of the road, regardless of the extent to which that road might be macadamised. He believed that it was doubtful whether the rails could be maintained only one inch above the general level of the road and advised that any significant projection above road level would be

Chapter XIV — The Jamaica Tramway Scandal, 1862-65

objectionable and might lead to accidents. He was not aware of similar rails having been used elsewhere for street or road railways and he thought that, in all probability, they had not been. The rails, laid on round sleepers, would be likely to break at or near the centre, sleepers laid on heavy wet clay with no ballast interposed would not function efficiently, decay of the bark of the sleepers would tend to loosen the rails, and the gauge would be unlikely to be maintained without cross-ties of wood or iron.

When confronted with Hawkshaw's report, Leahy had sufficient sense not to challenge directly the opinion of the leader of the civil engineering profession in Britain; instead, he complained that the queries put to Hawkshaw, had been "framed by men who had no professional or practical knowledge of such questions" and had left him no option but to criticise the scheme. Soon afterwards, the project was again condemned in the report of a Committee of the House of Assembly which had heard evidence over 13 days from 24 witnesses, many of whom asserted that the state of the public road was then "most fearful" and "beyond description". One regular road user said that the rails were from six to eight inches above the road surface and very dangerous to traffic, especially at night, while the Acting Colonial Engineer reported that the rails had been irregularly laid, and the sleepers were not properly ballasted, because there had been no proper direction of the work. In his evidence, Leahy had told the committee that it would be difficult for him to continue his connection with the tramway, having been unjustly and harshly treated by the Governor and his credit "in a monetary sense being seriously damaged by the proceedings"; besides, having regard to the clamour which had been raised against the construction of the line, he felt it best "for peace sake" to surrender his interest to the Government on equitable terms.

Leahy abandoned the tramway project and left for London in February 1864. Having met the Duke of Newcastle's private secretary at the Colonial office on 16 March, he wrote to the Duke requesting that he should be given access to everything Eyre had written about him so that he could properly defend himself. He claimed that it was well known that, if Darling had continued as Governor, he would not have been "sacrificed", as he had been, and went on to advance the absurd argument that he had been legally obliged to go ahead with the scheme once it was authorised by the Main Road Board.[19] The reply sent to Leahy simply reminded him that as the grounds for his removal were to be found in the proceedings of the Privy Council of Jamaica, which were known to him, it was unnecessary that he should be informed of the contents of Eyre's despatches on the subject.[20]

When the resignation of the Duke of Newcastle as Colonial Secretary and his replacement by Edward Cardwell was announced on 7 April, Leahy seized the opportunity to seek a rehearing of his case. Writing to Cardwell, he referred him to the plea made on his behalf by his brother, the Archbishop of Cashel and Emly, "who, I believe, is not entirely unknown to you" and who would not have interested himself in the case "were he not convinced of the justice of my cause". Asking that his case should be reconsidered, as the subject was of such vital importance to him, "both professionally and officially" but, at the same time, realising the difficulty of reversing the decision to dismiss him, he asked Cardwell "to kindly give me an appointment at the Cape or some other colony".[21] As Chief Secretary for Ireland from 1859 to 1861 when the national education system was the subject of intense discussion between the Catholic hierarchy and the Government, Cardwell would have met the Archbishop on a number of occasions, but this could hardly have been expected to lead him to set aside the blunt advice of his officials that Leahy was not eligible for re-employment in the public service; the reply which issued was couched in more diplomatic terms, expressing regret that an offer of re-employment could not be made.[22]

Meanwhile, in Jamaica, the Government refused to have anything further to do with the tramway project. The rails and sleepers were torn up by road users and thrown into the ditches at the sides of the road or turned to other uses, just as the unique rails and sleepers ordered by Leahy for the Cork & Bandon Railway, almost 20 years earlier, had been; relics of the tramway were still to be seen along the road 50 years later, according to Lord Sydney Olivier who was Governor of Jamaica from 1907 to 1913.[23] Instead of the tramway, an 11-mile extension of the Jamaica Railway Company's Kingston – Spanish Town line to connect it to Old Harbour was completed in July 1869; ten years later, the railway was acquired by the Government of the colony and was extended to Porus, a further 24 miles away, in 1885.

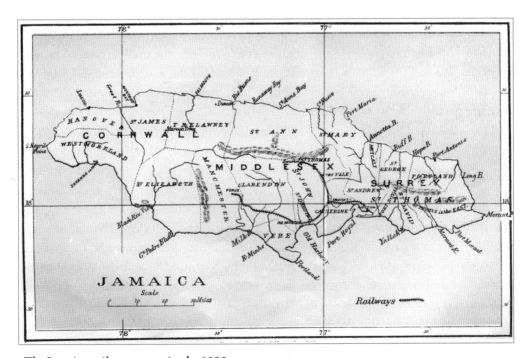

The Jamaica railway system in the 1880s.

Chapter XIV — The Jamaica Tramway Scandal, 1862-65

Political impact

The impact of the tramway scandal was felt for several years because, as Governor Eyre told the Colonial Secretary, "like everything else in Jamaica, the question has now been turned into a political one ... to be used as a ground of accusation against myself". George Price was determined to have Eyre removed from his post and, in a series of increasingly bitter and personalised submissions to the Colonial Secretary, he stepped up his efforts in the spring of 1864 to convince him that Eyre should carry primary responsibility for having allowed Leahy's fraudulent scheme to go ahead. Price was closely related to two members of the House of Lords – the Earl of Shrewsbury was his first cousin[24] and Lord Desart was the son of his sister-in-law[25] - but he seems to have failed to convince either of these that the matter deserved an airing in the House. However, when a third close relation - his brother-in-law, Lord Dunsany[26] - became an Irish Representative Peer in 1864, Price quickly persuaded him to raise the issue in the Lords and he did so on 3 June.[27] Moving a motion which called for presentation to the House of copies of all correspondence, reports and despatches relating to the tramway, Dunsany heavily criticised Eyre for not having intervened decisively at an early stage and for having allowed Leahy "whose duty it was ... to see that the work was honestly performed" to become contractor for the work. The papers called for by Dunsany, comprising no less than 356 foolscap pages of closely printed material, were published in the House of Lords Sessional Papers in July 1864 but, although Dunsany secured the presentation of an additional 141 pages of material in July 1865, he seems to have made no effort to initiate further debates on the subject.[28] The presentation of yet another batch of papers to the Lords in 1867[29] brought the total number of printed pages devoted to Leahy's tramway project to more than 500, a remarkable total by any standards for a relatively small project initiated by a single individual, and one which can have few parallels in the British Parliamentary Papers.

Price set out for London at the end of June 1864 aiming, with the aid of his relations in the House of Lords, to press for an inquiry into the tramway affair and threatening that "further agitation of the question in Parliament" and "future difficulties" for the Government could ensue if he did not get a hearing. But the bureaucrats in the Colonial Office stood firm and refused to meet him. In September, while staying with his sister-in-law, the Dowager Countess of Desart, at her house at Pembroke Road, Dublin, he made another effort to engage the Colonial Office in substantive correspondence on the subject and, after his return to Jamaica later that month, he continued to address detailed letters to the Colonial Secretary and to the local press arguing that there had been a "most culpable and persevering neglect of duty" by the Governor in relation to the scheme. This "perpetual and interminable correspondence", as Eyre described it, continued until April 1865 when Price left the colony to take up residence in England, but it failed to achieve his objective.

In search of fame and fortune: the Leahy family of engineers, 1780-1888

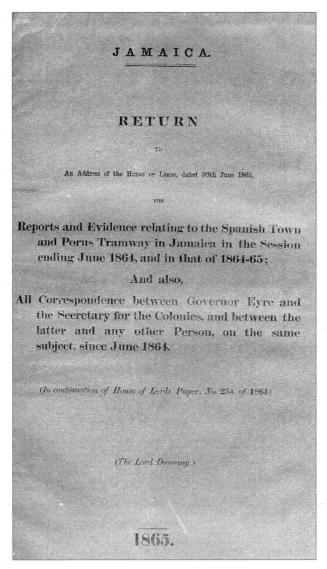

Cover page of the second of three substantial collections of reports and correspondence relating to the Jamaica Tramway presented to the House of Lords between 1864 and 1867 (House of Lords Sessional Papers 1865 Vol XL (257)).

Rebellion comes to Jamaica

Throughout 1864 and 1865, while the tramway debacle was still the subject of despatches between London and Jamaica, trouble of a more serious kind was brewing in the island. In fiery and inflammatory speeches, a radical coloured Assemblyman, George William Gordon, had been denouncing Eyre because of his attitude to measures which would improve the condition of Jamaica's poor and black population. Gordon, who had originally been a supporter of the tramway project, earning for himself the soubriquet *the Pecksniff of patriotism* in one of the island's newspapers,[30] was, from the time of his election in March 1863, severely critical of the waste of money on roads and bridges in the previous few years "notwithstanding all we have heard about the talent of the imported engineer, Mr Leahy". He engaged in one of his most violent outbursts in January 1864 when Eyre notified the Assembly of the appointment of a native of Jamaica, who held no formal engineering qualifications, to replace Leahy temporarily as Colonial Engineer. This was an illegal act, according to Gordon, and if the law was disregarded in this way by the Governor, it would lead to anarchy and bloodshed and the people would have to fly to arms and become self-governing.[31]

Rebellion broke out in Jamaica on 11 October 1865, when a protégé of Gordon's led hundreds of his supporters into Morant Bay – about 25 miles east of Kingston – setting fire to the courthouse where the vestry were meeting and killing a total of 18 militia men, officials and civilians. Eyre declared martial law and had the rebellion put down quickly and with great severity by units of the British army and navy. Gordon was hanged, over 400 others were shot or hanged, at least 600 men and women were flogged, and 1,000 dwellings were burnt down[32]. When details of these measures emerged in London,[33] a storm of protest arose, leaving the Colonial Office no option but to establish a Royal Commission whose report led to Eyre's recall in July 1866. For several years afterwards, the deposed Governor was at the centre of what amounted to an English intellectual civil war; there were several failed attempts by a group calling themselves the Jamaica Committee, led by John Stuart Mill, Thomas Huxley and others, to bring him to trial and, in opposition to them, the Eyre Defence Committee was set up, supported by such notable figures as Charles Dickens and the poet laureate Alfred (later Lord) Tennyson, and by many members of both Houses of Parliament and of the armed forces. Funds were raised on each side and, remarkably, in view of his own dismissal, Edmund Leahy and his wife were among those who publicly supported Eyre; *The Times* of 21 January 1867, in a large advertisement from the Eyre Defence and Aid Fund, listed them among those who contributed £1 each.

"the maggot in Price's brain"
The tragic events of October 1865 provided an opportunity for Price to renew his criticism of the handling of the tramway scheme. In a letter to the Prime Minister, Earl Russell, he made the bizarre and irrational claim that the "the dissentions and misfortunes in that island are due *solely* to the encouragement given to a transaction known in Jamaica as the Tramway Fraud" and the fact that Eyre had been rewarded by promotion to the rank of Governor, notwithstanding his role in the affair.[34] In attempting to justify this, Price argued that the fraud had diverted the attention of the Assembly between 1862 and 1865

and that it was on the tramway question that Gordon had spoken "most frequently and urgently in the Assembly and … placed it first in the list of grievances he complained of". In July 1866, claiming that the Jamaica Royal Commission had failed to elicit "the true cause of the discontent prevalent among all classes in Jamaica", he took his case to the public by publishing a book which he called *Jamaica and the Colonial Office: Who caused the Crisis*.[35] This dealt in great detail with what he described as the "tramway swindle" and repeated the argument that the handling of Leahy's scheme was the primary cause of the "general irritation and disaffection" which had culminated in the rebellion. Price's book had little impact at the time, and failed to attract attention in Parliament, as he had hoped, but it was followed and extensively quoted almost 70 years later by Lord Sydney Olivier, a former Governor of Jamaica, in his harshly critical book *The Myth of Governor Eyre*.[36] In turn, this damning biography, in which the version of events set out by Price was "everywhere accepted as holy writ", was followed by many subsequent writers.[37] An exception was W L Mathieson who, in *The Sugar Colonies and Governor Eyre 1849-66*,[38] suggested that it was absurd to adopt the line taken by Price. A more recent and more sympathetic biographer of Eyre also takes this view, describing the tramway affair as "the maggot in Price's brain" and, not unfairly, refers to him as a monomaniac who was obsessed with the tramway scandal.[39]

"my error, if any, was one of judgment only"

Leahy, like Price, sought to use the rebellion to justify reconsideration of the tramway affair. Writing to the Colonial Secretary in December 1865, and apologising for drawing his attention to such a personal matter in "the present deplorable state of Jamaica", he complained about the "inconsiderateness with which I feel I have been visited in depriving me of my late office of Chief Engineer in Jamaica"; this would never have occurred, he contended, "except under the disorderly state of things evoked just previously in the Island, which has since culminated in the late deplorable results, and which at present very much tend to prove my case".[40] Then, in a more conciliatory and contrite tone than he had previously adopted, he went on to ask that the past should be disregarded "on the grounds of my solemn assurance that my error, if any, was one of judgment only and also believing that I was not entirely unsanctioned by the Governor". On this basis, and because he had lost not only his office but also large sums of money, he asked Cardwell to allow his name to stand on his list for re-employment on some suitable occasion. Officials in the Colonial Office drafted a reply to Leahy indicating that the circumstances of his dismissal precluded him from being considered for further public employment but, due to the pressure of business in the aftermath of the rebellion, the letter was put aside; when it surfaced five months later, the view was taken that it was best left unanswered.

In his letter to the Colonial Secretary, Leahy had advised "from my experience of Jamaica, and as a member of its Legislature, that neither the late nor the present state of

Chapter XIV — The Jamaica Tramway Scandal, 1862-65

society there is at all suited to the actual form of government by a Council and a House of Assembly", adding that nothing, in his judgment, "would settle the island so well as a Military Government".[41] While this drastic remedy was not adopted, Eyre did persuade the Assembly to pass legislation which would sweep away the Legislative Council, the Assembly itself and the Executive Committee and allow for the substitution of the standard system of crown colony government. This was readily endorsed by legislation passed at Westminster, one MP remarking that "the only good law the Assembly passed was that by which it abolished itself".[42] Lord Dunsany availed of the second stage debate in the Lords to give the tramway affair another airing, quoting from some of the papers which had been presented to the House at his request, but his was a lone voice.

In 1867, after the Duke of Buckingham and Chandos had become Colonial Secretary, Leahy made another effort to secure re-employment in the colonial service; he told the Duke that he had "suffered a good deal of punishment for whatever amount of blame may be attached to the indiscretion of being connected with the construction of tramways ... whilst filling the office of Chief Engineer" and "whether it was right or wrong" he hoped that the view might be taken that he had suffered enough.[43] The reply was curt and unambiguous: "the Duke cannot hold out to you any prospect of re-employment".[44] Ten years later, and again in 1879, Leahy was still writing to successive Colonial Secretaries, demanding fresh enquiries into his case, and claiming arrears of salary and compensation "for services rendered and money expended in connection with the tramway". In a final letter in 1883, 20 years after work on the tramway had ceased, he requested that he should be allowed complete it "in accordance with contract". Unfortunately, the reaction of the Colonial Office to the letters of 1877 and 1879 and to this last extraordinary demand cannot now be established as the relevant papers have been "destroyed under statute".[45]

REFERENCES

1. Despatch No 604, dated 13 May 1863 to Lt Governor Eyre, *House of Lords Sessional Papers 1864,* Vol XIII (254) page 295
2. Douglas Hall, *Free Jamaica 1838 – 1865: An Economic History,* Yale University Press, 1959, pages 35, 92
3. *ibid*
4. BT 31/408/1566, Memorandum of Association of the Jamaica South Coast Railway Company dated 20 June 1859
5. J C Conroy, *A History of Railways in Ireland,* Longmans, Green and Co Ltd., London, 1928, pages 242 - 246
6. Jack Simmons and Gordon Biddle, *The Oxford Companion to British Railway History,* Oxford, 1997, pages 537 - 539
7. Edmund Leahy, *A Practical Treatise on Making and Repairing Roads,* John Weake, Architectural Library, London, 1844, page 85
8. Except where otherwise indicated, what follows, including quotations, is based primarily on the three volumes of documents relating to the Tramway, comprising some 500 closely printed foolscap pages, presented to the House of Lords: *House of Lords Sessional Papers 1864,* Vol XIII (254); *House of Lords Sessional Papers 1865* Vol XL (257); *House of Lords Sessional Papers 1867* Vol XIII (211).
9. Edward John Eyre (1815 – 1901) initially came to notice as an explorer in Australia, being the first white man to make the 1,000 mile crossing of the Great Australian Bight on the southern coast of the continent in 1841. He had served as Lieutenant Governor of New Zealand from 1846 to 1853 and as Governor of the Leeward Islands before his appointment to Jamaica in 1862.

In search of fame and fortune: the Leahy family of engineers, 1780-1888

10. George Price, *Jamaica and the Colonial Office: Who caused the Crisis*, Sampson Low, Son and Marston, London, 1866, page 34
11. William Henry Barlow (1812–1902), an English railway and consulting engineer, is best known as the designer of London's St Pancras train-shed and for his part in the design of the second Tay Bridge; in 1849, he had patented a very short-lived variation of Brunel's bridge rail, with the bottom flanges converted into prongs to hold the rail in place in ballast without using sleepers.
12. Lewis Sinclair, the surveyor who had carried out the tramway surveys for Leahy in 1861-62, expected to receive about £400 in fees for this work but, in 1863, had been paid only £50 for his services.
13. Geoffrey Dutton, *The Hero as Murderer*, London, 1967, page 227
14. CO 137/378, Despatch No 17 dated 20 January 1864 from Governor Eyre to the Duke of Newcastle; George B Pennell was appointed to replace Leahy in April 1864, CO 137/386, Despatch of 13 April1864 from the Colonial Office to Governor Eyre
15. CO 137/374, letters of 21 July and 6 August 1863 from Edmund Leahy to H W Austin, Government Secretary, Jamaica
16. CO 137/374, letter of 10 August 1863 from David Smith to the Executive Committee
17. CO 137/374, Despatch No 189 of 8 August 1863 from Governor Eyre to the Colonial Office
18. CO 137/377, letter of 12 September 1863 from the Board of Trade to the Colonial Office
19. CO 137/386, letter of 17 March 1864 from Edmund Leahy to the Duke of Newcastle
20. CO 137/386, letter of 30 March 1864 from the Colonial Office to Edmund Leahy
21. CO 137/386, letter of 7 April 1864 from Edmund Leahy to Edward Cardwell
22. CO 137/386, letter of 20 April 1864 from the Colonial Office to Edmund Leahy
23. Lord Olivier, *The Myth of Governor* Eyre, Hogarth Press, London, 1933, page 80
24. Elizabeth Lambart, mother of George Price, was the youngest daughter of Charles Lambart of Beaupark, Co Meath. His aunt, Frances Thomasine Lambart, had married the 2nd Earl Talbot on 28 August 1800 and was the mother of the 18th Earl of Shrewsbury and Talbot (1803-1868), a Rear-Admiral in the Navy and senior Earl in the peerage of England.
25. Price's eldest brother, Rose Lambert Price (1799-1826) married Catherine, widow of the second Earl of Desart in 1824; her son, John, born 12 October 1818, who had become third Earl of Desart in 1820, on the death of his father, became an Irish representative peer in 1846. He was Under-Secretary for the Colonies from February to December 1852 and died on 1 April 1865, aged 46.
26. George Price had married Emily Valentine Plunkett, sister of Edward Lord Dunsany (1808-89) on 18 October 1839; Dunsany took the Oath of Office as an Irish representative peer on 14 March 1864.
27. Hansard, Third Series, Vol CLXXV, col 1119
28. *House of Lords Sessional Papers 1864* Vol XIII (254); *House of Lords Sessional Papers 1865* Vol XL (257)
29. *House of Lords Sessional Papers 1867* Vol XIII (211)
30. *Colonial Standard*, 13 February 1865
31. *Colonial Standard*, 14 January 1864
32. *Report of the Jamaica Royal Commission 1866*, HC 1866 Vol XXX (3683)
33. *The Times*, 17 and 20 November 1865; Despatch No 251 dated 20 October 1865 from Eyre to Edward Cardwell in *Papers relating to the Disturbances in Jamaica (Part I)*, HC 1866 Vol LI (3594)
34. CO 137/398, letter of 14 December 1865 from George Price to Earl Russell; copy also attached to Despatch No 21 from Edward Cardwell to Governor Sir H Storks in *Further Papers relating to the Disturbances in Jamaica*, HC 1866 Vol LI (3594 I)
35. George Price, *Jamaica and the Colonial Office: Who caused the Crisis*, Sampson Low, Son and Marston, London, 1866
36. Lord Olivier, *The Myth of Governor Eyre*, Hogarth Press, London, 1933, pages 53 - 91; in a later book, *Jamaica: The Blessed Island* (Faber & Faber, London, 1936), Lord Olivier refers to Price as Captain George Price MP, confusing him with his son, Captain George Edward Price MP (1842-1926) who, in 1884, had contributed a preface to a pamphlet supporting the colonists' demand for the restoration of representative government in the island: *Jamaica: papers relating to proposed Change in the Form of Government*, London, 1884.
37. Geoffrey Dutton, *The Hero as Murderer*, London, 1967, pages 229, 235 and 401-402; for earlier sympathetic assessments of Eyre's conduct in Jamaica, see Hamilton Hume, *The Life of Edward John Eyre, late Governor of Jamaica*, London, 1867, and J A Froude, *The English in the West Indies*, London, 1888, pages 256-263.
38. W L Mathieson, *The Sugar Colonies and Governor Eyre 1849-66*, London, 1936

Chapter XIV — The Jamaica Tramway Scandal, 1862-65

39. Geoffrey Dutton, *The Hero as Murderer*, London, 1967, pages 237 and 402
40. CO 137/398, letter of 4 December 1865 from Edmund Leahy to Edward Cardwell, Colonial Secretary
41. ibid; having been absent from the Assembly for more than 12 months, Leahy's application for further leave of absence was rejected by the House of Assembly in November 1865 and his seat was declared vacated (*Jamaica Gleaner*, 8 November 1865).
42. Mr Crum-Ewing, Hansard, Third Series, Vol 181, col 921
43. CO 137/429, letter of 28 August 1867 from Edmund Leahy to the Duke of Buckingham and Chandos
44. CO 137/429, letter of 12 September 1867 from the Colonial Office to Edmund Leahy
45. CO 351/10, CO 351/11 and CO 351/13, Registers of (incoming) Correspondence, Jamaica, 1875-77, 1878-79 and 1883-84, entries relating to letters of 26 November 1877, 9 July 1879 and 17 January 1883; the Colonial Office practice of copying replies to letters relating to Jamaica into an Entry Book ceased in 1872.

CHAPTER XV

EDMUND LEAHY AND THE CASHEL ELECTION, 1868

"I feel it to be the proudest thing now in my life to be in a position to give you the character of so good a man".

Edmund Leahy, 1868[1]

"the corruptionist's native habitat"

By 1868, Edmund Leahy had spent more than 30 years of his life on the fringes of political systems, working with successive grand juries in Cork from 1834 to 1846 and, in the 20 years that followed, dealing with officials of the Ottoman Government, engaging with colonial governors in the Cape of Good Hope and in Jamaica, and corresponding with and meeting successive British Foreign and Colonial Secretaries and their senior officials in his efforts to gain support for various schemes. His only direct involvement in politics had arisen in March 1863 when he was elected to the Jamaica House of Assembly in an election marked by corruption and threats of violence. His final bout of political activity at the election for the borough of Cashel, County Tipperary, in 1868 was again characterised by rampant corruption and was to have costly results for the candidate he promoted at the election – a wealthy carpetbagger – and for Cashel itself which lost its direct representation at Westminster as a consequence of the corrupt election. Petitions resulting from the election were tried in 1869 and Election Commissioners were subsequently appointed to inquire into the corrupt practices which had been found to prevail; what follows, except where otherwise indicated, is based on the published reports and minutes of evidence of these inquiries.[2]

Leahy was a friend and business associate of Henry Munster, an English barrister who had provided legal opinions favourable to his claims against the Ottoman Government in the late 1850s and who, in return, had received a loan of some money from Leahy. By 1868, however, Leahy seems to have become dependent on Munster for employment and financial support. In the summer of that year, he was in Ireland on Munster's behalf, looking at an estate near Killenaule which was for sale, and conducting some other

Chapter XV — Edmund Leahy and the Cashel Election, 1868

business, when it became clear that a general election was likely to take place before the end of the year. He suggested to Munster, who was living temporarily in Sheffield at the time and had suffered heavy business losses in the previous few years, that he might become a candidate for the borough of Cashel and the latter agreed, even though he admitted that he had never been in Cashel, had "no particular interest in the place" and had no previous intention of offering himself as a candidate in any Irish constituency. Some personal antagonism between Munster and the sitting MP, James Lyster O'Beirne, a Dubliner who lived in London, seems to have influenced Munster's decision to oppose him.[3] O'Beirne had been associated with Overend, Gurney & Company, one of London's leading banking companies which crashed in May 1866 with liabilities amounting to £11 million, precipitating a commercial crisis and numerous company failures;[4] Munster, who was among those affected, was later to describe O'Beirne as one of a "gang of the most disreputable scoundrels that ever deluded any civilised society".[5]

The introduction of Henry Munster as a candidate at Cashel gave rise to difficulties for Leahy's brother, Patrick, who, as Archbishop of Cashel and Emly, took an active interest in Tipperary politics. He held O'Beirne in high regard as a man who had proved himself an extremely useful and clever member of Parliament "and whose whole parliamentary career justifies the confidence that he will acquit himself equally well for the time to come". In addition, it was common knowledge that bribery and corruption were rife at the 1865 election in Cashel and, anticipating that what had happened before would happen again, the archbishop warned Edmund "in the strongest terms I could use, against having anything to do with the election of Cashel – that he would be getting into a business out of which, perhaps, he could not get, without regretting he had ever got into it; and I warned him repeatedly, notwithstanding my personal regard for Mr Munster". But, notwithstanding these warnings, Edmund Leahy pressed ahead with plans to have his friend elected to represent the borough.

Prior to the introduction in 1872 of voting by secret ballot,[6] bribery and corruption were common features of contested parliamentary elections in Ireland. The borough constituency of Cashel, due to its relative poverty and small number of electors (203 in 1868), fitted neatly into the category of nineteenth-century parliamentary constituencies which has been described as "the corruptionist's native habitat".[7] From 1852 onwards, each of the five general elections in Cashel was contested and widespread corruption on these occasions made the constituency especially notorious by national standards.[8] Groups of electors literally auctioned themselves in return for communal or individual benefits and, by the 1860s, success at a Cashel election depended on cash and little else.[9] When bribery was condemned from the pulpit in 1859, the only result was that electors increased the price of their votes to take account of the fact that their souls could be damned.[10] At the following election in 1865, when J L O'Beirne was believed to have spent up to £3,000, his "generosity to local institutions and businesses made him electorally irresistible".[11] It must have been clear, therefore, from the outset that Leahy and his associates would have to challenge O'Beirne at his own game if they were to have any chance of unseating him.

"sham liberal" or "real friend"?

Edmund Leahy based himself in Cashel from mid-September onwards and quickly made all the necessary arrangements for the coming campaign. A local solicitor, Michael J Laffan, was engaged as election agent at a fee of 100 guineas but was directed in writing by Munster that as "Mr Leahy is an older man, and of much experience, I hope you will take his counsel on all matters". Together, Leahy and Laffan devised the campaign strategy, drafted an election address for publication in the local papers[12] and began the recruitment of a team of assistants to conduct the campaign. The address touched on the major issues of the day, including the land and education questions, and the disestablishment of the Church of Ireland, a subject on which Munster professed to have strong views; in one of his speeches he told the electors that "the most burning question" before them was "the pulling down of that monstrous iniquity which is to every Irishman, to every Catholic, the badge of slavery".[13]

When the candidate himself arrived in Ireland in mid-October, he was met at Thurles railway station by Leahy with a four-horse carriage and, according to the correspondent of *The Times*, "made a public entry into Cashel of the Kings ... accompanied by a great concourse of the populace, whom he addressed".[14] The *Freeman's Journal* took a dim view of Munster's decision to contest the election, describing him as "a sham liberal" and telling its readers that "it is really time for those invaders who seek by force of money to prove that Ireland is 'dear' to their hearts as well as to their pockets, to be informed that they are not wanted".[15] Similarly, the *Tipperary & Clare Independent* asked why Munster should "obtrude himself on Cashel" and seek to displace the sitting liberal MP whose "voice and vote were always for the Irish cause".[16] On the other hand, the *Cashel Gazette* lauded Munster's decision to contest Cashel, describing him as a "Roman Catholic gentleman of high social standing and princely fortune" who had generously supported charities in England; while warning, with extraordinary prescience, of the penalty of disfranchisement which could follow if bribery were to be engaged in, the *Gazette* openly called on the electors of Cashel to vote for Munster "as a real friend who has the power and the will to promote the best interest of the city".[17] It subsequently emerged, however, that the owner and editor of this journal, J Davis White, a Cashel solicitor, who had acted as O'Beirne's agent at the 1865 election, had turned against him because of an unpaid debt arising from that election, and that he had received a cheque for £50 for undertaking "the advocacy" of Munster's cause in 1868.[18] White made no apology for this; indeed, in a memoir which was published posthumously, he noted that the representation of Cashel "had for many years been a matter of bargain and sale" and he defended the electors who "only sold themselves, in preference to being sold by others, as was the custom".[19]

Chapter XV — Edmund Leahy and the Cashel Election, 1868

Campaign spending

Even before Munster's arrival, a liberal spending programme had been initiated in Cashel by his campaign team. A house at Abbey View was rented and furnished to accommodate him and his party and a campaign headquarters was set up in rented rooms at Corcoran's Hotel. Work on the construction of a six-mile-long private telegraph line between Cashel and Goold's Cross on the Dublin–Cork railway was in progress under Leahy's direction, allegedly for Munster's own convenience but with the hope that the Government would take over the line, which cost about £400, at a later stage; there were suggestions that this was merely to be the forerunner of a branch railway from Goold's Cross to Cashel.[20] An initial deposit of £5,000 was made in Munster's name at the National Bank in Cashel and Leahy arranged that the deposit receipt would be widely exhibited to electors and others to demonstrate Munster's financial status and serious intent. Houses, and rooms in houses, were rented from electors, ostensibly to provide rooms for committee meetings or to accommodate election workers and Munster's servants, although some of them were never used. As the campaign progressed, more furniture was purchased for the rented houses, and a carriage and pair of horses was acquired, one of them at a price of £55 even though the seller had bought it only a few days earlier for £14. Other purchases included a cow, a jennet which cost £20 (and which turned out to be such an obstinate and dangerous brute that it was later given away) and several new suits for the candidate who claimed that he had been "rather chaffed by Mr Leahy about my personal appearance which unquestionably was shabby". Munster became "an extensive customer" at various local shops, including a toy shop, 100 tickets costing £12 10s were taken for a concert to be given by a local music teacher, and up to £42 was spent on tickets for a theatrical performance in aid of the Society of St Vincent de Paul. Individual deserving cases of charity were relieved by suitable payments, beggars in the hotel yard received daily donations, spirits were delivered to the poor "as presents for Christmas," and a subscription of 20 guineas earned Munster the status of life governor of the County Infirmary.

"actuated by none but the purest motives"

According to one of his priests, money had first been spoken of at a Cashel election in 1852 when Patrick Leahy, then President of St Patrick's College, Thurles, had presented as a possible candidate a man who was wiling to pay up to £400 towards local charities. Whether this had anything to do with the strategy pursued by Henry Munster can only be a matter of conjecture but the fact is that he expended at least £1,000 on charitable and church-related purposes in Cashel during the course of the 1868 election. The largest single donation went towards the cost of bringing a Christian Brothers school to Cashel – a project about which Archbishop Leahy had been in touch with the Board of National Education for Ireland as far back as 1856 when he was Parish Priest in Cashel.[21] Initially, Munster's advisors had suggested that this contribution, which they saw as a "judicious investment", should be a conditional one, to be returned if he failed to be elected. However, Munster rejected this strategy: the contribution, he declared, "should not be looked upon as in any way relating to the election" and he directed that the payment should be made once Edmund Leahy was satisfied that suitable premises were available and that the school could be opened within nine months. A few days later, having inspected the old fever hospital which the Town Commissioners proposed to make available for the school, Leahy announced the donation at a public meeting and handed over a cheque for £500 to the parish priest, Archdeacon Quirke, who was at pains to declare publicly that the donation would not cause him to "raise my voice nor my little finger for Munster's political interest".

Archbishop Leahy himself, while he asserted that he had observed "the most strict neutrality" at the election and never sought, directly or indirectly, to influence the vote of any elector, accepted a cheque for £100 towards the building fund after he had shown Munster over his unfinished cathedral at Thurles; he believed that the offering was "actuated by none but the purest motives", and that it had been "nobly and charitably offered". The archbishop was also believed to have received a cheque for £5 which, as he recalled it, was "to get a few masses offered up for Mr Munster". Dean Walter Cantwell, parish priest of Fethard, was given a cheque for £200 to purchase a site for a new convent in the town but, unlike the others, he subsequently spoke publicly in favour of Munster at the hustings, describing him, according to one newspaper report, as a man of pure honour and of charity, a genuine Catholic, whom he believed to have been "sent by God to bless Cashel by his presence".[22] Another beneficiary was the library of St Patrick's College, Thurles, which received a donation of £20. Finally, and contrary to Edmund Leahy's advice that "it would be useless and injurious just at present to do much for those nuns and their convent", Munster directed that goods should be donated to a fundraising bazaar in aid of the Presentation Convent and arranged that a sum of £50 would also be spent there; half of this was spent by Leahy's wife who, with his son, was staying at Munster's rented house in Cashel. In addition, the convent received approximately £150 in a further series of donations.

As the campaign intensified, the local press was outspoken in its comments on the rival candidates. The *Cashel Gazette* attacked O'Beirne for having reneged on promises made in 1865, criticised his lack of attention to Cashel, commented favourably on Munster's intention to reside there, and spoke of the many advantages that would follow from having a resident MP. On the other side, criticism in the *Tipperary & Clare Independent* of Munster and his supporters intensified, one editorial describing him as "a downright political adventurer and hypocrite" who had "fallen into the hands of men who make politics a speculation for their own selfish ends" and who was dependent on "corruption and political prostitution to degrade the people of Cashel and oust Lyster O'Beirne". While this particular onslaught was the subject of an apology and retraction in the

Chapter XV — Edmund Leahy and the Cashel Election, 1868

following issue, the paper continued its strong support for O'Beirne, listing out his voting record in the House of Commons to back up his case for re-election.

Election meetings, at which Munster was usually accompanied by Leahy and his son, were reported extensively in the local press. At one such meeting on 2 November, Leahy spoke at length and in passionate terms about his friend, telling the attendance that since he had last been in Cashel, over 30 years earlier –

> I have been through a great many parts of Europe, Asia, Africa and America. I have seen, perhaps, more than most of you ... and I am able to tell you that, amongst all the good men I have seen, and all the good I have witnessed, I have met with none that I could recommend to you so good and deserving as the gentleman who now seeks to be your representative. I have been a long time away out of my native land. I have not changed one particle in heart, in soul and in love for the old creed and the old country – and I would sacrifice my life for them. ... I feel it to be the proudest thing now in my life to be in a position to give you the character of so good a man.[23]

And, as if that emotional appeal were not enough, Leahy concluded by announcing that Munster was willing, in addition to all of his other donations, to provide £1,000 towards the restoration of the ruined buildings on the Rock of Cashel, if they could be returned for use as a Catholic cathedral; it was hardly a coincidence that this was a major objective of Archbishop Leahy.

Direct payments for votes
Munster had been advised from the outset by Edmund Leahy that he would not succeed without recourse to direct payment for votes but he insisted throughout that he would not authorise bribery (and was later to deny that he had any knowledge of bribery at any stage of the campaign). In the belief that it would avoid a breach of electoral law, he himself had devised a scheme of paying retainers to electors – originally set at £5 but increased to £10 on Leahy's suggestion – in respect of temporary jobs amounting to mere sinecures, on condition that those who received these payments would not vote for either of the candidates.[24] However, as the election date drew near, and with the *Freeman's Journal* suggesting that Munster, who was referred to disparagingly as "the Tory candidate", had withdrawn entirely,[25] Leahy seems to have come to the conclusion that success was still not assured and so, in the ten days immediately preceding the poll, he arranged that a total of £1,100 would be spent or committed in direct bribes to individuals. Electors called by arrangement to Corcoran's Hotel to receive payments of £30 or £40 each in sealed envelopes, although in some cases only half notes were given out, on the understanding that the other half of the note would be made available after the election to those who had

met their commitments. In addition, drink and food were liberally dispensed at Munster's expense to electors and non-electors alike who visited the hotel, even though "treating" was an offence under electoral law;[26] bills for drink alone were later found to total £254 and one witness reported that wholesale drunkenness prevailed throughout the town in the later stages of the campaign.

In all, Munster was calculated to have spent over £3,800 in the effort to secure his election, although his official return of election expenses came to less than £1,000. While O'Beirne could afford to devote only very limited resources of his own to meet election expenses, he raised loans of £400 and received a grant of £500 from a fund which had been set up to support liberal candidates. This enabled him and his supporters to engage in bribery, but on a much more limited scale. In addition, he rented a large number of houses for use as committee rooms or tally rooms, and held an eve of poll dinner for some 50 electors in a local hotel which ended only with breakfast the following morning. A similar dinner, followed by dancing and singing, was arranged for Munster's supporters at Corcoran's Hotel, culminating in a breakfast at which one scandalised observer noted that meat was being eaten – even though it was a Friday!

"the corrupt practice of bribery had prevailed extensively"

Munster and O'Beirne, as well as six other candidates, were duly proposed and seconded at the hustings on 18 November and when the poll took place two days later, over 90% of the electors turned out. O'Beirne received 100 votes and Munster 84 but Leahy and Munster were determined to carry on the struggle to win Cashel. There were suggestions that Leahy met some of O'Beirne's supporters and offered them substantial sums if they would assist in assembling evidence to support an election petition. In addition, much of the money which had been promised in bribes, or was owed for goods and services, was withheld in an effort to influence the evidence of those who might be called as witnesses in the trial of a petition. In December, after O'Beirne had sued Munster for slander, claiming damages of £5,000, Munster lodged two election petitions under legislation which had introduced new procedures for questioning the result of an election;[27] he claimed that "the majority of good and lawful votes" had been given for himself and that O'Beirne's majority was only a "colourable" one, "made up of the votes of persons who had ... corruptly accepted meat, drink and entertainment ... and who had received or agreed for money, employment or other valuable consideration ... or who had been retained or employed as agents, for hire and reward" by O'Beirne. Baron Fitzgerald, a Judge of the Irish Court of Exchequer, was appointed to try the petitions and held public hearings in Cashel on five days in February 1869. Having listened to the evidence of up to 100 witnesses, the judge held that the election was void, that neither Munster nor O'Beirne had been duly elected and that "the corrupt practice of bribery had prevailed extensively" at the election. This latter finding made it obligatory that, following consideration by Parliament of Baron Fitzgerald's report,[28] Commissioners should be appointed under an Act of 1852[29] to inquire further into the corrupt practices, and this was done by Royal Warrant in June 1869. The three barristers who formed the commission sat for 22 days between 4 October and 20 November, and heard evidence from some 250 witnesses. Their proceedings were reported extensively in the local press[30] and even *The Times* carried daily reports, one of which noted that the Commissioners were "pursuing their inquiries with diligence and firmness ... dragging to light transactions which had been kept carefully concealed".[31]

Witnesses at the trial of the petitions and before the Commissioners demonstrated extraordinary lapses of memory and evasive answers were the order of the day. Others absconded even though they had received a summons to attend, and a few of those heavily involved on Munster's side had left Cashel in good time, encouraged or assisted to do so by Edmund Leahy, to avoid being served with a summons. Although a *subpoena* had been issued, Leahy himself did not attend the trial of the petitions although, according to Munster, he had met him in Cashel on the Sunday before the trial began. Similarly, while there was evidence that Leahy had been staying at Munster's house in Cashel on the day on which the Commissioners first sat, he did not attend to give evidence even though Munster had given an assurance that he would do so.

The absence of Edmund Leahy made it difficult for the Commissioners to piece together what exactly had been done in the course of the election but they succeeded in establishing that he had played a central role in managing the campaign. Witnesses told that he was seen "going here and there, night and day" and that he was Munster's chief adviser, an active canvasser and speaker on his behalf, and his companion on the canvass of electors. There was evidence that Leahy had arranged employment at the election for a number of individuals and free accommodation for others, including a man who knew very few people in Cashel but whose wife was related to Leahy. And, most important of all, there was direct evidence that it was Leahy who had supplied £350 in notes and £600 in gold coins to Patrick Laffan, a young medical student who had managed the payment of most of the bribes and retainers to individuals; the gold coins had been obtained at a bank in Dublin on foot of a cheque issued by Leahy and left for collection by Laffan at the home in Thurles of James McNamara, a Franciscan friar, who received a £5 cheque for facilitating the operation.[32]

When no reply had been received from Leahy to a letter from the Commission requesting his attendance, Munster stated that he understood that Leahy was going to Brussels in connection with a tramway project. Asked again a few days later if he knew where Leahy was, Munster replied that he did not, and could not understand why he had not appeared. When all the available witnesses had been heard, the Commission decided to adjourn to enable Leahy to attend but, when he was again called on 1 November he did not appear although he had acknowledged in a letter from Vienna that he had received the summons to do so. When Munster told the Commission that the latest letter he had received from Leahy spoke of travelling to Constantinople, the Commission decided to adjourn again to 20 November to give him another opportunity to attend. But, when the final session of the Commission opened on that day, there was still no appearance by Leahy. While showing utter contempt for the Commission in this way, Leahy had in fact travelled to Rome from where he was carrying on correspondence with the Colonial Secretary (as his "most obedient humble servant") about the possibility of Government support for telegraph schemes in the South Atlantic.[33]

"a great difference between influencing opinion and influencing opinion corruptly"
It was not possible for Archbishop Leahy to escape questioning by the Election Commissioners, as his brother had done, even though he cannot have relished the prospect of having to appear before them. While he had always publicly defended his right "as a citizen of this realm" to take an active part in politics, he and his clergy had been criticised two years earlier by a Select Committee of the House of Commons which found that "divers of the Roman Catholic Clergy exercised their influence upon their congregations in a manner calculated to prejudice the free choice of the electors" at a Tipperary by-election in 1866.[34] On that occasion, he had openly supported the liberal candidate, arranged that all of his priests would act together so as to prevent "scandalous divisions" or "public disedification", and urged all of them to call on the "the honest electors of the county" to vote for the liberal, provided the priests did so only in the chapel yards and "in their capacity of citizens and not as clergymen".[35]

The archbishop had remained strictly neutral at the Cashel election, even though he and his clergy had publicly declared their support for the two liberal candidates at the election for Tipperary County which was being held at the same time.[36] Nevertheless, when he took the witness stand on 22 October 1869, he was confronted with awkward and potentially embarrassing issues in the light of the evidence that had been given in the 16 sitting days before he appeared. It was clear by then that his own brother had masterminded the election strategy and the generous spending programme which was under investigation by the Commissioners. It was clear also that well over 25% of this spending had benefited the Church in one way or another. In addition, there had been allegations in the evidence of one of his own parish priests that Leahy had previously put forward as a candidate for Cashel a wealthy barrister who would be both willing and able to contribute up to £400 towards the charities of the town; and the same priest had openly admitted that he saw nothing wrong in telling the electors to take all the money Munster would give them because "they would be very great fools if they refused it". Finally, there was even suspicion in some quarters that the archbishop had been involved in the corrupt practices, and that he was implicated in some way in the provision of the £600 in gold coins which his brother had made available for the payment of bribes.

Having satisfied the Commissioners that he had no knowledge of the bribery or other

Chapter XV — Edmund Leahy and the Cashel Election, 1868

corrupt practices which had been engaged in and that he was not the source of any of the funds which had been used for these purposes, Archbishop Leahy was pressed by George Waters, Chairman of the Commission, to give his views on whether some of the election activities had involved bribery. A fascinating battle of wits and of words ensued between the skilled theologian and the experienced Queen's Counsel, himself a Catholic, centring on the question of how an act which clearly constituted bribery under electoral law might be judged in the context of moral law. The archbishop argued that an offer to contribute liberally to local charities had only been "a collateral and subordinate consideration" when speaking of the man whose candidature he had recommended in 1852; he went on to tell the Commissioners "as one knowing a little about morals and questions of morals" that there was nothing improper in what he had done in 1852 "and nothing at all in it bordering on corruption, or any interference with purity of election". In further examination on the question of what constituted bribery, he agreed that while donations of £1,000 for charitable purposes at the 1868 election might have influenced voters, "there is a great difference between influencing opinion and influencing opinion corruptly"; he went on to argue that no question of bribery arose where an honest voter, in discharging his "conscientious obligation" to opt for one of the two candidates fit to be elected, decided, as a subordinate consideration, to give his vote to the man who would promote the material prosperity of the area, or support its charities. And in a further reply which helped to support the justification offered by Munster for some of his election expenditure, he agreed that while it would be a low motive, it would be a legitimate one, for a candidate who did not wish to engage in bribery to give liberally to charities in order to counteract any imputation of meanness.

Cashel is disfranchised

The Election Commissioners completed their report in December 1869 with commendable speed, efficiency and economy,[37] and found that the election had been conducted in a corrupt manner on the part of each of the candidates. They held that a total of 77 electors had been guilty of corrupt practices, along with 22 others, including Edmund Leahy. They specifically addressed the question of whether the donation of

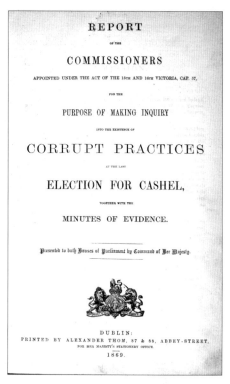

Cover page of the report of the Cashel Election Commissioners, 1869 (HC 1870 Vol XXXII).

In search of fame and fortune: the Leahy family of engineers, 1780-1888

£500 to the Christian Brothers was intended to induce voters to support Munster and had no hesitation in finding that the transaction constituted a corrupt practice within the meaning of the relevant legislation. When the report was published, Munster still insisted that he knew nothing of any corrupt or unlawful practices and claimed that these could only have arisen from the injudicious interference of over-zealous friends, counter to his own explicit instructions. And, loyal to the end, John Davis White's *Cashel Gazette* refused to publish what its proprietor considered to be a one-sided report and decried the fact that Munster's "reasonable explanations on oath, together with the lucid and able arguments of His Grace, the Most Rev Dr Leahy, were not sufficient to save him".[38]

Faced with the finding that corrupt practices had been engaged in at two successive elections in Cashel, the Government brought forward a Bill to disfranchise the borough, depriving it of the representation which it had enjoyed in parliament since 1585. Noting that "in one shape or another … fully one half of the constituency was implicated in the corrupt practices" at Cashel, the Solicitor General for Ireland suggested in the Commons that "the House might come to the conclusion that if the truth could have been fully elicited, a considerable percentage of the remaining half would have been added" to the list of the guilty.[39] There was no opposition to the Government's proposal and the Bill received the Royal Assent on 1 August 1870.[40]

Over a year later, Henry Munster, in a series of letters to Archbishop Leahy,[41] apologised for the fact that he personally had been compromised in the course of the election campaign but disclaimed all responsibility for "the lamentable results". He continued, however, to insist that he did not "connive at bribery in any shape", and that he had repeatedly warned Edmund Leahy, both generally and "in special reference to Your Grace's being his brother", that he was not to become implicated or involved in any scheme that might cause difficulty for the archbishop. He threw all of the blame for the illegal practices on Edmund and appeared to accuse him of attempting irregularly to draw £3,000 from his London bank. Clearly, a friendship and business relationship which had lasted for over ten years was then very definitely at an end.

"bent on getting a seat and indifferent to the amount he spends"
After the electoral disaster at Cashel, Henry Munster did not retire from the political arena, "a wiser but a poorer man," as one of the local newspapers had suggested that he might do;[42] instead, as *The Times* put it, he appeared "to be bent on getting a seat and indifferent to the amount he spends on his object".[43] Thus, while the Election Commissioners were still sitting at Cashel in November 1869, he reacted favourably to an invitation by a number of Mallow publicans to become a candidate at a by-election which was due to be held there early in 1870[44] arising from the appointment of the sitting member, Edward Sullivan QC, as Master of the Rolls. When the delegation returned to Mallow with "a glowing account of the charitable character of Mr Munster at Cashel," a

Chapter XV — Edmund Leahy and the Cashel Election, 1868

formal requisition was sent to him by 63 of the town's 204 electors, including 28 of the 48 publicans on the register. Munster arrived in Mallow in mid-January to find that there were two other candidates in the field - George Waters QC, who had just completed his work as chairman of the Cashel Election Commission and a conservative candidate, Major Laurence E Knox, the founder and proprietor of the *Irish Times* and whose own election to represent Sligo in November 1868 had been declared void following a petition.[45] Waters retired from the contest, believing that he had no chance of success, leaving Mallow to be "wooed by two candidates, neither of them unversed in the wiles of electioneering", as *The Times* put it.[46]

When the poll was taken on 3 February, Munster was declared the winner, having defeated Knox by 91 votes to 83. Knox immediately lodged a petition (with assistance from Waters) alleging bribery, corrupt treating and other corrupt practices by Munster and his agents and, in April, Mr Justice Michael Morris[47] sat for ten days at Mallow, often from 9.00am to 7.00pm, to hear evidence from both sides. It emerged that "the Cashel bribers", as the Judge described them, had been at work in Mallow, that houses and rooms had been rented, as in Cashel, that large amounts of porter, tea, sugar and bread had been purchased from publicans (some of whom were also grocers) by Munster's agents for distribution to the electors, and that the vast majority of the town's publicans had voted for him. Munster attempted to explain what the Judge called "his great and sudden popularity" by reference to the desire of the electors of Mallow to be represented by a Catholic, but the Judge believed that he could "account for it on much more material and local considerations"; thus, although direct bribery was not proven to have occurred, the Judge still had no hesitation in declaring Munster's election to be void on the grounds that he had been guilty, by his agents, of corrupt treating. Munster himself escaped personal censure but "if there were an Habitual Criminals Act for parliamentary candidates," declared *The Times*, "his position would have been embarrassing and the report of the learned Judge might have been somewhat different".[48]

On 10 May, within three weeks of the delivery of this judgment, a second by-election was held in Mallow. Major Knox, who had been a founding member of Isaac Butt's Home Government Association, was again nominated, this time standing on the Home Rule platform – the first candidate in the country to do so.[49] With Henry Munster legally disqualified from standing again for seven years, George Waters re-entered the lists as a liberal and defeated Knox by 93 votes to 85. Waters' appointment as a county court judge[50] led to yet another by-election on 7 June 1872 at which Munster's 23-year-old son, William Felix, describing himself as a liberal and a home ruler, presented himself as a candidate, telling the electorate that he was doing so "in fulfilment of a pledge" made to them by his father and in order "to reverse by your votes the decision that the last (sic) election was corrupt".[51] His opponent, the Home Rule candidate, John G McCarthy, had been in the field before Munster's arrival and was strongly supported by the local parish priest and

In search of fame and fortune: the Leahy family of engineers, 1780-1888

other clergy, by prominent politicians, including J F Maguire MP and A M Sullivan, and by editorials in the *Cork Examiner*.[52] McCarthy, however, was personally unpopular with the advanced nationalists of Cork[53] and, partly for this reason, Munster defeated him by 91 votes to 78, despite his youth and lack of knowledge of the area and of Irish affairs generally.[54] Henry Munster himself, long after his period of disqualification had come to an end, made a final effort to take an Irish seat in the Commons in February 1887 when, as a liberal unionist, he was heavily defeated by a nationalist candidate at a by-election in Donegal South.

The 1868 Cashel election seems to have marked the end of Edmund Leahy's inglorious involvement in political activity in Ireland. He was fortunate that disagreement arose in 1870 between the Home Office in London and the Lord Lieutenant and his legal advisers in Dublin as to who should take the initiative in instigating criminal prosecutions arising from the Election Commissioners' report;[55] but for this, Leahy could have had to answer charges which, if proven, would have left him facing the prospect of a long jail sentence. He was not listed among "the Cashel bribers" who were part of Henry Munster's team at the February 1870 by-election at Mallow and reports of the successful campaign there in 1872 by Munster's son make no reference to him. Archbishop Leahy, on the other hand, seems to have provided some support, moral or otherwise, on that occasion, earning him a letter of thanks and a cheque for the cathedral building fund from the newly-elected MP.[56]

REFERENCES

1. *The Cashel Gazette*, 7 November 1868, report of a speech by Edmund Leahy
2. *Copy of the Minutes of the Evidence taken at the Trial of the Cashel Election Petitions (1869)* HC 1868-69 Vol XLIX (121); *Report of the Commissioners appointed... for the purpose of making Inquiry into the existence of Corrupt Practices at the last Election for Cashel, together with the Minutes of Evidence*, HC 1870 Vol XXXII (c.9); see also CDA, Christopher O'Dwyer, *The Life of Dr Leahy 1806-1875*, thesis presented to the Faculty of Arts, St Patrick's College, Maynooth for the degree of MA, September 1970, chapter 13; and James O'Shea, *Priests, Politics and Society in Post-famine Ireland: A Study of County Tipperary 1850-1891*, Wolfhound Press, Dublin, 1983
3. According to the *Who's Who of British Members of Parliament, Vol I, 1832 – 1885* (ed Michael Stenton, Sussex, 1976), Lyster was born in Dublin in 1820, educated at TCD, and was engaged in commerce in London at the time of his election as an "independent liberal".
4. In a letter to *The Times*, published on 29 January 1869, O'Beirne claimed that his only connection with Overend, Gurney & Co had arisen between 1861 and 1863 when he had been appointed by the bank to superintend the Milwall Iron Works on the Thames, formerly the property of the well-known engineer and naval architect, C J Mare, at a salary of £1,500 a year.
5. *The Cashel Gazette*, 19 November 1868
6. Parliamentary and Municipal Elections Act, 1872, 35 & 36 Vict., c.33, commonly called the Ballot Act
7. K Theodore Hoppen, *Elections, Politics and Society in Ireland, 1832–1885*, Clarendon Press, Oxford, 1984, page 75; Cashel was listed as one of 68 UK constituencies (21 of them in Ireland) containing less than 400 voters in a return presented to Parliament in 1859, HC 1859 (Session 2) XXVI (77).

Chapter XV — Edmund Leahy and the Cashel Election, 1868

8. Kieran Devery, "The function of hotels in parliamentary borough elections in mid-nineteenth century Ireland", in *Irish History: A Research Yearbook No 2*, Four Courts Press, 2003, page 53; a petition was lodged claiming that general and extensive bribery had prevailed to secure the election of the liberal candidate, Sir Timothy O' Brien in 1857 but did not proceed to a hearing for technical reasons – see *Report from the General Committee of Elections on the Cashel Election Petition*, HC 1857-58 Vol XII (157).
9. K Theodore Hoppen, *Elections, Politics and Society in Ireland, 1832–1885*, Clarendon Press, Oxford, 1984, page 450; "National Politics and Local Realities in Mid-Nineteenth Century Ireland", in *Studies in Irish History*, University College Dublin, 1979, page 204
10. K Theodore Hoppen, *Elections, Politics and Society in Ireland, 1832–1885*, Clarendon Press, Oxford, 1984, page 77
11. R V Comerford, "Tipperary Representation at Westminster 1801- 1918", in *Tipperary: History and Society*, ed William Nolan, Geography Publications, Dublin, 1985
12. See, for example, *The Cashel Gazette*, 3 October 1868
13. *The Cashel Gazette*, 24 October 1868, report of an address on 17 October
14. *The Times*, 20 October 1868
15. *Freeman's Journal*, 28 October 1868
16. *The Tipperary & Clare Independent*, 26 September 1868
17. *The Cashel Gazette*, 3 and 17 October 1868
18. For a biography of White, see Denis G Marnane, "John Davis White of Cashel (1829-1893)", in *Tipperary Historical Journal*, 1994
19. Denis G Marnane, "John Davis White's Sixty Years in Cashel", *Tipperary Historical Journal*, 2004
20. A six-mile long branch line from Goold's Cross to Cashel was eventually completed and opened in December 1904.
21. CDA, 1856/9, letter of 20 November 1856 to Leahy from the Board of National Education for Ireland
22. *The Cashel Gazette*, 19 November 1868
23. *The Cashel Gazette*, 7 November 1868
24. Section 8 of the Representation of the People (Ireland) Act, 1868 (31 & 32 Vict., c. 49) provided that an elector who, for reward, had been retained, hired or employed by a candidate for the purposes of an election would not be entitled to vote at that election.
25. *Freeman's Journal, 13 November 1868*
26. Bribery etc Act, 1854, 17 & 18 Vict., c.102
27. The Election Petitions and Corrupt Practices at Elections Act, 1868, 31 & 32 Vict., c. 125, received the Royal Assent on 31 July 1868; see Cornelius O Leary, *The Elimination of Corrupt Practices in British Elections 1868 – 1911*, Clarendon Press, Oxford, 1962.
28. Hansard (Third Series), Vol CXCV, 29 and 30 April 1869, cols 1935, 1971
29. An Act for more effectual Inquiry into the Existence of Corrupt Practices at Elections for Members to serve in Parliament, 15 & 16 Vict., c.57
30. *The Cashel Gazette*, 9, 16, 23 and 30 October and 6 November, 1869; *Tipperary & Clare Independent, Tipperary Free Press*
31. *The Times*, 9 October 1868
32. Father James McNamara OFM was based in Thurles from the early 1830s until his death in 1881 and was held in high regard by Archbishop Leahy – see Patrick Conlan OFM, "The Franciscan House in Thurles" in *Thurles: The Cathedral Town, Essays in honour of Archbishop Thomas Morris*, Geography Publications, Dublin, 1989, pages 179-183.
33. CO 318/257, letters of 12 August and 18 December 1869 from Edmund Leahy to Earl Granville
34. *Minutes of Evidence taken before the Select Committee on the Tipperary Election Petition, with the Proceedings of the Committee*, HC 1867 Vol VIII (211)
35. Evidence of Dr Leahy in April 1867 to the *Select Committee on the Tipperary Election Petition*, HC 1867 Vol VIII (211)
36. *The Tipperary & Clare Independent*, 14 November 1868, advertisement in the name of P Leahy
37. Each Commissioner was remunerated on the basis of a daily fee of five guineas for each of the twenty-two sitting days, and for the additional seven days required to complete the report, with an allowance of £1 a day for subsistence; the entire proceedings cost only £1,220 – *Return of the Accounts of the Sums expended in connection with the Election Inquiries at Dublin, Sligo and Cashel*, HC 1872 Vol XLVII (105).
38. *The Cashel Gazette*, 12 March 1870
39. Hansard (Third Series), Vol CCII, 16 June 1870, col 309
40. An Act to disfranchise the Boroughs of Sligo and Cashel, 33 & 34 Vict., c.38
41. CDA 1871/79, letters of 9, 22, 24 and 28 December 1871 from Henry Munster

42. *Tipperary & Clare Independent*, 21 November 1868
43. *The Times*, 26 April 1870
44. This summary account of the proceedings at the Mallow election draws heavily on the judgment delivered at the conclusion of the trial of the subsequent election petition – see *Copies of the Shorthand Writer's Notes of the Judgments delivered by the Judges selected in pursuance of the Parliamentary Elections Act, 1868 for the Trial of Election Petitions since April 1869, etc*, HC 1872 Vol XLVII (268).
45. *Copy of Minutes of the Evidence taken at the Trial of the Sligo Borough Election Petition 1869*, HC 1868-69 Vol XLIX (85)
46. *The Times*, 26 April 1870
47. Morris had been liberal MP for Galway city from 1865 to 1867 before his appointment to the bench; he became Lord Chief Justice of Ireland in 1887 and was created Baron Killanin in 1900.
48. *The Times*, 26 April 1870
49. R V Comerford, *The Fenians in Context: Irish Politics and Society 1848-82*, Wolfhound Press, Dublin, 1985, page 190
50. Born in Cork in 1827, the second son of George Waters, a brewer and distiller, Waters was educated at TCD, called to the Irish Bar in 1849, became a QC in 1859 and came to notice in 1867 when he was one of those engaged in the defence of Fenian prisoners; he was county court judge for County Waterford for some 20 years before transfer to the Cavan–Leitrim circuit where he served until retirement at the end of 1904. He had for many years been President of the Society of St Vincent de Paul in Ireland before his death in Dublin on 21 April 1905 (*Who's Who of British Members of Parliament, Vol I, 1832–1885*, ed Michael Stenton, Sussex, 1976).
51. CE, 23 May 1872, election address of W F Munster
52. CE, 20 and 22 May 1872
53. Emmet Larkin, *The Roman Catholic Church and the Home Rule Movement in Ireland, 1870-74*, Gill & Macmillan, Dublin, 1990
54. W F Munster, born in France in 1849, a graduate of the University of London and a member of the Inner Temple, held the Mallow seat until the February 1874 general election at which he did not stand; he died on 11 April 1877 (*Who's Who of British Members of Parliament, Vol I, 1832–1885*, ed Michael Stenton, Sussex, 1976).
55. PRO, HO 45/8435, Sligo and Cashel Elections Commission (Ireland): proof for prosecution
56. CDA 1872/26, letter of 26 June 1872 from W F Munster, House of Commons, London, to Archbishop Patrick Leahy

CHAPTER XVI

EDMUND LEAHY
1864-88

Efforts to regain a county surveyor position
Edmund Leahy was just 50 years old when he returned to London from Jamaica in 1864 and, while nothing has emerged to indicate how he supported himself and his family in the immediately following years, the indications are that his efforts to resume practice as a civil engineer were not particularly successful.[1] Having failed to reverse his dismissal from the position of Colonial Engineer and Architect in Jamaica, or to secure alternative employment in the colonial service, he made a number of efforts in 1867 to secure a public service position in Ireland, relying heavily on his former contacts with individuals who were then prominent in the government of the country.

Leahy's first application[2] in April was addressed to the Under-Secretary at Dublin Castle, Major-General Sir Thomas Larcom, who had been in charge of administration at the Ordnance Survey Headquarters in the Phoenix Park, Dublin, during his own brief employment on the survey in County Donegal in 1833-34. Venturing "to seek a renewal of a long acquaintance" with Larcom, and explaining that he had since been engaged in "engineering pursuits ... as various as they have been widely diffused over most places in Europe, Asia and Africa", Leahy sought to re-enter the ranks of the Irish county surveyors from which he had resigned in 1846, or to obtain some other appointment in connection with public works. He mentioned that his "valued and kind friend", General Sir John Fox Burgoyne, and General Lord Strathnairn, Commander-in-Chief of the Forces in Ireland, whom he had met while working in Turkey in the 1850s, would both be glad to speak for him and added, in somewhat contradictory terms: "I shall say nothing of my brother, Dr Leahy, Archbishop of Cashel, as he feels disinclined to make any application to the Government for anyone connected with himself". Larcom replied that it was not in his power to accede to Leahy's application: the Board of Works had no engineering work in hands on which Leahy could be employed and, under the County Surveyors (Ireland) Act, 1862,[3] future vacancies for surveyors were to be filled on the basis of public competitions and examinations conducted by the Civil Service Commissioners.

Appealing immediately to Larcom for reconsideration of his case, Leahy claimed to be entitled to the benefit of a transitional section of the Act which allowed for the appointment to county surveyor positions, without further examination, of persons who had qualified for these posts under the arrangements which had been in place between 1834 and 1862.[4] He wrote simultaneously to Lord Strathnairn asking him to offer "a favourable word" on his behalf to Larcom and the General immediately obliged by arranging for his military secretary to make a submission in which Leahy was described as a civil engineer of great experience, very respectably connected, and whom Strathnairn had always found to be perfectly trustworthy.[5] In deciding, however, to rely on Strathnairn's influence to advance his case with Larcom, Leahy was not to know that relations between the two men had been strained almost to breaking point in the previous year because of serious differences of opinion as to the best disposition of Strathnairn's 23,000 troops in anticipation of a possible Fenian rising.[6] In the event, his application to resume a county surveyor position was again rejected in May 1867.[7] A few months later, Leahy wrote again to Larcom applying for an appointment as Inspector of Prisons and advised that his friend, Captain M E Archdall, MP for Fermanagh, would support his application. In addition, he submitted testimonials which he had just received from Burgoyne and Strathnairn who both attested to his services to them in Turkey in the period immediately before the Crimean War.[8] Leahy was told initially that his application would be brought to the notice of the Lord Lieutenant, but was disappointed to learn in December that "there is no probability of such an appointment becoming vacant".[9]

The Irish Industrial Society

Throughout 1867, and in parallel with his efforts to secure a position in the public service in Ireland, Leahy was also attempting to promote a scheme for creating enhanced employment opportunities for the youth of the country - and, of course, a position for himself. His scheme was remarkably similar to a proposal to establish the "Dargan Industrial College" which had been adopted by a Testimonial Committee in July 1853 as a means of honouring the prominent railway contractor who had financed that year's Dublin Industrial Exhibition. The proposed Dargan College was to provide "for the instruction in the practical arts of industry of young men of genius whose humble birth or limited means might offer a barrier to advancement, and for the general diffusion of industrial knowledge throughout the land", but the scheme did not proceed and the funds collected were allocated instead to help fund the establishment of the National Gallery in Dublin.[10]

The new organisation planned by Leahy was to be called the *Irish Industrial Society*. In letters to the Chief Secretary, Lord Naas (soon to become sixth Earl of Mayo), in June 1867 he explained that the purpose was "to instruct the Irish youth of both sexes in the best modes of applying their labour to the Mechanical Arts, particularly to those having relation to the working of Metals, Iron, Wood, Clay, Stone, Hemp, Flax, Animal Products such as Leather, Hair, Wool etc, and generally as to their application to ordinary uses".[11] To demonstrate the need for such a scheme of technical instruction, he pointed to the fact that "thousands who yearly leave Ireland have only their labour, unskilled and untaught, to rely upon – they are therefore of necessity forced into the lowest and least remunerative occupations, with little chance of advancement to the better positions which would be open to them, had they a knowledge of the mechanical arts". He went on to suggest that his proposed society should aim to make "skilled labourers of the untaught poor", to develop "the mechanical intelligence at present lying dormant amongst the masses ... like a mine of untold wealth", and to secure that "with a little culture ... the millions of Irish heads and hands, now incapable of any but the poorest occupations ... might soon be very considerably increased in value".

To achieve all of this, Leahy's plan was that the new society (of which he was to be honorary secretary) would set up "a Chief Institution in Dublin, with subordinate representatives in the several counties" to provide the necessary instruction to "scholars of the national and other schools". Accepting that Lord Naas, because of "the numerous engagements incidental to his position", would not be able to take a very active part in promoting the scheme, Leahy told him that "I propose to take a good deal on myself and at once to make a beginning" by asking major figures in the political field and in business to act as patrons. Asked by one of his officials if he knew anything of this man, Lord Naas wrote "I don't know anything whatever about him" – and that was enough to put an end to that particular scheme. Apart from one further letter in 1879 offering suggestions to the Chief Secretary for measures to relieve distress, the exchange also seems to have marked the end of Leahy's efforts to begin a new public service career in Ireland.[12]

Leahy's correspondence with Lord Strathnairn

Leahy's correspondence with Lord Strathnairn in 1867 throws interesting light on some rather surprising opinions of the man – described as "one of the greatest enigmas of British military history" – who, as Lieutenant-General Sir Hugh Henry Rose, had come to Ireland in July 1865 to take command of the army and to arrest the spread of Fenianism.[13] Created Baron Strathnairn in 1866 and promoted full general early in 1867, Strathnairn's flying columns were still scouring the disaffected areas of Tipperary, Cork and Waterford following the Fenian rising of March 1867 when Leahy first sought and obtained his support for an engineering appointment in Ireland. But while the General was vigorously putting down any traces of insurgency, his personal view was that extreme and violent measures would never pacify Ireland; instead, the country's historical evils should be dealt with, starting with a solution of the church question – not by disestablishing the Church of Ireland, but by "concurrent endowment" which would improve the position of the Catholic Church. He expressed this view at great length in a letter of 16 March 1867 to his superior, the Duke of Cambridge, Commander–in-Chief of the British Army, while complaining that "the Roman Catholic clergy, high and low, who condemn Fenians, go hand in hand with them at the Elections". By way of illustrating this latter point, and having in all probability noted a report in *The Times* that very day of Archbishop Leahy's pastoral letter denouncing the Fenian rising (a criminal enterprise, equal in folly, according to Leahy, to that of 1848), he went on to tell the Duke that the archbishop had preached in favour of Captain White (a National Association candidate at the Tipperary by-election in 1866) at the Cathedral in Thurles and that one of his priests had brought "a column of voters" to Thurles to support White.[14]

Against this background, when he received another letter from Edmund Leahy in July requesting a testimonial as to his services at the time of the Crimean War, Strathnairn decided to ask him, in a postscript to his reply:

> to give me an introduction to your brother, the Archbishop of Thurles (sic). I think one cause of affairs not going on so favourably in Ireland as elsewhere are social separations, not differences. For instance, myself and many more whom I know are most anxious to know and be on the most friendly relations with the Roman Catholic Clergy, and of course, particularly with its high dignitaries – but we never have the opportunity, the means or the advantage or the qualification of becoming acquainted with them.[15]

It seems unlikely that Strathnairn's initiative led to any substantive contact between him and the archbishop who was among the more nationalist of the Irish bishops in the 1860s and an active campaigner for disestablishment of the Church of Ireland.[16] The papers of Dr Leahy in the Cashel Diocesan Archives and Strathnairn's papers in the British Library[17] do not include any correspondence between the two men and it seems that whatever hopes Strathnairn may have had of reaching an understanding with the archbishop were quickly dissipated. By November 1868, he had come to the view that the Bishop of Kerry, David Moriarty, who was "of a tolerant and Christian frame of mind",[18] was the most likely of the Catholic hierarchy to reach an accommodation with the Government, while "the Cardinal Cullens" – an expression which was probably intended to include prelates like Leahy – "will create and perpetuate … the bitter fruits of religious strife or sectarian intolerance".[19] And to make any kind of working relationship even more unlikely, Strathnairn thought it necessary one year later to advise the Lord Lieutenant that bringing to justice one of Leahy's parish priests (Fr Ryan of Cashel) and other unnamed Catholic clergymen who "excited and encouraged impulsive populations to treason and assassination" would be "one of the best guarantees … for the preservation of the Public Peace".[20]

Telegraph cable schemes
Having failed, even with the support of his contacts in high places, to secure employment in Ireland, and complaining that he was disadvantaged by the fact that the valuable references from Burgoyne and Lord Strathnairn which he had submitted to the Chief

Secretary's office in Dublin had not been returned to him,[21] Leahy resumed his earlier practice of submitting fanciful schemes for the support of the Foreign Office and the Colonial Office. Nine years after he had first sought government assistance for proposals to lay submarine telegraph cables, and a year after the Atlantic cable had been laid from Valencia to Newfoundland, he sought the assistance of Lord Stanley, then Foreign Secretary, in facilitating negotiations which he proposed to undertake with the governments of countries in North and South America and in Cuba "with the view of an extended system of telegraphic communication in that quarter". In September 1867, without making any enquiries about Leahy's credentials, his associates, or the financial resources available to him, Stanley provided him with an introductory letter to the various British representatives in those countries and told them that he had "no objection to your presenting Mr Leahy to the proper authority of the country in which you reside, leaving it to him to explain his plans and solicit the concurrence of that Government, but making it clearly understood that her Majesty's Government are in no way responsible either for the plan itself or for the execution of it".[22]

Two years later, in 1869, Leahy sought the patronage and support of Earl Granville, the Colonial Secretary, for a more elaborate scheme which would "present many advantages to so many British Colonies now without any means of telegraphic connection with this country".[23] Enclosing for reference a copy of Lord Stanley's letter, he reported that he had for some years been "engaged in establishing Telegraphic Communication throughout the entire of the West Indian Islands, the British and other South American Colonies and States, and thence by Submarine Cable from Brazil to Sierra Leone, Gambia, Morocco and Gibraltar to Falmouth" and claimed that this line would give "the West Indies and South America the shortest and best connection with Europe, free from accidents from Icebergs or Fisheries to which the existing Atlantic Cables are so liable". In assessing Leahy's proposal, a Colonial Office official noted that the colonies in the West Indies were free to make any arrangements they wished to facilitate telegraphic communication, provided that no exclusive privileges were given to any company, and that several of the colonies had already decided to subsidise the construction of cables. He noted also that Sir Charles Bright[24] of the Ocean Telegraph Company was already in the field and had visited most of the colonies, but still proposed that a reasonably helpful reply should be issued to Leahy. A more senior official, however, who obviously had a longer and better memory, advised that "if this is the Civil Engineer of Jamaica, of the same name, who was dismissed for misconduct, I think the answer should be a dry acknowledgement of receipt" and added that Lord Stanley would hardly have written his introductory letter in 1867 "if he had known the circumstances".[25] A simple acknowledgement was duly issued and when Leahy wrote again from Rome in December 1869 requesting a substantive reply,[26] the response was simply to send him a duplicate. By then, a cable had been laid from Florida to Cuba and one of Bright's companies – the West India and Panama Telegraph Company – had

Chapter XVI — Edmund Leahy, 1864-88

begun to lay more than 4,000 miles of submarine cables linking the various islands in the West Indies and connecting them with Cuba and the South American mainland; this scheme was successfully completed in 1872 and nothing has emerged to suggest that Leahy had any connection with it.

Undeterred by the fact that his 1869 proposals had effectively been ignored, and taking no account of the fact that some long lines of undersea cable were already being laid by well-established companies without subsidies of any kind,[27] Leahy wrote again to Earl Granville in June 1870 on behalf of *The British, African and South American Telegraph Company* of which he claimed to be the engineer, and asked for an indication of "how far we can reckon upon the support and co-operation of the Government" in the enterprise.[28] In addition, believing that the settlement at the Gambia, where the company intended to have "one of its chief stations", was likely to be transferred to France, he asked that any inter-governmental agreement should reserve for his company the privilege "of landing the Telegraph Cables there and maintaining them in and through the colony free from hindrance or obligations other than our Government would impose". Once again, Leahy's past history was the deciding factor in determining the response to these latest requests: "he is the wrong man in the wrong place" was the comment of one official, while another added that it would not, under any circumstances be thought proper to comply with his request. A letter of 4 July 1870 stating that the Colonial Secretary was unable to entertain his proposals effectively ended this series of efforts by Leahy to obtain official support for his attempts to develop a career in the rapidly expanding business of building a global communications network.[29]

Apart from letters which related to the loss of his position in Jamaica, the Colonial Office Daily Registers of Correspondence for the years 1871 to 1888[30] record only one further communication from Edmund Leahy – a letter of 15 February 1879 in which he submitted proposals for connecting Aden and Natal by a 5,000-mile-long submarine telegraph cable.[31] Similar proposals from a number of companies had been under consideration by the Government since 1873 but all of them were contingent on the provision of subsidies or guarantees towards the investment of well over one million pounds which would be involved.[32] An agreement was eventually made on 30 July 1879 between the Crown Agents for the Colonies and the Eastern Telegraph Company Ltd under which a cable was to be laid from Durban to Zanzibar and Aden and subsidised to the extent of £2,500 a year.[33] It seems very unlikely, however, that this was the scheme about which Leahy wrote in February 1879 or that he had any real involvement in the project which was successfully completed in the following year.

Possible tramway schemes
According to the evidence given by his friend and business associate, Henry Munster, to the Cashel Election Commissioners in 1869, Leahy had been planning to go to Australia

in mid-1868 "because he did not see his way to sufficient employment here".[34] Munster had strongly advised him not to go, because when he had previously been in a hot climate (in Jamaica) it had "nearly killed him", and he therefore came to an understanding with Leahy to supply him with funds for his personal purposes and to engage him in connection with some development projects of his own.[35] At a later stage, according to Munster's evidence, Leahy went to Brussels to examine that city's tramways "he being engaged in a scheme in which, to a certain extent, I agreed to co-operate – laying down tramways in Dublin" and involving "the construction of a patent omnibus of his".[36] A week later, Leahy was apparently in Vienna and intending to travel on to Constantinople[37] and in December he was staying at a hotel in Rome[38] where his brother, Archbishop Leahy, was attending the First Vatican Council. However, while the first horse-trams ran in Brussels in 1869, in Constantinople in 1871 and in Dublin in 1872,[39] there is no evidence that Leahy played any part in bringing any of these schemes into operation.

"in high spirits" in New York

Edmund Leahy was listed as a civil engineer in the Trades and Commercial sections of the *Post Office London Directory* for each of the years 1866 to 1871[40] and although this suggests that he was seeking professional engagements in the area in those years, no details of any such engagements have emerged. The fact that he was not listed in the Directory for any of the years 1872 to 1881 is *prima facie* evidence that he was not normally resident in London during those years. His brother Patrick, whose practice it was to list the addresses of family members in his Ordo, recorded an address for Edmund at 245 East 13th Street, 2nd Avenue, New York, in 1871[41] and, in December that year, Henry Munster (the Cashel election candidate) reported to the archbishop that he had learned through a solicitor that Leahy had written from New York "in high spirits and resolved to settle there".[42] But how long Leahy may have remained in New York, or what business may have taken him there, has not been established. He was not included in the 1880 Census for New York City and his name does not appear in any available directory at the address noted by his brother.[43]

Return to London

Leahy was back in London at the time of the 1881 Census of England.[44] He resumed his listing as a civil engineer in the Trades and Commercial sections of the *Post Office London Directory* for 1882 but was not listed in the Directory for any subsequent year. By then, he was 72 years of age and, with no record of achievement in the English engineering world in the previous two decades, it would have been very difficult for him to attract worthwhile commissions in competition with the growing number of well-educated younger men who then made up the profession. By 1885, he had taken up a position as manager of the Finsbury Park Turkish Baths in North London,[45] one of about 100 establishments of this kind which were set up in the city from 1860 onwards. Located at

Chapter XVI — Edmund Leahy, 1864-88

149 Fonthill Road, the baths managed by Leahy had opened in July 1883 and were reported to cater for both ladies and gentlemen.[46] They had a short life, surviving only until 1891.[47] By then, Edmund himself had passed on.

REFERENCES

1. Leahy was listed as a civil engineer in the Trades and Commercial sections of the *Post Office London Directory* for the years 1866 to 1871.
2. NAI, CSORP 1867/7297, letter of 17 April 1867 from Edmund Leahy
3. 25 & 26 Vict., c. 106
4. NAI, CSORP 1867/8223, letter of 26 April 1867 from Edmund Leahy
5. NAI, CSORP 1867/7666, letter of 26 April 1867 from Edmund Leahy and letter of 29 April 1867 from the Military Secretary to Lord Strathnairn
6. Leon Ó Broin, *Fenian Fever*, Chatto & Windus, London, 1971, pages 98-109
7. Another former county surveyor, Thomas Turner (1820-91), also applied unsuccessfully in the 1860s to be reappointed to a county surveyor post but was ultimately appointed to a post in County Dublin in 1883.
8. NLI, Earl of Mayo papers, Ms 11,154, letter of 26 August 1867 from Edmund Leahy to Sir Thomas Larcom and letter of 22 October 1867 from Leahy to the Earl of Mayo
9. ibid, letter of 20 December 1867 from Edmund Leahy, incorporating text of letter of 19 December from the Chief Secretary's Office
10. Peter Somerville-Large, *1854- 2004:The National Gallery of Ireland*, National Gallery of Ireland, 2004
11. NLI, Earl of Mayo papers, Ms 11,153, letters of 5 and 7 June 1867 from Edmund Leahy to Lord Naas
12. NAI, CSORP 1879/19262, letter of November 1879 listed in Chief Secretary's Office Register but not extant
13. ODNB, biographical note by Brian Robson; since Leahy had met him in Constantinople in the early 1850s, Rose had served with distinction as British Commissioner at the French headquarters during the Crimean War; subsequently, as commander of the Central India Field Force in 1857-58, he had played a major part in suppressing the Indian mutiny and served as Commander-in-Chief of the army in India from 1860 to 1865.
14. BL, Rose Papers, Add 42823, letter of 16 March 1867 from Strathnairn to the Duke of Cambridge
15. BL, Rose Papers, Add 42824, letter of 16 July 1867 to Edmund Leahy
16. See Chapter XVIII below
17. BL, Rose papers, Add 42796 - 42838
18. Moriarty is of course remembered for his statement in February 1867, following a minor outbreak of Fenian violence in Kerry, that "eternity is not long enough, nor hell hot enough, to punish such miscreants"; unlike Leahy, Moriarty had expressed support for a form of "concurrent endowment" scheme, as favoured by Strathnairn, instead of disestablishment.
19. BL, Rose Papers, Add 42825, letter of 3 November 1868 to the Countess of Portarlington
20. BL, Rose Papers, Add 42826, letter of 30 November 1869 to Earl Spencer
21. NLI, Mayo papers, Ms 11,169, letter of 28 December 1867 from Edmund Leahy to the Earl of Mayo
22. CO 318/257, copy of letter of 21 September 1867 signed by Lord Stanley and attached to letter of 12 August 1869 from Edmund Leahy to Earl Granville
23. CO 318/257, letter of 12 August 1869 from Edmund Leahy to Earl Granville
24. Charles Tilson Bright (1832 -1888) was appointed engineer–in–chief to the Atlantic Telegraph Company in 1856 at the age of 24 and was knighted in 1858 when it appeared that the cable laid that year was working satisfactorily; he was subsequently engaged in cable-laying in the Mediterranean, in the Persian Gulf and in the West Indies.
25. CO 318/257, internal minutes endorsed on letter of 12 August 1869 from Edmund Leahy to Earl Granville
26. CO 318/257, letter of 18 December 1869 from Edmund Leahy to Earl Granville
27. Charles Bright, *Submarine Telegraphs: Their History, Construction and Working*, London 1898
28. CO 87/98A, letter of 28 June 1870 from Edmund Leahy to Earl Granville
29. CO 415/1, letter of 4 July 1870 from the Colonial Office to Edmund Leahy
30. CO 382/24 to 382/42, Colonial Office Daily Registers of Correspondence, 1871 to 1888
31. CO 382/33, letter No 2503, Colonial Office Register of Correspondence, 1879
32. CO 882/4/1, Memorandum regarding proposed telegraphic cable from Aden via Mauritius and Natal, to the Cape, 26 February 1877

33. TNA, TS 18/245
34. *Report of the Cashel Election Inquiry Commission, Minutes of Evidence*, HC 1870 Vol XXXII (c. 9), page 316
35. ibid, pages 316 and 387
36. ibid, pages 215 and 392
37. ibid, page 393
38. CO 318/257, letter of 18 December 1869 from Edmund Leahy to Earl Granville
39. Michael Corcoran, *Through Streets Broad & Narrow: A History of Dublin Trams*, Midland Publishing, Leicester, 2000
40. *Post Office London Directory*, 1866 to 1871
41. CDA, Ordo of Archbishop Leahy for 1871
42. CDA 1871/79, letter of 9 December 1871 from Henry Munster to Archbishop Patrick Leahy
43. Research conducted in 1998-99 by Mrs Sheila Macauley
44. Census of England 1881, RG 11, Folio 0027/81, page 9
45. *Islington Street Directory, 1886; Kelly's London Suburban Directory, 1888*
46. *The Times*, 4 July 1883
47. By 1892, Nos 143-151 Fonthill Road were occupied by the London Co-Operative Supply Stores Ltd; the site is occupied today by the sometimes controversial North London Mosque.

CHAPTER XVII

THE LAST YEARS OF THE LEAHY FAMILY

Edmund
After their return from Jamaica in 1864, Edmund Leahy, with his wife, Juliet, and son Gerald, lived at 21 Cardington Street, London, close to Euston Station. While Edmund struggled throughout the 1860s to recover from his financially disastrous years in Jamaica and to create new employment opportunities for himself, his brother, Patrick, had established himself as one of the leading figures in the Catholic Church in Ireland. As Archbishop of Cashel and Emly, he worked closely with Cardinal Cullen, Archbishop of Dublin, on a variety of issues, including the controversial national education and university issues on which he had been active, in conjunction with Archbishop Slattery, since the mid-1840s. When he was about to travel to London in November 1865 for a meeting with the Home Secretary and other Ministers to discuss these issues, he wrote to Cullen, who was already in London, telling him that he could be contacted at 21 Cardington Street, "where my only brother resides".[1] After a few years, however, the archbishop seems to have given up the practice of lodging with his brother, possibly because Edmund's new residence at Heaton Road, Peckham Rye, a developing suburb in south-east London, would have provided less convenient lodgings; in February 1868, when he was being pressed by Cullen to travel to London again for another meeting with Ministers about the university question, he found it necessary to write to the Cardinal asking him to recommend a suitable place to stay![2] Edmund and his family were still living at Heaton Road in April 1871, when the census was taken[3] but he moved to New York later that year.[4]

The Leahy sisters
During the 1860s, Archbishop Leahy also kept in touch with his unmarried sisters whom he was continuing to support financially. Helena and Anne, who had been in London since the early 1850s, were living in Kensington in 1861 when a census was taken;[5] they both described themselves as "fundholders" in completing the return. The sisters moved soon afterwards to Ramsgate where they were joined by Susan after her return from Trinidad in 1863. In November of that year Helena died at the age of 49.[6] Anne and Susan then moved to Margate, a few miles away, and the archbishop visited them there in 1870.[7]

Chapter XVII — The last years of the Leahy Family

They were back again in London by the mid-1870s when at least one of them was living at 20, Upper Phillimore Place, off Kensington High Street.[8] She was joined there for a brief period around 1880 by her nephew, Albert William Denis Leahy - eldest son of Denis Leahy - who was already in practice as a surgeon.[9]

The Ryan Family

Margaret Leahy, the only sister who married, led a peripatetic life. She was married in the early 1840s to Patrick Ryan,[10] who had been in practice as a solicitor in Lower Gardiner Street, Dublin, since 1839. Ryan was subsequently in practice at East Main Street, Thurles,[11] and the family was living there in 1843.[12] From 1848 to 1852, the Ryans lived at 12 Marino Crescent, Dublin,[13] but by 1854 they had moved again to Sea View House, Ballyard, Tralee,[14] where Patrick Ryan died "after a lingering illness" in May 1855, aged 47.[15] Margaret continued to live in Tralee with her children at least until November 1857[16] but she too came to live in London in 1858 before moving in 1865 to Cork where she was still living in 1871.[17] She also appears to have lived in Boulogne at one stage when she and her family were being supported by her brother, Patrick.[18]

Margaret had at least four children, two boys and two girls. Her eldest son, James Patrick Ryan, born in 1843, studied medicine at the Catholic University Medical School in Dublin, and gained Licentiates of the Royal College of Surgeons (Ireland) and of the Royal College of Physicians (Ireland) in 1865. He worked initially in Los Angeles and in Chile and was an ambulance surgeon in the Franco-Prussian war of 1870-71. He practiced in the 1870s in Killarney where his address, according to the official Medical Register, was "The Palace",[19] presumably the residence of David Moriarty who was Bishop of Kerry from 1856 until his death in 1877. Ryan emigrated to Australia in 1873 and worked as an ophthalmic surgeon in Melbourne until his death on 19 June 1918.[20] Margaret's second son, George, born in 1845, also qualified as a medical doctor, having gained Licentiates of the Royal Colleges of Surgeons and Physicians, Edinburgh, in 1867. He practiced for a short time in Dublin before joining the Army Medical Service with which he served until 1899 when he had attained the rank of Lieutenant Colonel. In retirement, he lived with his wife, Louisa, at Ryde, in the Isle of Wight, where he died on 5 October 1904.[21]

In June 1873, at the Catholic University Church in Dublin, Archbishop Patrick Leahy officiated at the double wedding of his nephew, James Patrick Ryan, and his niece, Mary Elizabeth (Minnie), then aged 24.[22] James married a widow, Jane Mary Ward, daughter of James M Tidmarsh JP, a Limerick merchant, while Minnie married John Matthew Galgey, son of a Cork merchant, William Galgey, who had lived at Clifton Terrace, Cork, immediately adjoining the Leahys' former residence at Bruin Lodge, since 1845.[23]

Archbishop Leahy

Throughout the 1860s, Archbishop Patrick Leahy played a major role in ecclesiastical affairs and in political affairs at both national and local level. A major objective of his was achieved in 1869 when the Cathedral of the Assumption in Thurles opened for worship, although it was not until after the archbishop's death that the building was completed. Leahy had announced his plans in 1862 to replace what had come to be known as the Big Chapel, built in 1807, and entered into building contracts in 1865.[24] With his attachment to everything Roman, the archbishop clearly had a large influence on the design of his new cathedral and it has been said of him that he was "at least partly his own architect".[25] It has been suggested also that he was "conversant with architectural principles thanks to his own family background"[26] and that "he must have had more than the average layman's knowledge of building matters".[27]

Cathedral of the Assumption, Thurles (Lawrence 6923, courtesy of National Libraray of Ireland).

Chapter XVII — The last years of the Leahy Family

While it is reasonable to assume that the future archbishop would have acquired some general knowledge of these matters from his father and that he may occasionally have assisted, perhaps during Maynooth holiday periods, in the family's surveying and other activities, there is nothing to suggest that family members had any particular influence on the plans for the new cathedral. All the indications are that neither the archbishop's father nor his brothers had any experience or competence in architecture and, in any case, they were rarely together after 1847. In 1867, however, when little work had been done on the interior of the new building, Henry Munster (later the Cashel election candidate), heard Edmund Leahy speak of a letter which he had received from the archbishop "talking of the style in which the cathedral was to be built and speaking of Owen Jones's book". Jones was an English architect, best known for his interior decoration work, and it is likely that the archbishop's letter was referring to his most influential and important book, *The Grammar of Ornament*,[28] published in 1856. Munster, who had a copy of the book, directed Edmund to send it to his brother[29] but whether the interior of the Cathedral (which for many is its best feature) was actually influenced by Jones's ideas is a matter for expert assessment.

Another high point in Patrick Leahy's later years as archbishop was the important role he played in the first Vatican Council which began in Rome in December 1869. According to an often-quoted letter from Cardinal Cullen to his secretary in Dublin, Dr Leahy, who was an ardent supporter of Papal infallibility, made a magnificent 90-minute speech at the Council in May 1870 in which "he tore Dr McHale [Archbishop of Tuam] to pieces and he extinguished the Primate of Hungary".[30] An English bishop felt that Leahy had made "one of the most clear, solid and luminous speeches yet heard in the Council" while an American bishop told his secretary that Leahy's speech "threw Cardinal Cullen into the shade".[31]

Patrick Leahy's final years were marred by a series of illnesses and he died of heart disease at his residence in Thurles on 26 January 1875, leaving a gap in the ranks of the Roman Catholic Hierarchy which, according to *The Times,* could not easily be filled.[32] The *Cork Examiner* noted that his sister, Mrs Ryan, was with him when he died[33] but other extensive newspaper reports of the archbishop's death, elaborate obsequies and burial in his cathedral made no mention of the presence of particular family members and did not list the chief mourners.[34] However, Mark Tierney's biography of Archbishop Croke, who was appointed to succeed Leahy in June 1875, records that when Croke arrived in Thurles to take up his new office, he found that much of Leahy's furniture and belongings had been taken away from the Palace by his relatives.[35]

A Leahy household in Kensington

When the 1881 census was taken, the Leahy household at Upper Phillimore Place, Kensington, had expanded. Anne Leahy was listed as head of the household which also included her sisters Susan and Margaret (Ryan), and her brother Edmund; all four described themselves as "annuitants".[36] Edmund's wife and son were not with him in Kensington at that stage and were not recorded in the Census at any other address in England. While Edmund was said by his sister to be a widower in 1888,[37] the index to the deaths registered in England and Wales in the years subsequent to 1871 (when he was living with Juliet and Gerald in London) does not contain an entry for either of them. Thus, Juliet's fate and that of her stepson remains a mystery. So too does the fate of Margaret Ryan of whom no record has been traced after 1881.

Edmund, Anne and Susan Leahy went to live in an attractive end-of-terrace three-storey over basement house at 111 Abingdon Road, off Kensington High Street, sometime after 1881. Susan died of bronchitis at that address, aged 63, on 7 April 1883 and her death certificate records the informant as Edmund Leahy of the same address, who was present when she died.[38] A death notice in *The Times* of 26 April 1883 read: "On the 11th April at 111 Abingdon Road, Kensington, SUSAN LEAHY, sister of the late Archbishop of Cashel, RIP".

Edmund's tragic end

Having taken up the position of manager at the Finsbury Park Turkish Baths in the mid-1880s,[39] Edmund moved to nearby 145 Fonthill Road. This was a rapidly developing residential area about five miles north of the centre of London and adjoining Finsbury Park which had been laid out by the Metropolitan Board of Works in the 1870s. But Leahy did not live long to enjoy the amenities of the area. On Monday, 5 March 1888, a London evening newspaper reported that:

> The body of a man, whose name was only ascertained this morning, was found last night about half-past nine. The inspector at the Edgware-road Station of the Metropolitan Railway had his attention called to something lying on the metals on the City side near to the end of the platform, and on his proceeding there to ascertain what it was he found it to be a man who had been run over by a passing train. He summoned assistance, and on an examination being made it was found that life was extinct, the deceased having both legs crushed, besides other injuries. The police searched the body, which was that of a very respectably-dressed man, aged about 65, and on his clothing being searched, £6 in money and a railway ticket from High-street, Kensington, to King's-cross, were found. The body was removed on the ambulance to the Marylebone mortuary.[40]

It was quickly established that the body was that of Edmund Leahy and arrangements were made to hold an inquest which was reported as follows in *The Paddington Times*:

Chapter XVII — The last years of the Leahy Family

On Wednesday, Dr. Danford Thomas held an enquiry at the Ossington Coffee Tavern, Paradise Street, Marleybone, touching the death of Edmund Leahy, whose dead body was found on Sunday evening on the line at the Edgware-road Station of the Metropolitan Railway. It appeared that the deceased was a civil engineer, and was 75 years of age, and resided at 145, Fonthill-road, Finsbury Park. On Sunday last, as was his custom, he spent the day with his sister, Annie Leahy, at 111 Abingdon-road, Kensington. He was then in the best of health and spirits. He left the High-street Station, Kensington, by the train which leaves there at nine minutes past 8 in the evening, and in less than half-an-hour afterwards his dead body was found near the tunnel on the line at the station mentioned. The evidence seemed to shew that the deceased got out of the train when it arrived at the Edgware-road Station to obey a call of nature, and was about to get into the Hammersmith train which followed a few minutes afterwards, and which usually stops here for coaling, the carriages being partly in the tunnel. He was by some means knocked down by this train and run over by the Kensington train, which came next, arriving at the station at 8.30. The deceased's legs were severed in three places. The jury returned a verdict of accidental death. On the body was found £7-16s-8½d, a railway ticket, an ivory rule, keys, &c.[41]

A report in *The Borough of Marylebone Mercury*[42] added the information that it was Leahy's regular practice on Sunday evenings to travel on the Metropolitan Railway (the underground) from High Street Kensington, not far from his sister's house, to King's Cross station, from where a street tramway ran towards his residence in Fonthill Road.[43] Arthur Charles Croker, a retired army officer who had known Leahy for 30 years, gave evidence that he had seen him off by train from High Street Kensington, having booked to King's Cross. Given that he had left the train prematurely at Edgware Road, the question of possible suicide was addressed at the inquest, bearing in mind the fact that at least one suicide was occurring on the railways each week in the 1880s.[44] Annie Leahy, however, said in evidence that she had no reason whatever to believe that Edmund had contemplated suicide and this was accepted by the coroner; there was no indication that Leahy was depressed and the fact that he was carrying the equivalent of £650 sterling in today's

Extract from certificate of the death of Edmund Leahy in March 1888 at Edgware Road station on the Metropolitan Railway (RD Marylebone, Vol 1a, page 494).

money would suggest that he was certainly not in poor financial circumstances. The cause of death was certified as "Found dead; Shock and Exhaustion; Injuries to legs and head; Run over by Train; Accidental".[45]

Edmund Leahy was buried on 8 March 1888 in a single grave in St Mary's Roman Catholic Cemetery on the Harrow Road in north-west London where his sister Susan had been interred five years earlier.[46] The cemetery immediately adjoins the larger and better known General Cemetery of All Souls, Kensal Green, the final resting place of some of the major figures of Victorian engineering, including Sir Marc Isambard Brunel and his son, Isambard Kingdom Brunel, Joseph Locke, Thomas Brassey and Sir John Rennie, and of literary figures such as Thackeray, Dickens and Trollope.[47] Unfortunately, the headstone on the Leahy grave had almost fallen when last viewed in August 2005 and the inscription, if there was any, was then illegible. And even the grave itself may soon disappear from view as the authorities at St Mary's appear to have embarked on a policy of supplementing their existing restricted burial space by the simple expedient of dumping earth to a depth of several metres over the older graves. By coincidence, the Leahy grave immediately adjoins that of Dr Edmund O'Leary (1843-83), another native of Tipperary and half-brother of the prominent nationalist, John O'Leary.

The grave of Edmund Leahy at St Mary's Roman Catholic Cemetery, London, and, on the left, the grave of Dr Edmund O'Leary, half-brother of the nationalist, John O'Leary.

The last survivor
Anne Leahy was probably the last survivor of the eight children born to Patrick and Margaret Leahy between 1806 and 1823. At the time of the 1901 Census of England, she was living "on her own means" at 22 Spencer Square, Ramsgate, where she and some of

Chapter XVII — The last years of the Leahy Family

her sisters had resided in the 1860s.[48] When she died at that address on 8 February 1905, her death certificate gave her age as 86 and described her as a woman "of independent means".[49] Interestingly, however, the certificate added that Anne was "Daughter of Patrick Leahy, a Civil Engineer (deceased)". The informant was her 60 year old niece, Mary E Galgey, who lived with her husband John, a merchant and candle manufacturer, at Oakhurst, Rushbrooke Road Lower, Queenstown (Cobh), County Cork.[50] The Galgey's only child had died by the date of the 1911 Census.

REFERENCES

1. DDA Cullen papers, 327/1 (109), letter of 24 November 1865 from Archbishop Patrick Leahy to Cullen
2. DDA, Cullen papers, quoted in Emmet Larkin, *The Consolidation of the Roman Catholic Church in Ireland 1860-1870*, Gill and Macmillan, Dublin, 1987, page 513
3. Census of Population, parish of Camberwell, borough of Lambeth, RG 10 732
4. CDA, Ordo of Archbishop Leahy for 1871; CDA 1871/79, letter of 9 December 1871 from Henry Munster to Archbishop Patrick Leahy
5. Census of England 1861, RD Kensington RG09/2/83/54
6. Helena (Ellen) Leahy's death certificate (RD Thanet, Vol 2a, page 450) describes her occupation as "Daughter of Patrick Leahy, a Civil Engineer (deceased)" and records her death on 28 November 1863 at 2 Spencer Street, Ramsgate; a death notice in the *Cork Examiner* of 9 December 1863 referred incorrectly to her as the "youngest daughter" of the late P Leahy, County Engineer for Cork.
7. CDA, Ordo of Archbishop Leahy for 1870, entry for 27 July
8. *Post Office London Directory*, 1876 to 1882 lists "Miss Leahy" as the occupant
9. *Post Office London Directory*, 1880
10. Patrick Ryan, the first son of James Ryan and his wife, Elizabeth Stokes, of Listowel, County Kerry, was admitted to King's Inns in 1833 where he qualified as a solicitor (E Keane and others (eds), *The King's Inns Admission Papers, 1607-1867*, Irish Manuscripts Commission, 1982).
11. *Thoms Directory*, 1845-1855; *Pettigrew and Oulton Dublin Almanac*, 1839-1844; *Slater's National Commercial Directory of Ireland*, 1846
12. CDA, Slattery Papers, 1843/2, letter of 16 February 1843 from Archbishop Slattery to Dr Patrick Leahy
13. *Thoms Directory*, 1849, 1850 and 1851
14. *Thoms Directory*, 1854
15. CE, 1 June 1855, births, marriages and deaths column
16. CDA 1857/22, letter from Bishop David Moriarty of Kerry to Archbishop Patrick Leahy, 6 November 1857 ("I saw Margaret in Tralee on Wednesday. She and the children are well")
17. CDA 1858/30A, letter of 23 June 1858 from Edmund Leahy to Patrick Leahy; Ordo of Archbishop Leahy, 1864, 1865 and 1871
18. CDA, O'Carroll Diary, entry for 22 May 1863, quoted in Christopher O' Dwyer, *The Life of Dr Leahy, 1806-1875*, MA thesis, Maynooth, 1970; the typescript of the diary prepared by Monsignor Feehan in 1997 does not contain this entry but this is probably explained by Monsignor Feehan's statement that, when he worked on the diary, several of the pages were badly repaired, and some sections were missing, including pages 106-112 which cover the months of May and June 1863.
19. *The Medical Register for the UK, 1873 - 1876*
20. *The Medical Register for the UK, 1865-1918; The Irish Medical Directory, 1874-1878; The Medical Directory, 1912*; Laurence M Geary records that "Irish-trained doctors were largely responsible for the introduction of ophthalmology to Victoria" - see "Australia felix: Irish doctors in nineteenth-century Victoria" in Patrick O'Sullivan (ed), *The Irish in the new Communities*, Leicester University Press, 1992.
21. *The Medical Directory, 1896; Commissioned Officers in the Medical Service of the British Army, 1660-1960*, London, 1968, page 1868
22. General Register Office, Dublin, register of marriages, 1873; *Cork Examiner*, 16 June 1873, notice in marriages column
23. *Aldwell's General Post Office Directory of Cork*, 1845

24. James O'Toole, "The Cathedral of the Assumption: an outline of its history", in *Thurles: the Cathedral Town*, Geography Publications, Dublin, 1989, pages 121-125
25. ibid
26. Kevin Whelan, "The Catholic Church in County Tipperary, 1700-1900", in *Tipperary: History and Society*, ed William Nolan, Geography Publications, Dublin, 1985, page 250
27. James O' Toole, "The Cathedral of the Assumption: an outline of its history", in *Thurles: The Cathedral Town*, Geography Publications, Dublin, 1989, page 120; an obituary in the *Limerick Reporter & Tipperary Vindicator* (29 January 1875) went much further, claiming that "few better understood chemistry, geology, the science of steam, the practical details of mechanism, experimental philosophy, engineering, metallurgy etc than he. … He was an accomplished mapper and able architect and engineer also".
28. Owen Jones, *The Grammar of Ornament*, Day and Son, London, 1856
29. Evidence of Henry Munster, before the Cashel Election Commissioners, 21 October 1869, page 312
30. Peadar Mac Suibhne, *Paul Cullen and his contemporaries*, Vol V, Leinster Leader, Naas, 1977, page 108
31. Emmet Larkin, *The Roman Catholic Church and the Home Rule Movement in Ireland 1870-1874*, Gill and Macmillan, Dublin, 1990, page 15
32. *The Times*, 28 January 1875
33. CE, 27 January 1875
34. CE, 4 February 1875; *Cashel Gazette & Tipperary Reporter*, 30 January and 6 and 13 February 1875; *Tipperary Advocate*, 30 January and 6 February, 1875; *Limerick Reporter & Tipperary Vindicator*, 29 January 1875
35. Mark Tierney, *Croke of Cashel*, Gill & Macmillan, Dublin, 1976, page 80.
36. Census of England 1881, RG 11, Folio 0027/81, page 9
37. Evidence of Annie Leahy at the inquest on Edmund Leahy, *The Borough of Marylebone Mercury*, 10 March 1888; Edmund's marital status was not recorded at the 1881 Census.
38. Death Certificate issued by the General Register Office, RD Kensington, Vol 1a, page 65
39. *Islington Street Directory, 1886; Kelly's London Suburban Directory, 1888*
40. *The Globe and Traveller*, 5 March 1888
41. *The Paddington Times*, Saturday, 10 March 1888
42. *The Borough of Marylebone Mercury*, Saturday, 10 March 1888
43. London's Metropolitan Railway, the world's first underground railway, running from Paddington via Edgware Road and King's Cross to Farringdon Street, opened in January 1863; by 1884, when it came to form part of the Circle Line, this first section of the underground was being used by more than 500 trains each day.
44. *General Report to the Board of Trade upon the Accidents which have occurred on the Railways of the United Kingdom during the year 1888*, HC 1889 Vol LXVII (c.5836); *Returns of Accidents and Casualties as reported to the Board of Trade by the several Railway Companies in the United Kingdom during the three months ending 31st March 1888*, HC 1889 Vol LXXXVII (c.5402)
45. Death Certificate issued by the General Register Office, RD Marylebone, Vol 1a, page 494
46. Grave No 6154; St Mary's Cemetery comprises about 29 acres and was opened in 1858.
47. For an authoritative and comprehensive survey of London's burial grounds in the Victorian era, see James Stephens Curl, *The Victorian Celebration of Death*, Sutton Publishing Limited, 2000
48. Census of Population 1901, civil parish of Ramsgate, borough of Ramsgate, RG 13 826
49. Death Certificate issued by the General Register Office, RD Thanet, Vol 2a, page 685
50. Census of Ireland 1901 and 1911, DED Queenstown No 2 Urban , Townland of Ringmeen (part of)

CHAPTER XVIII

CONCLUSION

Apart from a small number of letters dating from the period 1855 to 1863, no private papers of members of the Leahy family have come to light. There are no known images, photographic or otherwise, of any member of the family except the Archbishop, and he is also the only one whose personality, appearance, attitudes and personal habits are noted in contemporary accounts. While the present study must therefore fall short of a full biography, it is possible to draw some conclusions on the basis of the substantial volume of official and business letters and other documents which are available.

Patrick Leahy senior – a competent all-rounder
Patrick Leahy senior might be described as a competent all-rounder in the first decades of the nineteenth century when the demarcation between professions was not as sharp as it was to become as the century progressed. As a land surveyor and cartographer, he prepared estate maps and maps of urban areas, including Waterford City and Clonmel, to a high standard but, in conjunction with these assignments, he was willing and able to offer advice on drainage, planting, land valuation, geology and mineralogical development, and to undertake infrastructural projects. He described himself as a civil engineer by 1818 and was associated not only with canal projects in the years 1809 to 1811 but also, in 1826, with the planning of one of the first railway schemes promoted in Ireland. In County Tipperary, where most of his early work was carried out, he appears to have been well thought of by the titled families and landed gentry - an essential requirement for any surveyor hoping to make a living at the time. When seeking a Government appointment in 1823, he was able to offer references from the Earl of Llandaff, the Earl of Clonmel, Lord Waterpark and Lord Norbury, the notorious hanging judge who had estates in Tipperary,[1] and his applications for a county surveyor post in 1817 and again in 1834 were similarly well supported. However, in spite of his good connections, he consistently failed in his attempts in the 1820s to gain recognition at national level, to win important commissions from the public authorities, or to establish himself as a significant figure in the emerging civil engineering profession.

Leahy's appointment in 1834 as county surveyor for the East Riding of Cork gained him a salaried post of considerable status for the first time in his life. While there were complaints about the inadequacy of the surveyors' official salary, there were opportunities to earn substantial additional income from private practice provided that official duties were not neglected and that conflicts of interest did not arise. It is likely that if he had taken up duty as county surveyor in the East Riding, without having his son Edmund assigned to the corresponding post in the West Riding, Leahy's previous experience would have enabled him to carry out his duties with reasonable efficiency and to the satisfaction of the grand jury. However, the family relationship between the two surveyors, and the presence with them in Cork of Denis and Matthew Leahy, provided opportunities and inducements which they obviously found to be irresistible. It made it possible for them to develop a very large private practice and to effectively subvert the provisions of the Grand Jury Acts under which county surveyors were prohibited from earning fees for designing projects that were to come before their own Grand Jury for sanction.

In all of this, Edmund Leahy appears to have been the dominant personality and the driving force in the family's various enterprises. That he and his father succeeded for over ten years in keeping at bay those members of the Cork Grand Jury who strongly disapproved of their conduct was remarkable, but the excesses of the younger man in his railway promotion activities in 1844-46 meant that there could be only one conclusion. If the relationship between Patrick and the Grand Jury had not already been eroded by more substantial controversies arising from the family's private practice activity, it seems unlikely that the complaint which ended his official career in 1846 would have been considered serious enough to warrant instant dismissal; instead, it provided a convenient legal basis to terminate a relationship which, already, had irretrievably broken down. That Patrick Leahy, at the age of 68, was forced to travel to South Africa to make a living, and that he should die there two years later and be buried in an unmarked grave, brought a sad end to the career of a proud man who had some lasting achievements to his credit, especially in respect of his maps.

Edmund Leahy
Of the four sons of Patrick and Margaret Leahy, Patrick and Edmund were clearly the most talented. Both of them were able communicators, and they were men of initiative, drive and determination. While many of his contemporaries, described Archbishop Leahy as "remarkable for his dignified bearing and uniform courtesy",[2] the diary of his critic, Rev James O'Carroll, records that he was a vain, proud and "would-be great man" with a pompous manner demonstrating his "pride and consequence".[3] Whichever of these descriptions was true of the archbishop, it seems clear that the image conjured up by O'Carroll would fit Edmund Leahy. Like many who practiced engineering in the first half of the nineteenth century, he lacked neither self-assurance nor self-confidence. But unlike many of the others, there was little to justify Leahy's self-importance or indeed, the arrogance, conceit and vanity which his conduct often demonstrated. To claim in 1859,

when he was 35 years old, that he had never been a subordinate in his life was strictly accurate, given that he had become an Irish county surveyor at only 20 years of age, had held that position until 1846 and had been working on his own account for the subsequent 13 years. On the other hand, to claim in 1863 that he was an experienced consulting engineer was misleading in the extreme: his career to that date bore no relationship to those within the engineering profession whose entitlement to that designation was hard-earned and widely recognised.

Leahy was clever and talented in many respects and hardly deserves the appellation "blundering buffoon" which Lee ascribed to him in discussing his engineering in 1844-45 of the Cork & Bandon railway.[4] He was, however, a willing participant in the excesses of the railway mania, and even in later years, his conduct was characterised by serious misrepresentation and deception. He promoted his various schemes with great skill, winning initial support from influential people and convincing gullible individuals that he had experience and skill and a record of achievement which he clearly did not have. His ambition far exceeded his ability. Like some of the other Victorian engineers and entrepreneurs, he continued throughout his life to put forward some improbable and far-fetched schemes which he had little prospect of ever carrying into effect. He was quick to attempt to enter new fields – railways in the 1840s, undersea telegraph cables in the late 1850s and urban tramways in the late 1860s – but left no lasting monument in any of them.

But was Edmund Leahy also a swindler like his contemporary John Sadleir (1813-56), "an Irishman from an obscure rural parish in west Tipperary whose wealthy connections, ruthlessness, ambition and speculative nature made him dream of vast wealth and high political office"?[5] There are undoubtedly similarities: like Leahy, Sadleir - "the Prince of Swindlers" - rose to prominence at a relatively early age by founding the Tipperary Joint Stock Bank in 1839. Like Leahy, he became heavily involved in railway promotion and development in the 1840s. But, with the possible exception of the tramway affair in Jamaica, Leahy's misdeeds were not in the same league as the corruption and fraud which marked Sadleir's career. Sadleir was known to Rev Patrick Leahy who was among the attendance when the Catholic Defence Association was launched in August 1851 with Sadleir as one of three joint treasurers,[6] and who lodged money sent to him by Edmund at Sadleir's bank in 1852. It is surprising, however, that Edmund seems never to have involved Sadleir in any of his schemes, although the latter was prominent in business and social life, and a Member of Parliament, when Leahy was attempting to build his new career in London in 1847-49.

Family attitudes

It was crucial for Patrick Leahy's survival in the business of land surveying and engineering in the early decades of the nineteenth century that he should be accepted by the gentry and major landowners as a reliable and loyal individual, without obvious associations with any of the national movements of the period. His support for the establishment, the forces of law and order and the Government were never in doubt. His new surveying methods were offered to the Government in 1821 primarily as objects of great value to the Army and the Navy. His 1834 proposals for mapping the baronies of Tipperary were put forward on the basis that they would assist police work in various ways and help to defeat conspiracies and combinations such as those engaged in by "a gang of desperate conspirators headed by the Keoghs" in the 1820s.[7] He took pride in the fact that, in the presence of Lord Norbury in the Privy Council Chamber, he was examined "on a very singular question" which had arisen in County Tipperary and which, he claimed, could only be decided on his evidence.[8] And in Cork, in 1836, he proclaimed his objective of ensuring that the entrances to the city would not be inferior to those of any other city in *our Empire*.

Edmund Leahy, like his father, was a life-long supporter and defender of the establishment on whose patronage he relied. He was, by any standards, an imperialist rather than a nationalist: his letters to British Ministers and officials from 1848 onwards are littered with references to *our Empire* and *our foreign possessions* and, for him, Ireland by the 1850s had become a "wretched country". He described the colonists at the Cape as "degenerate" individuals who failed to do their duty as loyal British subjects. For Jamaica, even after the excesses of the authorities in putting down the rebellion of 1865, he urged the establishment of a Military Government to "settle" the island, and subscribed to the fund set up to support the deposed Governor. To deal with a Kaffir uprising on the borders of the Cape Province, he advocated the use of artillery, even though opinion in Ireland, as expressed in the *Nation* newspaper a few years earlier, would have wished for a Kaffir success on the basis that the English had no more business at the Cape than in Ireland.[9] And even in countries outside the Empire, Leahy was quick to denigrate the natives and to adopt racist attitudes: the Greeks, for example, were bandits and scoundrels, fit only to be hunted through the mountains like wild boar, and the Turks were criticised for their "Levantine venality".

While the arguments which Edmund advanced for his various road, railway and other schemes, especially in Ireland, often centred on the creation of employment and wealth for local communities, there was never an indication that he had any real concern for the ordinary people of the areas in which he served, either in Ireland or abroad. By contrast, Archbishop Leahy, who was among the more nationalist of the Irish bishops in the 1860s and 1870s,[10] was willing to go further than many of his episcopal colleagues in leading and supporting crusades on matters which affected his flock, believing that it was his duty and that of his clergy to work for the temporal as well as the spiritual welfare of his people. He became increasingly concerned in the 1860s about the decline in the number of small tenant farmers and the alarming increase in emigration from rural areas, and believed that

In search of fame and fortune: the Leahy family of engineers, 1780-1888

"this wicked Anti-Catholic, Anti-Roman, Anti-Irish, Anti-everything-dear-to-us Government is looking on with delight, seeing that the direct, the certain effect of these laws [the Landlord and Tenant Acts] is to root out our Catholic people".[11] He initiated petitions in 1863 to promote tenants' rights and the disestablishment of the Church of Ireland and was among a small number of bishops present in December 1864 at the launch in Dublin of the National Association which was to campaign on these issues and in favour of denominational national education. In 1863, he strongly opposed a proposal by Archbishop Cullen to send an address on behalf of the Irish bishops to the Prince of Wales on the occasion of his forthcoming marriage, arguing that the proposal was quite uncalled for "while our poor countrymen are, many of them, starving or on the brink of starvation and others in thousands are flying from our shores to any part of the world where they can find the necessaries of life and all the while a wicked heartless Government is laughing with delight";[12] this was just one of a number of occasions on which he was not afraid to disagree strongly with Cullen, indicating that he was not "entirely Cullen's man" as Norman has described him.[13] Edmund Leahy, in the same year, was careful to send a subscription of three guineas from Jamaica to the Prince Consort Testimonial Fund which had been set up after the death of Prince Albert at the end of 1861.[14]

Serving the Empire

In the nineteenth century, Irish people from almost every section of society who might be constrained at home had access to the social and economic opportunities which the British Empire provided. Lawyers, doctors and engineers from Ireland accounted for significant proportions of those recruited for service in the colonies, particularly in the second half of the century.[15] Graduates of the School of Engineering, established at Trinity College Dublin in 1841, gained positions of authority and influence in colonial territories from the 1850s onwards when a Trinity diploma or degree was recognised as a passport to a rewarding career in an expanding imperial economy.[16] By 1861, more than half of the Trinity engineering graduates were working abroad, "the advance guard ... of the army of TCD men who were to provide the British Empire for nearly a century, in numbers far exceeding the proportion of the total roll of graduates of the British Isles, with administrators, jurists, doctors, missionaries and engineers".[17]

Before the age of the graduate engineer, however, and before the era of competitive examinations for public service positions, relatively few Irish engineers seem to have attempted, like the Leahys, to move from a career in Ireland to employment in the colonial service. But the Leahys' success in gaining such posts must be seen in context. By 1850, the patronage traditionally exercised by British Colonial Secretaries had diminished considerably and it had become the practice in the great majority of colonies for offices, other than that of Governor, to be filled from the colonists on the recommendation of the Governor.[18] Local selection was generally not possible, however, in colonies with tropical

Chapter XVIII — Conclusion

climates, such as Trinidad, where a pool of suitable candidates could not be found, and persons fit for particular offices had to be sent out from England.[19] When Matthew and Denis Leahy applied successively for the position of Colonial Engineer in Trinidad, they seemed to have faced little or no competition, probably because of the relatively small salary attaching to the post and the inhospitable environment which the island offered to Europeans - and which was to cost them both their lives. Edmund's original appointment in Jamaica was to a subsidiary position in what was then a depressed and turbulent colony, and he failed twice to gain the well-paid and much sought-after post of Colonial Engineer at the Cape of Good Hope. Thus, while the three brothers' appointments in the West Indies may appear at first sight to have been a remarkable achievement, it can also be seen as evidence that none of them was able to gain sufficient professional engagements of a more desirable kind to support themselves and their families in England, even though the 1850s and 1860s were, for the most part, years of economic prosperity and rapid development there.

What is remarkable however, is the extent of the access to senior figures in the Colonial Office and Foreign Office which some of the Leahys were able to achieve almost immediately after their move to London in 1848. Even after it had become a separate organisation in 1854, detached from the War Office, the Colonial Office was a small organisation, always under pressure to deal not only with business and correspondence originating in Britain but also with the numerous despatches from colonial governors to which a response had to be ready for issue by the next mail-packet. And yet, Secretaries of State and senior officials found time, over a period of 20 years or more, to grant interviews to Edmund Leahy, to read letters and lengthy submissions from him and personally to direct the form of the reply. Edmund, however, pressed matters too far at times, and seems to have been regarded as a nuisance by more than one Colonial Secretary: Earl Grey, for example, in 1851, directed that only a short formal reply should be sent to one of Leahy's submissions "considering what a troublesome man Mr Leahy is inclined to be"[20] and he responded to another submission by telling Leahy that it would require more consideration than he had time to devote to it.[21] On the other hand, Lord John Russell, as Foreign Secretary, caused his officials and the embassy at Constantinople to devote a considerable amount of time and effort between 1859 and 1864 to the pursuit with the Turkish Government of Leahy's claim for payment for his railway plans.

Lack of consultation and co-ordination between different branches of the British administration in the Victorian era is evident from the manner in which proposals from Edmund Leahy were dealt with. The fact that the Leahys had left Ireland in 1847 with serious questions to answer regarding the conduct of their official duties and the promotion of their railway schemes does not seem to have had any adverse effects on the consideration by the Colonial Office and Foreign Office of proposals submitted by Edmund in the following few years. The wildly exaggerated claims which he made in 1858 about his performance and achievements as a county surveyor in Ireland were never checked but seem to have been taken at face value when his applications for employment in the colonial service were being assessed. In September 1867, four years after he had been dismissed on foot of his role in the Jamaica tramway affair, the Colonial Office told Leahy unambiguously that he had no prospect of further employment in their sphere of operations[22] - and yet Lord Stanley, the Foreign Secretary, in the same month, provided Leahy with an introductory letter to the various British representatives in the West Indies and the Caribbean encouraging them to present him to the proper authorities so that he could promote one of his telegraph cable schemes.[23] Two years later, when a senior Colonial Office official who remembered Leahy's dismissal from his post in Jamaica saw a copy of the letter, he remarked that Stanley would hardly have written it "if he had known the circumstances".[24] But even this experienced and well-informed official had no way of knowing that the man with whom he was then in correspondence was, effectively "on the run" from the Cashel Election Commissioners who could obtain no definite information as to his whereabouts.

Exploiting the sources for engineering biography
While sons often followed their fathers into engineering careers in the nineteenth century, and two or more brothers working together in engineering was not unusual, there can be few other cases in which a father and three sons worked together, as the Leahys did in the 1830s and 1840s. However, their contribution to infrastructural development in Ireland was quite limited and their careers are of interest primarily because of the range of projects they attempted and because of their persistence in attempting to establish themselves in the years after 1847 in Europe, South Africa and the West Indies. Some of their contemporaries in Ireland, including a number of those others who became county surveyors in 1834, have left more substantial endowments of roads, bridges, other public works and buildings but, in most cases, the names of these men and details of their projects remain hidden in the archives.

The fact that so much of the Leahys' work, and so many of the schemes they proposed, involved public authorities in Ireland has allowed a great deal of information on their activities to be gleaned from surviving public records in the National Archives of Ireland. These same records contain a vast amount of interesting information about contemporaries of the Leahys. The Chief Secretary's Office Registered Papers and Official Papers, for example, include for each year numerous letters from surveyors and engineers, as well as letters from other correspondents, about a variety of infrastructural and architectural projects. The Letter Books and Minute Books of the Commissioners of Public Works and their predecessors record in great detail the progress of a large number of schemes carried out at local level with the aid of grants and loans from the Government, and illustrate the contributions made to these works by individual engineers and

Chapter XVIII — Conclusion

architects; the extensive collection of OPW Architectural and Engineering Drawings provides a valuable supplement to the written record in many cases. In time, sources like these will no doubt be exploited more fully to fill out the story of the physical development of nineteenth-century Ireland and of the men who were responsible for planning and carrying out that development.

The large numbers of Irish-born engineers who worked in the British colonies or elsewhere in the world in the nineteenth century have attracted even less attention in published work than those who remained at home. For these individuals, the Colonial Office papers relating to the different colonies, as well as the records of the Foreign Office, can provide valuable personal and other career information. Registers of Incoming Correspondence, Letter Books, and volumes of Dispatches to and from Governors, Ambassadors and Consuls are held at The National Archives in London (formerly the Public Record Office) for most colonies and for countries outside the Empire; the fact that these holdings are included in an on-line catalogue (PROCAT) and that individual volumes generally contain an index of names and subjects makes it possible to access relevant materials relatively quickly. These and similar sources in other repositories deserve to be better known and more extensively used to tell the story of this particular category of nineteenth-century Irish emigrants who contributed substantially to development in all five continents.

REFERENCES

1. John Toler (1745-1831), 1st Earl of Norbury, a native of Tipperary and a leading opponent of Catholic Emancipation, earned his reputation as a hanging judge in the aftermath of the 1798 rebellion.
2. ODNB, G C Boase, "Leahy, Patrick (1806-1875)", rev D Mark Tierney
3. CDA, O Carroll Diary, entries for 2 June and 24 September 1862
4. Joseph Lee, "The Construction Costs of Early Irish Railways 1830 -1853", *Business History*, Vol 9, 1967, pages 98 - 103
5. James O' Shea, *Prince of Swindlers: John Sadleir MP 1813 – 1856*, Geography Publications, Dublin, 1999, page viii
6. ibid, pages 187-189
7. NAI, CSORP 1834/903, letter of 21 February 1834 from Patrick Leahy; this may be related to a case, reported in *The Times* of 2 September 1816, in which a Special Commission at Clonmel under Lord Norbury convicted Patrick Keogh of destroying the dispensary at Bansha and the adjoining house; Keogh was taken from Clonmel to Ballagh, where the dispensary formerly stood, and was executed there.
8. NAI, CSORP 1823/6319, memorial of 31 July 1823 from Patrick Leahy
9. *The Nation*, 16 October 1847, page 860
10. Leahy was not, of course, a supporter of the more extreme nationalist movements: he deplored and was contemptuous of the Fenian rising which, he wrote in a pastoral letter, equalled in folly that of 1848 (*The Times*, 16 March 1867).
11. Letter of 27 March 1863 to Rev Tobias Kirby quoted in Emmet Larkin *The Consolidation of the Roman Catholic Church in Ireland, 1860-1870*, Gill and Macmillan, Dublin, 1987
12. DDA, Cullen papers, letter of 5 March 1863 from Leahy to Dr Cullen
13. E R Norman, *The Catholic Church and Ireland in the Age of Rebellion*, Longmans, Green and Co, London, 1965, page 13
14. CO 351/6, letter of 29 April 1863
15. Alvin Jackson, "Ireland, the Union, and the Empire, 1800-1960", in Kevin Kenny (ed), *Ireland and the British Empire*, Oxford University Press, 2004; David Fitzpatrick, "Ireland and the Empire", in Andrew Porter (ed), *The Oxford History of the British Empire, Volume III, The Nineteenth Century*, Oxford University Press, 1999

16. R C Cox, *Engineering at Trinity*, School of Engineering, TCD, Dublin, 1993
17. R B McDowell and D A Webb, *Trinity College Dublin, 1592-1952, An Academic History*, Cambridge,1982
18. *Report from the Select Committee on Official Salaries,* HC 1850 Vol XV (611), evidence given in June 1850 by the Colonial Secretary, Earl Grey
19. ibid
20. CO 48/322, letter of 25 November 1851 from CO to Edmund Leahy
21. CO 48/322, letter of 8 December 1851 from CO to Edmund Leahy
22. CO 137/429, letter of 12 September 1867 from CO to Edmund Leahy
23. CO 318/257, letter of 21 September 1867 signed by Lord Stanley, attached to letter of 12 August 1869 from Edmund Leahy
24. CO 318/257, minute endorsed on letter of 12 August 1869 from Edmund Leahy

APPENDIX I

FAMILY CORRESPONDENCE IN THE PAPERS OF ARCHBISHOP LEAHY

On the initiative of the late Dr Thomas Morris, Archbishop of Cashel and Emly, the papers of the archbishops of Cashel and Emly were catalogued and calendared by Dom Mark Tierney OSB in the 1960s. Copies of the calendars, with notes and indexes, are available at the National Library of Ireland. Special List No 171 relates to Dr Leahy's papers which may be viewed on microfilm at the Library (Pos. Nos. 6005 – 6010). While it has been stated that Dr Leahy took great care to preserve his papers intact (Mark Tierney, "Cashel Diocesan Archives", in *Thurles: the Cathedral Town: Essays in honour of Archbishop Morris*, ed William Corbett and William Nolan, Geography Publications, Dublin, 1989, pages 267 -275), this applies primarily to official papers and to the years after his episcopal ordination. Thus, while more than 900 items, mainly incoming correspondence, are included in the calendar of his papers, only 60 of these relate to the period before June 1857. Moreover, relatively few items of family correspondence have survived, mainly letters from Edmund and Matthew Leahy in the years 1855 to 1863. There are no letters from Denis Leahy or from any of the sisters whom the archbishop was supporting financially for so long.

CDA 1855/1: Letter of 8 April 1955 – Easter Monday – from Edmund Leahy, Constantinople, to The Very Rev Doctor Leahy VG, Vice President, Catholic University, Stephen's Green, Dublin

My Dear Pat,

I wish you a happy Easter and trust that this time of general rejoicing will find you free from that bad complaint under which you have been suffering.

I enclose you <u>Second</u> of Exchange for the £100 which 1 sent you on 25th March and if the <u>First</u> has come to hand safely, of course, the <u>Second</u> now forwarded, will be useless or unnecessary.

There is no news here – everything proceeds as usual, but it's supposed that the Allies will commence bombarding Sebastopol on the 11th inst. from the new line of forts and entrenchments which have been just completed. Whether the bombardment will lead to an assault on the town, remains to be seen. In the meantime fresh troops arrive every day on both tides, promising a bloody fight whenever the final struggle is made for Sebastopol.

Appendix I

Mat is very well, thank God, but the country is infested with banditti – in fact the Greek is naturally a bandit – Our Police have already shot 15 of them and hunt them night and day thro' the mountains – like wild boar. In the last affair 8 of the police were killed, they killing just an equal number of the robbers. Eventually we will clear the country of these scoundrels, but it is very disagreeable to be obliged to adopt such measure. These scoundrels of Bandits commonly bless themselves and make the sign of the cross before coming to the fight; in short such a combination of fanaticism, ignorance and villany could not I believe be found elsewhere. The Turks are bad enough, but those rascally Greeks are about the worst class of men of God's creation. Their cunning and quickness, which some people call natural talent, are not to be surpassed, but their vices surpass all other qualities.

I am anxious to hear from you as I have not had a letter for many posts – please send the enclosed to Juliet. I will send you more money in a week of two; but in the meantime if it was not giving you too much trouble I would like you to give Mr. James Keane of Cork £25 on my account and request him to let you know how I stand in his books after that payment. I will also ask you to remit a sum to Doctor Griffith of the Cape, as soon as I send you another remittance, <u>but not before,</u> for the purpose of having a Tomb build over our poor father…..

(Letter ends abruptly)

CDA 1855/2: Letter of 30 April 1855 from Edmund Leahy, Constantinople, to Very Rev Patrick Leahy

My dear Pat,

Herewith I send you First of Exchange for One Hundred and Fifty Pounds which you can apply in such way as you consider most useful for the girls and Juliet. Send Juliet £12 independent of the regular sum of £5 a month and the £10 of which I wrote to you in a former letter. I wish to give her the present £12 to dispose of for her wants such as clothes for herself and Gerald and if she wishes to giver her mother anything for the expense she has been to her – but you need not mention these things in your letter, only send her the £20 and don't say anything more than that; besides it, you will remit the £5 a month as previously arranged.

These two sums or debts which I wish to pay through you if you have no objection to the trouble viz. £50 to Troughton & Simms, the mathematical instrument makers of Fleet Street, London; and £12 to Sir Harry Smith, the late governor of the Cape of Good Hope. His address is Lieutenant General Sir Harry G. Smith, Bart., G.C.B., Horse Guards, London but I write to him myself by this post as well as to Troughton & Simms enclosing orders on you for the above sums and when you hear of them please to honour my drafts. Sir Harry's is a debt of honour – when we were in difficulties at the Cape, our poor Father wrote to him and Sir Harry very kindly enclosed him a check for £12 without our poor Father's even asking for it. If it was not inconvenient to you to drop him a line asking what bank you would lodge that sum to his credit, it would perhaps be the handsomest manner of acting. You need not write to the London people (T & S) as they will no doubt forward my order but which Sir Harry might not like to do. It is a pleasure to repay the latter.

I am sorry indeed to find by your last letter that your health is so much impaired but I hope that your fears are not well founded and by a little relaxation and the change of climate all may be well again. I would much like you to decide on meeting me in Italy and if you do, I may be able to meet you in Paris or perhaps London. I think you had better decide on this trip at once and write me immediately. The Turkish Government are nearly decided on the Railway and other works of the kind, and we are such good friends (particularly Rashid Pasha, the Grand Vizier to whom I may say I entirely owe our present independence) that I think they will not entrust those things to anyone from me – should they do so, I shall be obliged to go to London to negotiate those affairs on their behalf. In short, things look very satisfactory, Thank god.

Mat is very well and there is no danger whatever of him – it is unnecessary to have any apprehension of accident from these Robber Gangs who infect our country – I mean our mines. You are right about "Pelion". I sent Mat your note. I hope we shall have our Furnaces in full work before a month, and I expect they will yield a good income. I have some idea of sending Lord Carlisle a present of some Gold and Silver Bars from there as he took some interest in our affairs when here. If you see him, say our works are going on well. We have now about 1,000 men at work, besides the English workmen we brought out.

There is no news from Sebastopol – the siege drags itself slowly along without providing any suitable effect on the Russians. Lord Stratford is there at present, but is expected here today.

I cannot say that I entirely approve of the Girls staying in LONDON – I fear they are not able to contend with the rogues of London; and if they would occupy themselves in a respectable way in or near Dublin, I think it would be best for them

Sir C. Fox treated me badly, but I don't intend having any crossness with him. I have written to Mr. Brassy who is a better and more reasonable man and for whom I have a great respect. Fox's man, Mr. Stokes, who was here did me as much harm as he could but Thank God I have disappointed him.

I had a letter from Denis last week and I fear he is not doing much good. If the Railway etc succeeds, I can do something for him.

It is a pity your health should fail you just when prosperity begins to dawn upon us. I hope in God you will be able to shake off the complaint under which you are suffering and I think by adopting my advice of coming away to the climate of Italy for a while you will do more in that way to restore your bodily health than in any other manner.

My dear Pat

Your affectionate Brother

Edmund Leahy

Please forward the enclosed to Juliet under cover addressed to
The Lady Arthur Lennox,
21 Obington Square,
Brompton,
London.

Appendix I

CDA 1856/12: Letter of 21 December 1856 from Edmund Leahy, Pelion Mines, Zaghora, Volo to The Very Reverend Dr Leahy VG etc. etc. etc., Catholic University, St Stephens Green, Dublin

My dear Pat,

I wrote you by last post and have since suffered a good deal in mind by the accusation you brought against me on account of the poor girls. I sent you previously to receiving your last, a request to give them and Juliet each £20 for Christmas, which I hope they have received by this time. As far as my means justify I think I have not been ungenerous to anyone having a family claim on whatever I have. There is a difference however between generosity and folly and unless you would wish me to be worse than foolish, you certainly would not desire me to bring any of them here, at least if you understood (which you clearly do not) the difficulty of our position in this country. If you would do me justice, I think you would be one of the first to admit that I have not been unmindful of a brother's regard for his poor sisters, but although you appear to entertain a different opinion and although I have a great respect and regard for everything you tell me, yet my dear Pat, I cannot in my conscience accuse myself of neglecting our poor sisters. I am easy in my mind respecting such a charge. I cannot say that I am quite so blameless towards my unfortunate wife and it is hardly fair to throw those poor sisters between us whenever I show any disposition to do her some justice.

I do not see the connection clearly between your holding or resigning Cashel and my refusing to commit the folly of bringing one of our sisters to this country. They are all in England and you are in that wretched country Ireland, and although your relative places might remain unaltered, you might well hold Cashel if it was worth having, without bringing them or any of them with you there. I never asked you to take one of them. Of course if you found that you could do so, I would feel very happy, but if otherwise, you would never find me the least disposed to upbraid you or the least disposed to continue as heretofore to give them all the pecuniary aid in my power.

Your last letter has hurt me much more than perhaps you intended or wished, but if you still think that I deserve your condemnation for omitting to do my duty to the poor girls, I beg you to explain wherein I am at fault and I will endeavour to set myself right. You might surely charge Denis as well as me, for after all he is as much a brother as I am. For him too I would have been glad to do some good although I fear very much that any effort I could make for him would not be very successful.

Our affairs and connections here are far from being agreeable. I can tell you that I am not sleeping on a bed of roses, and am far from sure that I am not near a good deal of trouble. My life has been remarkably full of difficulties of all sorts and perhaps for that reason I am not so easily discouraged or frightened by misfortune. Still, I would be very much discouraged if you often wrote me in the like strain to your last.

Wishing you a happy Christmas and hoping that you are quite recovered from the dangerous fall you had.

I am my dear Pat

Your affectionate brother,

Edmund Leahy

Don't forget the usual address – Hotel l'Angleterre, Constantinople.

continued

Would you kindly enclose 1/4p worth of postage stamps in the accompanying letter to be forwarded to Sir H. Smith, etc. etc.

I wish very much I could persuade you to pass a little time with us on Mount Pelion. If you would, we will make an excursion to the vale of Tempé and all the classic quarters of Thessaly; which I am sure would do you much good in health and mind. E.L.

CDA 1856/12: Letter of 21 December 1856 from Matthew Leahy, Pelion Works, Zaghora, Thessaly, to Archbishop Patrick Leahy

My dear Pat,

It is a long time since you have sent me a line – you write frequently to Edmd. I wrote you a good while since from Constantinople when I mentioned about Denis's coming here to send me a couple of prayer books in English but you never said a word about them in you letter to Edmd. If there was an opportunity of sending them I would be very glad to have them. I said an ordinary Prayer Book, a Missal and perhaps "Chattenor" for a psalm reading each day. These would be well for others too.

We are very lonely here for the want of newspapers and I therefore hope you may be able to send us some soon. It is a great loss to us also not to have any of the other periodicals of the day – transactions of the different societies etc., which would enable one, though far removed from such societies, to know what they are doing and what else may be going on in the scientific world. It is the more necessary that we should be well acquainted with these things as we don't know how such knowledge may affect our present or future business. Will you therefore see if you can choose some one or two such publications as you may think most useful to us under such circumstances.

Denis I think has no notion of coming here. Even if he came, I am afraid he would not do much good for himself or us. However, he is a fool in his old days. If he came here, he certainly could do little more than act as a clerk, at least for some time, as he knows nothing of the works going on here and less of any foreign language.

The works are going on far better than when last I wrote you, some two or three months ago, but they require great care, supervision and constant fatigue on our part. I told you then that the works yielded about £400 per month, now they give about £1,000 and there is no knowing what the progress in this may be a few months hence but mines are not always to be reckoned upon as one does not know for certainty what the ground may give.

I am sorry you have made up your mind not to come and see us. You would like the trip and how delighted I would be to see you!

I hope you will remember to send us the Times and "Galligans Messenger". As for the poor girls coming here, how could they come or <u>live</u> here. No one to speak to all the day and sometimes for days. Every morning we go away down the mountain <u>1500</u> ft. and don't come back till night. The ground is too steep and rocky to ride and we return quite tired out. Sometimes we go off as we do tomorrow to see some of

Appendix I

the mines and don't return for three or four days. The country is beset with robbers and blaguards and we never stir without eight or ten guards all around as well as ourselves.

If you are writing to the girls or Denis, tell them I would write to them this post but have not time. Next post I will. Those boxes in the Custom Ho at London are very well worth £6 or much more. They contain very valuable works and most useful to us just now if we had them. There were in them a splendid work on railways published by "Mathieu" of Paris, Mosely's Mechanics of Engineering, Hall's Calculus, a splendid work of Buchanan Mechanics published by George Ressnie – and many other such. If you can you ought send us these. There was but one sofa left there – the other we gave as present to the steward. Remember me to Mrs Whelply.

Both of us quite well.

Your a.f.b.

Matthew Leahy.

CDA 1859/35: Letter of 26 November 1859 from Edmund Leahy, Spanishtown, Jamaica to Archbishop Patrick Leahy

My dear Pat

I have just received your letter in time to reply by return mail. You have not favoured me with many letters and I regret that the only one I have had for months should be so unreasonable.

You proposed to hold some money safely for me in <u>my own name</u> in the Bank of Ireland, and I sent it you in that understanding. You however applied it to your own use, and now when you <u>imagine</u> (for it is only imagination after all) that I was about to call upon you to pay a portion, you write me not only a passionate but an abusive letter.

Learn now what you do not appear to know, that your name is in no way, directly or indirectly, connected with Austin's Bills. The fellow's letter is vaguely worded, but as I have already said you nor your name has any existence on those Bills.

Mat's wants seem to require some aid, and I will thank you to let him have what his wants require debiting me with the amount.

Your affectionate brother
As ever
Edmund Leahy

Will you have the kindness not to delay forwarding this letter to Mat, for I am sure he will be very weary until he hears from me.

CDA 1860/12: Letter of 11 March 1860 from E Leahy, Spanish Town, Jamaica, to Archbishop Patrick Leahy

My Dear Pat,

I got your note of the 15th February last Tuesday.

I wrote you last, whilst smarting under your accusation of drawing a bill upon you, in favour of Austin, without your consent. Had you confined your abuse to the condemnation of such an act I would have simply assured you of a falsity and there would have been an end to the matter but it was too bad not only to assume as fact, that which was false; but in addition to insinuate, or more, that I had made a fool of you, or a tool of you, or something worse. You now write that you "will not say what I think of it" (my letter) "or what it gave me to feel – or what to think of your future". Enough. Men do greatly change in time.

Men do change indeed, or else you would not write the above. After a knowledge of the falseness of the accusation you brought against me, I might, I think, in common justice, expect an excuse or an apology, in place of aggravation. Am I to conclude that you still persist in flinging that false charge against me? You surely must know that you were in error on the 15th February when penning the lines I copy on the other page; for the bill itself was sent out here a month previously and an action now pends in the Law Courts of this island upon it. I almost fear to suppose by your act – else how could Austin find my address immediately after the date of your first letter?

My present means do not enable me to pay so large a bill and therefore I must let it take its course and abide the consequences; but if my present embarrassment and humiliation in my new office here, is in any way going to your writing to Austin, I can only say it is giving me a poor recompense for my endeavours to relieve you from difficulties. You know you offered to pay £100 to Austin when you were or said you were in difficulties; but I paid it myself out of my limited means in hand, sooner than cause you any inconvenience; although at the same time you owed me a large sum of money. The recollection is painful to me - but I hope you have not assisted Austin or in any way enabled him to cause my present difficulties.

You say you wish to know how much of my money you have. As well as I can at present find out, you owe me about £650. Allowing for all the payments you made to Juliet and Denis for me, and half your payments to the Girls.

You appear to forget that besides the larger sums, I sent you also many smaller ones – some for £250 – some for £50. And so on – not one of each, but many. Besides I think £300 or thereabouts, when you were appointed Archdeacon, which you required for expenses on taking charge. Of this latter, however, I intended you should repay yourself the bill I drew upon you from the Cape. But allowing for this, I find you owed me about £650 when I left England for here. What you have since given to Matt, and half the support we agreed upon for the Girls, should be taken out of it.

I can easily understand, that in the present circumstances of Ireland, and your diocese in particular, your wants may surpass your means. But you ought at least, bear me in mind; for although I would not

Appendix I

inconvenience you, yet you would be in error, if you supposed I was not in need – "hard up" would express my state better than anything else.

Matt's extravagance is unwarrantable. You ought endeavour to get him a <u>first</u> class professional place like inspector of railways. In the meantime you ought to give him a reasonable amount for support and charge to me. You say you suffer. <u>Have I not?</u>

Your affectionate brother,

E Leahy

APPENDIX II

MEMORANDA OF A VISIT TO THE SITE OF THE RUINS OF THE ANCIENT CITY OF SIZICUS IN ASIATIC TURKEY

BY E. LEAHY, C.E. 1857[1]

Communicated by Sir Roderick I. Murchison.

Left Constantinople for Panorma,[2] in my steamship *Star* on Saturday Morning, 23rd May, at 9 o'clock A.M. Panorma is about 70 miles S.W. by W. from the Seraglio Point, and is situated on the Asiatic shore of the Sea of Marmora. Population about 4000, one half being Christian and the other half Mahometan. The place is remarkable chiefly for its proximity to the site of the ancient city of Sizicus, and for some quarries of handsome red marble found in an insulated deposit of limestone at the southern suburbs of the town.

I was accompanied by Dr. and Mrs. Sarell, Mr. Philip Sarell, of the British Embassy, and the Greek Archbishop or "*Despot*" of Sizicus, and after a pleasant run of about nine hours we anchored opposite the town of Panorma at 6 o'clock P.M.

Next day the whole population of the town turned out to welcome the "Despot" (the general name in the East for bishop), who had not seen his flock for the last three years, being too much occupied with "important business" in Constantinople.

The Archbishop's influence procured us the best horses and saddles in the town, and, accompanied by him, we all started at 10 o'clock A.M. for the ruins of Sizicus, where we arrived in about an hour, the distance being only 5 miles. These ruins are situated on the N.E. end of the isthmus separating the peninsula of Artaki from the mainland.

Comparatively few traces of Sizicus now exist above ground; even the name would in all probability have long since been unknown in the country but for the creation of an archbishopric of the same name.

The most remarkable ruins are an aqueduct and some sarcophagi; the latter are, indeed, in such good preservation that they cannot well be called ruins. Close on the sea-shore are two of these tombs; they were lately uncovered, being only 3 to 4 feet below the surface, and are in fine preservation. The covering-lid of each sarcophagus is hewn out of one block of white marble, of which there are extensive beds in the adjacent island of Marmora, and each of those blocks must have weighed upwards of 20 tons. The interior of each sarcophagus was divided into two stories by a thick flagged floor, inserted in and supported by the side-walls. In the lower story were found eight, and in the upper seven, human skeletons. In the general outline the figure of the Egyptian sarcophagus in the British Museum is alike to those of Sizicus. The recesses of the architraves of the latter do not appear to have been quite finished, but the workmanship of the mouldings is excellent. One sarcophagus is ornamented with an "egg and dart" moulding, running quite around in full relief, and wrought as finely as anything of the sort, ancient or modern, within my knowledge.

The whole breadth of the Isthmus of Sizicus is covered with broken columns and massive walls faced with square blocks of black granite, and backed with rubble masonry set in lime cement. The walls are distinctly traceable across the isthmus, from sea to sea, at the junction of the peninsula of Artaki; and appear to have served as the line of fortifications for Sizicus facing the continent. The city extended from those walls into the peninsula of Artaki, and at a distance of about a mile there still remain the ruins of a large aqueduct, in many parts over 100 feet high.

Strabo represented the peninsula as an island, and there is a tradition amongst the present inhabitants that the sea formerly ran across the isthmus, and that ships passed and repassed; but, if ever such a communication existed, there is no trace of it to be found at present.

The chart of this coast (Sea of Marmora), published and compiled in 1830-31 by our Admiralty, from surveys purporting to have been made by French, Spanish, and English, is not as accurate as could be wished. This chart shows a considerable inlet or canal cutting the isthmus almost across, whilst in fact there is no inlet whatever. The isthmus maintains a regular breadth of about a mile, without any indent or projection from sea or land. There is a sort of marsh in the middle which becomes a lake in wet weather.

Another and more serious error of the chart is the rock which is represented *above water* 1 mile S.S.E. of Mola Island, at the entrance of the Bay of Panorma, and in the direct course from Constantinople to Panorma. There is no rock visible in that situation; but exactly in the same position there is a rock *having four feet water over it,* and is consequently very dangerous. Many vessels have been lost upon it, owing possibly to an over-confidence in the chart, and a consequent belief that *no dangerous* rock existed *under* water.

The Admiralty charts, prepared from the surveys of British officers, receive universal confidence for accuracy, and, in truth, they defy all criticism both for accuracy and clearness. An error of the sort just alluded to is the more dangerous because of this confidence, as foreigners unable to read English rely upon the Admiralty's well-known stamp, and discover perhaps when too late that the surveys were duly noticed in the title of the chart as not being made by the Admiralty officers. This chart of the Sea of Marmora ought to be corrected and republished.

It is possible that the rock alluded to, which is now under water, might have sunk since the time the surveys were made, and it is also possible that the same cause which depressed the rock might have elevated the isthmus of Artaki. This supposition would be partly consistent with the representations both of the chart and of the Strabo, but still not quite reconcilable with the comparatively recent date of the Admiralty chart.

At the same time, it may be well to remark that the earthquake of March, 1854, with which we in these localities were visited, caused many changes in the country between Artaki and Mount Olympus, so much so that the courses of some rivers near Brusa were entirely altered, and have remained so since.

Edmund Leahy.

REFERENCES

1. As published in *Proceedings of the Royal Geographical Society*, London, vol 2, 1858, pages 376 – 378; the original manuscript by Edmund Leahy is held in the archives of the Royal Geographical Society JMS/9/123.
2. Written as Pandermo in the original manuscript; now Bandirma

Bibliography

NEWSPAPERS

Borough of Marylebone Mercury
Cape of Good Hope and Port Natal Shipping and Mercantile Gazette
Cape Town Mail
Cashel Gazette
Cashel Gazette and Tipperary Reporter
Clonmel Advertiser
Cork Constitution
Cork Examiner
Cork Southern Reporter
Dublin Evening Post
Freeman's Journal
Globe and Traveller (London)
Irish Railway Gazette and General Commercial Advertiser
Jamaica Gleaner
Kerry Evening Post
Kerry Examiner
Limerick Reporter & Tipperary Vindicator
Paddington Times (London)
The Times
Tipperary & Clare Independent
Tipperary Advocate
Tipperary Free Press

PERIODICALS

Illustrated London News
Irish Builder and Engineer
Journal of the Irish Railway Record Society
Proceedings of the Institution of Civil Engineers
The Engineer
Transactions of the Institution of Civil Engineers of Ireland
Trinidad Royal Gazette

MANUSCRIPT SOURCES

NATIONAL ARCHIVES OF IRELAND
Census of Ireland, 1901, 1911
Chief Secretary's Office Registered Papers
Chief Secretary's Office Unregistered Papers
Council Office Papers

Bibliography

Letter Books of the Directors General of Inland Navigation, Southern District
Minute Book of the Commissioners for the Improvement of the Bogs in Ireland
Minute Books of the Directors General of Inland Navigation
Official Papers
OPW Architectural and Engineering Drawings
Ordnance Survey Letter Registers
Ordnance Survey Monthly Returns
Public Works Letter Books
Relief Commission papers

NATIONAL LIBRARY OF IRELAND
Earl of Mayo papers
Irish Road Maps
Larcom papers
Longfield Map Collection
William Smith O'Brien papers

THE NATIONAL ARCHIVES (PRO) LONDON
Board of Trade papers
Colonial Office papers
Foreign Office papers
Home Office papers
War Office papers

BRITISH LIBRARY
Journals of Charles Blacker Vignoles
Rose papers (Papers of Field-Marshal Lord Strathnairn)
Layard papers

OTHER REPOSITORIES
Archives of the Geological Survey of Ireland, Dublin
Archives of the Institution of Civil Engineers, London
Archives of the Institution of Engineers, Dublin
Birmingham City Archives
Cashel Diocesan Archives
Cork Archives Institute
Dublin Diocesan Archives
House of Lords Record Office, London
London Metropolitan Archives
National Archives of South Africa
Royal Geographical Society, London
South Tipperary County Museum, Clonmel
Tipperary Libraries, Thurles
Waterford City Archives

In search of fame and fortune: the Leahy family of engineers, 1780-1888

BRITISH PARLIAMENTARY PAPERS
Estimate of the Sum that will be necessary to complete the Surveys of the Districts, HC 1812 Vol V (94)
Inland Navigation: Papers relating thereto, July 1812, HC 1812 Vol V (366)
Third Report from the Committee on Inland Navigation in Ireland, HC 1812-13 Vol VI (284)
Reports of the Commissioners Appointed to Enquire into the Nature and Extent of the Several Bogs in Ireland and the Practicability of Draining and Cultivating them, HC 1810 Vol X (365) (First Report); HC 1810-11 Vol VI (96) (Second Report); HC 1813-14 Vol VI (130) (Third Report); HC 1813-14 Vol VI (131) (Fourth Report)
Report of the Board of Civil Engineers which sat at No 21 Mary Street, Dublin, from 23rd December 1817 to the 19th January 1818, HC 1818 Vol XVI (2)
Report from the Select Committee on the Survey and Valuation of Ireland, HC 1824 Vol VIII (445)
Eighth Report of the Commissioners of Irish Education Inquiry, HC 1826-27 Vol XIII (509)
Report from the Select Committee on Grand Jury Presentments, Ireland, HC 1826-27 Vol III (555)
Fifth Report of the Inspectors General on the General State of the Prisons of Ireland, HC 1827 Vol XI (471)
Seventh Report of the Inspector General on the General State of the Prisons of Ireland, HC 1829 Vol XIII (10)
Tenth Report of the Commissioners of the Irish Fisheries, HC 1829 Vol XIII (329)
Report on the Southern District in Ireland, HC 1829 Vol XXII (153)
Accounts of all sums advanced .....between 1819 and 1829 for public works in Ireland and for the employment of the Poor, HC 1829 Vol XXII (317)
Accounts of all Sums advanced by the Public to Grand Juries in Ireland during each of the last Ten years for Mail Coach Roads and Prisons, HC 1829 Vol XXII (327)
Report of the Select Committee on the State of the Poor in Ireland (Summary Report), HC 1830 Vol VII 667)
Eighth Report of the Inspectors General on the General State of the Prisons of Ireland, HC 1830 Vol XXIV (48)
Report on the Roads made at the Public Expense in the Southern Districts of Ireland by Richard Griffith, Civil Engineer, HC 1831 Vol XII (119)
Tenth Report of the Inspectors General on the General State of the Prisons of Ireland, HC 1831-32 Vol XXII (152)
Reports from the Select Committee on Public Works, Ireland, HC 1835 Vol XX (329) (573)
Annual Report of the Ecclesiastical Commissioners for Ireland 1834, HC 1835 Vol XXII (113)
First Report of the Commissioners of Municipal Corporations in Ireland, HC 1835 Vol XXVII (23)
Appendix to the First Report of the Commissioners of Municipal Corporations in Ireland, Part I, HC 1835 Vol XXVIII
Third Annual Report from the Board of Public Works in Ireland, HC 1835 Vol XXXVI (76)
Thirteenth Report of the Inspectors General on the General State of the Prisons of Ireland, HC 1835) Vol XXXVI (114)
Abstracts of Grand Jury Presentments in 1834, HC 1835 Vol XXXVII (220)
A Return of the Names of Each Engineer Employed under the Commissioners of Public Works in Ireland ... in the years 1832, 1833 and 1834, HC 1835 Vol XXXVII (536)
Report from the Select Committee on County Cess (Ireland), HC 1836 Vol XII (527)
Annual Report of the Ecclesiastical Commissioners for Ireland 1835, HC 1836 Vol XXV (130)
Abstracts of Grand Jury Presentments in 1835, HC 1836 Vol XLVII (119)
Third Report from the Select Committee on Fictitious Votes, Ireland, HC 1837 Vol XI (480)

Bibliography

First Report of the Commissioners of Inquiry into the state of the Irish Fisheries, HC 1837 Vol XXII (77)
Second Report of the Commissioners of Inquiry into the state of the Irish Fisheries, HC 1837 Vol XXII (82)
Abstracts of Grand Jury Presentments in 1836, HC 1837 Vol LI (110)
Annual Report of the Ecclesiastical Commissioners for Ireland to the Lord Lieutenant 1836, HC 1837-38 Vol XXVIII (53)
Report upon the experimental improvements on the Crown Estates at Pobble O'Keeffe in the County of Cork for 1837, HC 1837-38 Vol XLVI (69)
Return of all Sums of Money voted or applied either by way of Grant or Loan in Aid of Public Works in Ireland since the Union, HC 1839 Vol XLIV (540)
Abstracts of Grand Jury Presentments in 1838, HC 1839 Vol XLVII (654)
Abstracts of Grand Jury Presentments in 1839, HC 1840 Vol XLVIII (41)
Return of the Number of County Surveyors and of their Deputies and Clerks, 1834-1839, and of the sums presented for their salaries, office expenses, etc. HC 1840 Vol XLVIII (291)
Ninth Annual Report from the Board of Public Works in Ireland, HC 1841 Vol XII (252)
Valuations for Poor Rates, Ireland, Local Reports, Second Series, Part I, HC 1841 Vol XXII (308)
Reports relative to the Valuations for Poor Rates and to the Registered Elective Franchise in Ireland, HC 1841 Vol XXIII (326)
Abstracts of Grand Jury Presentments in 1840, HC 1841 Vol XXVII (143)
Report of the Commissioners appointed to Revise the Several Laws under or by virtue of which moneys are now raised by Grand Jury Presentment in Ireland, HC 1842 Vol XXIV (386)
Tenth Annual Report from the Board of Public Works in Ireland, HC 1842 Vol XXIV (384)
Abstracts of Grand Jury Presentments in 1841, HC 1842 Vol XXXVIII (90)
Eleventh Annual Report of the Board of Public Works in Ireland, HC 1843 Vol XXVIII (467)
Abstracts of Grand Jury Presentments in 1842, HC 1843 Vol L (146)
Select Committee on Townland Valuation of Ireland, HC 1844 Vol VII (513)
Report of the Commissioner for inquiring into the Execution of the Contracts for Certain Union Workhouses in Ireland, HC 1844 Vol XXX (562)
Return of the Number of Days appointed by the Sheriff for Transacting the Fiscal Business in each County in Ireland, HC 1844 Vol XLIII (130)
Abstracts of Grand Jury Presentments in 1843, HC 1844 Vol XLIII (194)
Return of Numbers of Churches and Chapels built, rebuilt or enlarged in each Diocese of Ireland since September 1833, HC 1844 Vol XLIII (279)
Returns of the Expenses of the Ecclesiastical Commissioners for Ireland for each of the last six years, HC 1844 Vol XLIII (319)
Copies of Correspondence between Her Majesty's Government in Ireland and the Grand Juries assembled at Summer Sessions 1844, on the subject of providing additional accommodation for Pauper Lunatics, HC 1844 Vol XLIII (603)
Evidence taken before the Commissioners appointed to Inquire into the Occupation of Land in Ireland, HC 1845 Vol XXI (657)
Report of the Commissioner appointed to enquire into the Execution of the Contracts for Certain Union Workhouses in Ireland, with Copy of Treasury Minute thereon, 20 March 1845, HC 1845 Vol XXVI (170)
Thirteenth Annual of the Board of Works in Ireland, HC 1845 Vol XXVI (640)
Second Report of the Commissioners appointed to Inquire into the State and Condition of the Tidal and Other Harbours, Shores and Navigable Rivers of Great Britain and Ireland, HC 1846 Vol XVIII (692)

Return of the Number of Inmates in Irish Workhouses, HC 1846 Vol XXXVI (297)
Third Report from the House of Lords Select Committee on Colonisation from Ireland, HC 1849 Vol XI (86)
Correspondence with the Governors of the Cape of Good Hope and Ceylon, respecting the Transportation of Convicts to those Colonies, etc HC 1849 Vol XLIII (217)
Report from the Select Committee on Official Salaries, HC 1850 Vol XV (611)
Copies or Extracts of Despatches relative to Convict Discipline and the Employment of Colonial Convicts in the formation and Improvement of Roads at the Cape, HC 1850 Vol XXXVIII (104)
Return of the White and Coloured Population of the Colony of the Cape of Good Hope, HC 1852 Vol XXXIII (124)
Returns of the several Sums of Money advanced by the Treasury for the Improvement of Tralee Harbour and for opening a Canal there, etc, HC 1852-53 Vol XCIV (781)
Metropolitan Commission of Sewers - Accounts, in Abstract, of Receipts and Expenditure during the Year 1854 etc, Schedule of Contracts for the performance of Special Works, HC 1854-55 Vol LIII (175)
Report of the Commissioners for Inquiring into the Erection of District Lunatic Asylums in Ireland, HC 1856 Vol LIII (9)
Returns of the Dates of Commissions issued by the Irish Government for the examination of Candidates for the Office of County Surveyor in Ireland, HC 1856 Vol LIII (335)
Report from the Select Committee on County & District Surveyors etc (Ireland), HC 1857 Session 2 Vol IX (270)
Report of the Metropolitan Board of Works for the year to 30 June 1857, HC 1857 (Sess 2) Vol XLI (234)
Returns by the Engineer of the Metropolitan Board of Works ... regarding the Ranelagh Sewer, HL 1857-58 Vol XVIII (221) (265)
Report from the General Committee of Elections on the Cashel Election Petition, HC 1857-58 Vol XII (157)
Returns of all the Small Constituencies which contain less than 400 Electors, HC 1859 (Session 2) Vol XXVI (77)
Report for the Year 1859 ... on the State of Her Majesty's Colonial Possessions, HC 1861 Vol XL (2841)
Report for the Year 1860.... on the State of Her Majesty's Colonial Possessions, HC 1862 Vol XXXVI (2955)
Report for the Year 1861.... on the State of Her Majesty's Colonial Possessions, HC 1863 Vol XXXIX (3165)
Names of County Surveyors in each County in Ireland and their Assistants, etc. HC 1863 Vol L (277)
Reports and Evidence relating to the Spanish Town and Porus Tramway in Jamaica, HL 1864 Vol XIII (254)
Report for the Year 1862 ....on the State of Her Majesty's Colonial Possessions, HC 1864 Vol XL (3304)
Report for the Year 1863 ....on the State of Her Majesty's Colonial Possessions, HC 1865 Vol XXXVII (3423)
Reports and Evidence relating to the Spanish Town and Porus Tramway in Jamaica, HL 1865 Vol XL (257)
Report of the Jamaica Royal Commission 1866, HC 1866 Vol XXX (3683)
Papers relating to the Disturbances in Jamaica (Part I), HC 1866 Vol LI (3594)
Further Papers relating to the Disturbances in Jamaica, HC 1866 Vol LI (3594 I)
Minutes of Evidence taken before the Select Committee on the Tipperary Election Petition, with the Proceedings of the Committee, HC 1867 Vol VIII (221)
Further Correspondence respecting the Spanish Town and Porus Tramway in Jamaica, HL 1867 Vol XIII (211)

Bibliography

Copy of Minutes of the Evidence taken at the Trial of the Sligo Borough Election Petition 1869, HC 1868-69 Vol XLIX (85)

Copy of the Minutes of the Evidence taken at the Trial of the Cashel Election Petitions (1869), HC 1868-69 Vol XLIX (121)

Report of the Commissioners appointed... for the purpose of making Inquiry into the existence of Corrupt Practices at the last Election for Cashel, together with the Minutes of Evidence, HC 1870 Vol XXXII (c.9)

Return of the Accounts of the Sums expended in connection with the Election Inquiries at Dublin, Sligo and Cashel, HC 1872 Vol XLVII (105)

Copies of the Shorthand Writer's Notes of the Judgments delivered by the Judges selected in pursuance of the Parliamentary Elections Act, 1868, for the Trial of Election Petitions since April 1869, etc, HC 1872 Vol XLVII (268)

Returns of Accidents and Casualties as reported to the Board of Trade by the several Railway Companies in the United Kingdom during the three months ending 31st March 1888, HC 1889 Vol LXXXVII (c.5402)

General Report to the Board of Trade upon the Accidents which have occurred on the Railways of the United Kingdom during the year 1888, HC 1889 Vol LXVII (c.5836)

LEGISLATION

An Act for Granting to His Majesty the Sum of Five Hundred Thousand Pounds for promoting Inland Navigation in Ireland, 40 Geo III c. 51 (Ir)

An Act to amend the Laws for improving and keeping in Repair the Post Roads in Ireland, 45 Geo. III, c. 43

An Act to provide for the more deliberate Investigation of Presentments to be made by Grand Juries for Roads and Public Works in Ireland, and for accounting for Money raised by such Presentments, 57 Geo III, c.107

An Act to suspend, until the End of the present Session of Parliament, the Operation of an Act made in the last Session of Parliament, to provide for the more deliberate Investigation of Presentments to be made by Grand Juries for Roads and Public Works in Ireland, and for accounting for Money raised by such Presentments, 58 Geo III, c.2

An Act to provide for the more deliberate Investigation of Presentments to be made by Grand Juries for Roads and Public Works in Ireland, and for accounting for Money raised by such Presentments, 58 Geo III, c.67

Weights and Measures Act, 1825, 6 Geo. IV, c.12

An Act for Making and Maintaining a Navigable Cut or Canal ... and for otherwise improving the said Harbour of Tralee, 9 Geo. IV, c.cxviii

Representation of the People Act, 1832, 2 & 3 Wm IV c. 45

Composition for Tithes Ireland, Act, 1832, 2 & 3 Will. IV, c. 119

An Act to alter and amend the laws relating to the Temporalities of the Church in Ireland, 3 & 4 Wm. IV, c.37

Grand Jury (Ireland) Act, 1833, 3 & 4 Wm IV, c.78

Grand Jury (Ireland) Act, 1836, 6 & 7 Wm. IV, c.116

Cork & Passage Railway Act 1837, 7 Wm. IV & 1 Vict., c. cviii

Poor Relief (Ireland) Act, 1838, 1 & 2 Vict., c 56

An Act to Consolidate and Amend the Laws for the Regulation of Grand Jury Presentments in the County of Dublin, 7 & 8 Vict., c.106

In search of fame and fortune: the Leahy family of engineers, 1780-1888

An Act for making a Railway from Cork to Bandon, 8 & 9 Vict., c.cxxii
An Act for more effectual Inquiry into the Existence of Corrupt Practices at Elections for Members to serve in Parliament, 1852, 15 & 16 Vict., c.57
Grand Jury (Ireland) Act,1853, 16 & 17 Vict., c. 136
Corrupt Practices Prevention Act, 1854, 17 & 18 Vict., c.102
County Surveyors (Ireland) Act, 1861 24 & 25 Vict., c.63
County Surveyors (Ireland) Act, 1862, 25 & 26 Vict., c. 106
Representation of the People (Ireland) Act, 1868 (31 & 32 Vict., c. 49
The Election Petitions and Corrupt Practices at Elections Act, 1868, 31 & 32 Vict., c. 125
An Act to disfranchise the Boroughs of Sligo and Cashel, 33 & 34 Vict., c.38
Parliamentary and Municipal Elections Act, 1872, 35 & 36 Vict., c.33

DIRECTORIES

Aldwell's General Post Office Directory of Cork, 1845
Commissioned Officers in the Medical Service of the British Army, 1660-1960
Cork Post Office General Directory, 1844-1845
Islington Street Directory
Kelly's London Suburban Directory
Pettigrew and Oulton Dublin Almanac
Post Office London Directory
Slater's National Commercial Directory of Ireland, 1846
The Irish Medical Directory
The Medical Directory (UK)
The Medical Register for the UK
The Medical Who's Who
The Quarterly Indian Army List
Thoms Directory

BOOKS AND OTHER PUBLICATIONS, PRE 1900

Bandon Navigation Committee, *Bandon Navigation: Report of the Proceedings of the Committee appointed 17 December 1841*
Bright, Charles, *Submarine Telegraphs: Their History, Construction and Working*, London 1898
Carbery, Lord, *Observations on the Grand Jury System of Ireland with Suggestions for its Improvement*, London 1831
Chatterton, Lady Georgina *Rambles in the South of Ireland during the Year 1838*, Vol I, London 1838
Clarke, Hyde, "The Imperial Ottoman Smyrna and Aidin Railway", in *The Levant Quarterly Review*, Vol II, No III, January 1861
Close, Col Sir Charles, *The Early Years of the Ordnance* Survey, David & Charles Reprints, 1969,
Coldwell's Tables for Reducing Irish Money into British Currency, second edition, printed by T Coldwell, 21 Batchelor's Walk, Dublin, 1826
Coldwell's Tables of the Weights and Measures, Dublin, 1826
Country Gentleman, *Hints on the System of Road Making*, Dublin, 1829
Francis, John, *A History of the English Railway*, London, 1851
Froude, J A, *The English in the West Indies*, Longmans, Green and Co., London, 1888
Gardner, W J, *A History of Jamaica from its discovery by Christopher Columbus to the present time*, London, 1873

Bibliography

Grieg, William, *Strictures on Road Police* (sic), Dublin, 1818
Helps, Arthur, *The Life and Labours of Mr Brassey*, London, 1872
Hume, Hamilton, *The Life of Edward John Eyre, late Governor of Jamaica*, London, 1867
Illustrated Official Handbook of the Cape and South Africa, London, 1893
Jones, Owen, *The Grammar of Ornament*, Day and Son, London, 1856
Kane, Sir Robert, *The Industrial Resources of Ireland*, Dublin, 1844
Keating, Rev M J, *Suggestions for Revision of the Irish Grand Jury Law* (nd)
Lardner, Rev Dionysius, *The Steam Engine familiarly explained and illustrated*, (5th edition), London 1836
Leahy, Edmund, *A Practical Treatise on Making and Repairing Roads, Illustrated by engravings and tables*, John Weale, Architectural Library, London, 1844
Leahy, Edmund, *Leahy v. The Ottoman Government: A Statement of the Claim of Edmund Leahy, Esq., CE, upon the Ottoman Government, in the matter of the Belgrade and Constantinople Railway*, I R Taylor, London, (?) 1859
Leahy, Patrick, Civil Engineer etc, *New and General Tables of Weights & Measures with ample calculations and suitable examples under each head*, Bentham and Hardy, Dublin, 1826
Lewis, Samuel, *A Topographical Dictionary of Ireland*, London, 1838
Markham, Captain A H, "The Arctic Campaign of 1879 in the Barents Sea", in *Proceedings of the Royal Geographical Society*, New Series, Vol II, 1880
Mitchel, John, *Jail Journal*, M H Gill, Dublin, 1913
Mullins, M B, Memoir of Barry Duncan Gibbons, *TICEI*, Vol VII, 1864
Mullins, M B, Presidential Address, *TICEI*, Vol VI, 1863
Murphy, Walter, *Remarks on the Irish Grand Jury System*, Cork, 1849
Neville, John, "Grand Jury Laws and County Public Works, Ireland", *Dublin University Magazine*, March 1846, Vol XXVII, No CLIX
Newenham, Thomas, *A View of the Natural, Political and Commercial Circumstances of Ireland*, 1809
Nimmo, Alexander, "Report of Alexander Nimmo, Civil Engineer MRIA, FRSE etc On the Proposed Railway between Limerick and Waterford, addressed to the Secretary of the Hibernian Railway Company", *Dublin Philosophical Journal*, No III, February 1826
Parnell, Sir Henry, *A Treatise on Roads*, (2nd edition), London, 1838
Personal Recollections of English Engineers, by a Civil Engineer, London, 1868
Price, George, *Jamaica and the Colonial Office: Who caused the Crisis*, Sampson Low, Son and Marston, London, 1866
Russell, W H, *The War* (2 vols), London 1855, 1856
Smith, Adam, *The Wealth of Nations Books IV-V*, Penguin Classics, 1999
Stanley, William Ford, *Surveying and Levelling Instruments Theoretically and Practically Described*, London, 1890
Stephenson, Sir Rowland Macdonald, *Railways in Turkey: Remarks upon the practicability and advantage of railway communication in European and Asiatic Turkey*, John Weale, London, 1859
Tables for reducing Irish Money into British Currency, Simms and McIntyre, Belfast, 1826
Tables of Currency, Coins, Weights and Measures etc, J Charles, Dublin, 1826
Townsend, Horatio, *A General and Statistical Survey of the County of Cork*, Cork, 1815
Wrottesley, George, *The Life and Correspondence of Field Marshal Sir John Fox Burgoyne* (2 vols), London, 1873
Young, Arthur, *A Tour in Ireland 1776-1779* (2 vols), London, 1780

BOOKS AND OTHER PUBLICATIONS, POST 1900

Andrews, J H, *A Paper Landscape: The Ordnance Survey in Nineteenth –Century Ireland* (2nd edition), Four Courts Press, 2002

Andrews, J H, "David Aher and Hill Clements's Map of County Kilkenny 1812-24", in W Nolan and K Whelan (eds), *Kilkenny: History and Society*, Geography Publications, Dublin, 1990

Andrews, J H, "Notes for a Future Edition of Brian Friel's *Translations*", in *The Irish Review*, No 13, Winter 1992-93

Andrews, J H, *Plantation Acres*, Ulster Historical Foundation, 1985

Andrews, J H, "Road Planning in Ireland before the Railway Age", *Irish Geography*, Vol 5, No 1, 1964

Andrews, J H, "The Longfield Maps in the National Library of Ireland: An Agenda for Research", *Irish Geography*, Vol 24 (1), 1991

Boase, G C, "Leahy, Patrick (1806-1875)", rev D Mark Tierney, *ODNB*, 2004

Broderick, David, *The First Toll-Roads: Ireland's Turnpike Roads 1729-1858*, Collins Press, Cork, 2002

Brodie, Antonia and others (eds), *Directory of British Architects 1834- 1914*, Continuum, London, 2001

Bull, E, "Aided Irish Immigration to the Cape 1823-1900", in *South African-Irish Studies*, Volume 2, 1992

Burke, Rev William P, *History of Clonmel*, Clonmel, 1907

Canning, Rev Bernard J, *Bishops of Ireland 1870-1987*, Donegal Democrat, Ballyshannon, 1987

Carbery, Lord, *Observations on the Grand Jury System of Ireland with Suggestions for its Improvement*, London 1831

Carter, Ernest F, *An Historical Geography of the Railways of the British Isles*, Cassell, London, 1959

Foley, Con, *A History of Douglas*, second revised edition, Cork, 1991

Colvin, Howard, *A Biographical Dictionary of British Architects, 1600-1840* (3rd edition), Yale University Press, 1995

Comerford, R V, *The Fenians in Context: Irish Politics and Society 1848-82*, Wolfhound Press, Dublin, 1985

Comerford, R V, "Tipperary Representation at Westminster 1801- 1918", in *Tipperary: History and Society*, ed William Nolan, Geography Publications, Dublin, 1985

Compton, Piers, *Cardigan of Balaclava*, London, 1972

Condon, James, "Mid-Nineteenth Century Thurles - The Visual Dimension", in *Thurles: The Cathedral Town, Essays in honour of Archbishop Thomas Morris*, Geography Publications, Dublin, 1989

Conlan, Patrick OFM, "The Franciscan House in Thurles" in *Thurles: The Cathedral Town, Essays in honour of Archbishop Thomas Morris*, Geography Publications, Dublin, 1989

Conroy, J C, *A History of Railways in Ireland*, Longmans, Green and Co Ltd., London, 1928

Cooke, Brian, *The Grand Crimean Central Railway*, Cavalier House, Cheshire, 2nd edition, 1997

Corcoran, Michael, *Through Streets Broad & Narrow: A History of Dublin Trams*, Midland Publishing, Leicester, 2000

Cox, R C, *Engineering at Trinity*, School of Engineering, TCD, Dublin, 1993

Cox, R C, "Robert Mallet, Engineer and Scientist", in *Mallet Centenary Seminar Papers*, Institution of Engineers of Ireland, Dublin, 1982

Cox, R C, *The School of Engineering, Trinity College Dublin: A Record of Past and Present Students 1841-1966* (4th edition), 1966

Craig, Maurice, *The Architecture of Ireland from the earliest times to 1980*, (revised edition) Batsford, London, 1989

Bibliography

Crampton, R J, *A Short History of Modern Bulgaria*, Cambridge, 1987
Creedon, C, *Cork City Railway Stations 1849-1985* (3rd edition), Cork, 1986
Creedon, C, *The Cork, Bandon and South Coast Railway,* Vol I, 1849-1899, Cork, 1986
Crossman, Virginia, *Local Government in Nineteenth-Century Ireland*, Institute of Irish Studies, Belfast, 1994
Curl, James Stephens, *The Victorian Celebration of Death*, Sutton Publishing Limited, 2000
Delany, V T H & D R, *The Canals of the South of Ireland*, David & Charles, Newton Abbot, 1966
Department of the Environment , Dublin, *Reports on inspection, assessment and rehabilitation of masonry arch bridges and of concrete bridges*, Dublin, 1988 and 1990
Devery, Kieran, "The function of hotels in parliamentary borough elections in mid-nineteenth century Ireland", in *Irish History: A Research Yearbook No 2*, Four Courts Press, 2003
Dickason, G B, *Irish Settlers to the Cape*, A A Balkema, Cape Town, 1973
Dickson, David, *Old World Colony: Cork and South Munster 1630-1830*, Cork University Press, 2005
Donnelly, James S Jr, *The Land and People of Nineteenth-Century Cork*, Routledge & Kegan Paul, London, 1975
Eden, Peter (ed), *Dictionary of Land Surveyors and Local Cartographers of Great Britain and Ireland, 1550-1850*, Kent, 1979
Fitzpatrick, "David, Ireland and the Empire", in Andrew Porter (ed), *The Oxford History of the British Empire, Volume III, The Nineteenth Century*, Oxford University Press, 1999
Flanagan, P J, "The Nore Viaduct, Thomastown", *JIRRS* Vol 9, No 52, June 1970
Fogarty, Rev Philip, "St Patrick's College, Thurles", MS in St Patrick's College Thurles
Ford, Joseph C, *Handbook of Jamaica, 1910*, Government Printing Office, Jamaica, 1910
Foster, R F, "An Irish Power in London": making it in the Victorian metropolis, in Fintan Cullen and R F Foster, *"Conquering England": Ireland in Victorian London*, National Portrait Gallery, London, 2005
Freely, John, *Inside the Seraglio*, Penguin Books, London, 2000
Friel, Brian, *Translations*, Faber and Faber, London, 1981
Geoffrey Dutton, *The Hero as Murderer*, London, 1967
Gold, Mick, "The Doomed City", in Paul Kerr et al, *The Crimean War*, Boxtree, London, 1997
Griffiths, A R G, *The Irish Board of Works, 1831-1878*, Garland Publishing, New York, 1987
Hall, Douglas, *Free Jamaica 1838 – 1865: An Economic History*, Yale University Press, 1959
Haslip, Joan, *The Sultan: The Life and Times of Abdul Hamid II*, Weidenfeld and Nicholson, London, 1958
Hasluck, Frederick William, *Cyzicus: being some account of the history and antiquities of that city, and of the districts adjacent to it*, Cambridge University Press, 1910
Herries Davies, Gordon L, *Sheets of Many Colours: The Mapping of Ireland's Rocks 1750-1890*, Royal Dublin Society, 1983
Hickey, Patrick, *Famine in West Cork: The Mizen Peninsula, Land and People, 1800 – 1852*, Mercier Press, Cork, 2002
Hoppen, K Theodore, "National Politics and Local Realities in Mid-Nineteenth Century Ireland", in *Studies in Irish History*, University College Dublin, 1979
Hoppen, K Theodore, *Elections, Politics and Society in Ireland, 1832–1885*, Clarendon Press, Oxford, 1984
Horner, Arnold , "Napoleon's Irish Legacy: the Bogs Commissioners, 1809-14", in *History Ireland*, Vol 13, No 5, September/October 2005

In search of fame and fortune: the Leahy family of engineers, 1780-1888

Irish, Bill, *Shipbuilding in Waterford: A historical, technical and pictorial study*, Wordwell Ltd, Bray, 2001
Jackson, Alvin, "Ireland, the Union, and the Empire, 1800-1960", in Kevin Kenny (ed), *Ireland and the British Empire*, Oxford University Press, 2004
James, J G, "The Evolution of Wooden Bridge Trusses to 1850", *Journal of the Institute of Wood Science*, Vol 9, No 3, June 1982
Johnson, Stephen, *Lost Railways of Co. Cork*, Stenlake Publishing Ltd, 2005
Jordan, Thomas E, *An Imaginative Empiricist – Thomas Aiskew Larcom(1801-1879) and Victorian Ireland*, Edwin Mellen Press, New York, 2002
Keane, E and others (eds), *The King's Inns Admission Papers, 1607-1867*, Irish Manuscripts Commission, 1982
Kelly, Liam, "A History of the Port of Tralee", in *Blennerville: Gateway to Tralee's Past*, Tralee, 1989
Kennedy, J M, *A Chronology of Thurles, 580 – 1978*, (revised and updated), Cavan, 1979
Kerr, Ian J, *Building the Railways of the Raj, 1850-1900*, Oxford India Paperbacks, Delhi, 1997
Kinross, Lord, *The Ottoman Empire*, The Folio Society, London, 2003
Lanigan, Anne, "The Workhouse Child in Thurles, 1840-1880", in *Thurles: The Cathedral Town, Essays in honour of Archbishop Thomas Morris*, Geography Publications, Dublin, 1989
Larkin, Emmet, *The Consolidation of the Roman Catholic Church in Ireland, 1860-1870*, Gill and Macmillan, Dublin, 1987
Larkin, Emmet, *The Roman Catholic Church and the Home Rule Movement in Ireland, 1870-74*, Gill & Macmillan, Dublin, 1990
Lee, Joseph, "The Construction Costs of Early Irish Railways 1830 -1853", *Business History*, Vol 9, 1967
Lincoln, Colm, *Steps and Steeples, Cork at the turn of the Century*, O' Brien Press, Dublin, 1980
Loeber, Rolf, *An Alphabetical List of Irish Architects, Craftsmen and Engineers from the 15th to the 20th century*, Irish Georgian Society, 1973
Lohan, Rena, *Guide to the Archives of the Office of Public Works*, Stationery Office, Dublin, 1994
Mac Suibhne, Peadar, *Paul Cullen and his contemporaries*, Vol V, Leinster Leader, Naas, 1977
MacLysaght, Edward, "Survey of Documents in Private Keeping, First Series", *Analecta Hibernica*, No 15, 1944, page 389
Marnane, Denis G, "John Davis White of Cashel (1829-1893)", *Tipperary Historical Journal*, 1994
Marnane, Denis G, "John Davis White's Sixty Years in Cashel", *Tipperary Historical Journal*, 2004
Marsh, Philip, *Beatty's Railway*, New Cherwell Press, Oxford, 2000
Mathieson, W L, *The Sugar Colonies and Governor Eyre 1849-66*, London, 1936
Mc Neill, D B, "Waterford & Central Ireland Railway", JIRRS Vol 13, No 74, October 1977
McCracken, Donal P, "The Odd Man Out: The South African Experience", in Andy Bielenberg (ed), *The Irish Diaspora*, Pearson Education Limited, 2000
McDowell, R B and D A Webb, *Trinity College Dublin 1592-1952: An Academic History*, Cambridge, 1982
McDowell, *The Irish Administration, 1801-1914*, Routledge & Kegan Paul, London, 1964
McMorran, Russell and Maurice O'Keeffe, *A Pictorial History of Tralee*, Tralee, 2005
Moody, T W and others (eds), *A New History of Ireland*, Vol 9, Maps, Genealogies, Lists, Oxford, 1984
Murphy, David, *The Arctic Fox, Francis Leopold McClintock, Discoverer of the fate of Franklin*, Collins Press, Cork, 2004
Murphy, Donal A, *The Two Tipperarys*, Relay Publications, Nenagh, 1994

Bibliography

Newman, Jeremiah, *Maynooth and Georgian Ireland*, Kenny's Bookshops and Art Galleries, Galway, 1979

Nolan, David M, *The County Cork Grand Jury, 1836-1899*, MA Thesis, UCC, 1974

Nolan, William (ed), *Tipperary: History and Society*, Geography Publications, Dublin, 1985

Nolan, William and K Whelan (eds), *Kilkenny: History and Society*, Geography Publications, Dublin, 1990

Nolan, William, "A Public Benefit": Sir Vere Hunt, Bart. and the Town of New Birmingham, Co. Tipperary, 1800-18, in *Surveying Ireland's Past: Multidisciplinary Essays in Honour of Anngret Simms*, Geography Publications, Dublin, 2004

Nolan, William, "Literary Sources", in *Irish Towns: A Guide to Sources*, Geography Publications, Dublin, 1998

Norman, E R, *The Catholic Church and Ireland in the Age of Rebellion*, Longmans, Green and Co, London, 1965

Ó Broin, Leon, *Fenian Fever: An Anglo-American Dilemma*, Chatto & Windus, London, 1971

Ó Dúgáin, An Bráthar Seán, "The Christian Brothers in Thurles", in *Thurles: The Cathedral Town, Essays in honour of Archbishop Thomas Morris*, Geography Publications, Dublin, 1989

O'Callaghan, Antoin, *Of Timber, Iron and Stone: A Journey through time on the Bridges of Cork*, Cork, 1991

O Donoghue, Brendan, "Bandon's Second Railway", in *Bandon Historical Journal*, No 10, 1994

O Donoghue, Brendan, "The Office of County Surveyor: Origins and Early Years", in *TICEI*, Vol 117, 1992

O'Donoghue, D J, *History of Bandon*, Cork Historical Guides Committee, 1970

O'Dwyer, Christopher, *The Life of Dr Leahy, 1806-1875*, MA thesis, St Patrick's College, Maynooth, 1970

O'Dwyer, Christy, "The Beleaguered Fortress: St Patrick's College, Thurles, 1837-1988", in *Thurles: The Cathedral Town, Essays in honour of Archbishop Thomas Morris*, Geography Publications, Dublin, 1989

O'Dwyer, Frederick, *The Architecture of Deane & Woodward*, Cork University Press, 1997

O'Dwyer, Frederick, "The Architecture of the Board of Public Works 1831-1923", in *Public Works, The Architecture of the OPW, 1831-1987*, Architectural Association of Ireland, 1987

O'Dwyer, Frederick, "The Foundation and Early Years of the RIAI", in *150 Years of Architecture in Ireland*, RIAI, 1989

O'Keeffe Peter and Tom Simmington, *Irish Stone Bridges: history and heritage*, Irish Academic Press, 1991

O'Keeffe, Peter J, *Ireland's Principal Roads 1608-1898*, National Roads Authority, Dublin, 2003

O'Keeffe, Peter J, "Richard Griffith: Planner and Builder of Roads", in *Richard Griffith Centenary Symposium Papers*, RDS 1980

O'Keeffe, Peter J, *Roads in Ireland: Past and Ahead*, An Foras Forbartha, 1985

O'Leary, Cornelius, *The Elimination of Corrupt Practices in British Elections 1868 – 1911*, Clarendon Press, Oxford, 1962

O'Shea, James, *Priests, Politics and Society in Post-famine Ireland: A Study of County Tipperary 1850-1891*, Wolfhound Press, Dublin, 1983

O'Shea, James, *Prince of Swindlers: John Sadleir MP 1813 – 1856*, Geography Publications, Dublin, 1999

O'Sullivan, Patrick (ed), *The Irish in the New Communities*, Leicester University Press, 1992

In search of fame and fortune: the Leahy family of engineers, 1780-1888

O'Toole, James, "The Cathedral of the Assumption: an outline of its history", in *Thurles: the Cathedral Town, Essays in honour of Archbishop Thomas Morris*, Geography Publications, Dublin, 1989
Olivier, Lord Sydney, *Jamaica: The Blessed Island*, Faber & Faber, London, 1936
Olivier, Lord Sydney, *The Myth of Governor Eyre*, Hogarth Press, London, 1933
Ponting, Clive, *The Crimean War*, Chatto & Windus, London, 2004
Power, Patrick C, *History of South Tipperary*, Mercier Press, Cork, 1989
Prunty, Jacinta, *Maps and Map-Making in Local History*, Four Courts Press, Dublin, 2004
Quane, Michael, "The Free School of Clonmel", JCHAS Vol LXIX, No 29, January – June 1964
Ruddock, Ted, biographical note on Alexander Nimmo, in *A Biographical Dictionary of Civil Engineers in Great Britain and Ireland, 1500- 1830*, Thomas Telford Publishing, London, 2002
Ryan, Father Francis, "St Patrick's College, Thurles", *Capuchin Annual*, 1960
Rynne, Colin, *The Industrial Archaeology of Cork City and its Environs*, Stationery Office, Dublin, 1999
Shee, Elizabeth and S J Watson, *Clonmel: An Architectural Guide*, An Taisce, Dublin, 1975
Shepherd, Ernie, *The Cork Bandon & South Coast Railway*, Midland Publishing, Leicester, 2005
Sherlock, Philip, *West Indian Nations - A New History*, London, 1973
Simmons, Jack and Gordon Biddle, *The Oxford Companion to British Railway History*, Oxford, 1997
Skempton, A W and others (eds), *A Biographical Dictionary of Civil Engineers in Great Britain and Ireland, Volume 1: 1550-1830*, Thomas Telford Publishing, London, 2002
Somerville-Large, Peter, *1854- 2004:The National Gallery of Ireland*, National Gallery of Ireland, 2004
South Africa Official Yearbook, 1979
Steele Gordon, John, *A Thread Across the Ocean*, Simon & Schuster, London, 2002
Stenton, Michael (ed), *Who's Who of British Members of Parliament, Vol I, 1832 – 1885*, Sussex, 1976
Talbot, Fred A, *Cassels Railways of the World* (2 vols) London nd
The Dictionary of Land Surveyors and Local Map-makers of Great Britain and Ireland, 1530-1850 (2nd edition), The British Library, 1997
Tierney, Mark, OSB, *Calendar of the Papers of Dr. Leahy, Archbishop of Cashel,1857-1875*, NLI Special List 171
Townsend, Horatio, *A General and Statistical Survey of the County of Cork*, Cork, 1815
Vaughan, W E and A J Fitzpatrick, *Irish Historical Statistics: Population, 1821-1921*, Royal Irish Academy, Dublin, 1978
Walker, Brian M (ed), *Parliamentary Election Results in Ireland, 1801-1922*, Royal Irish Academy, Dublin, 1978
Walsh, Malachy, "Cork Bridges", *The Engineers' Journal*, August, 1981
Watson, J G, *The Institution of Civil Engineers: A Short History*, Thomas Telford Ltd., London, 1988
Webb, Sidney and Beatrice, *English Local Government: Statutory Authorities for Special Purposes*, Longmans, Green and Co, London, 1922
Whelan, Kevin, "The Catholic Church in County Tipperary, 1700-1900", in Nolan, William (ed), *Tipperary: History and Society*, Geography Publications, Dublin, 1985
Williams, R A, *The Berehaven Copper Mines*, The Northern Mine Research Society, Sheffield, 1991
Wynne Jones, Joan, *The Abiding Enchantment of Curragh Chase – A Big House Remembered*, 1983

Index

Abbeyfeale, 133
Aberdeen, Earl of, 182
Adare Parish, 27 (note 3)
Aden, 160
Adrianople (Edirne), 175, 185-186
Aher, David, 8, 11, 17
Aidin, 187
Albert Bridge, 45, 50 (note 85)
Albert, Prince, 297
Alexandria, 160
Alfred, Prince, 214
Allihies Copper Mines, 80, 95 (note 63)
Allihies, 80-82
Allow River, 89
Anahala Bog, 80
Anderson, Charles Frederick, 37
Anderson, Elizabeth, 202, 213-214, 217-218
Andrews, J H, x-xi, 6, 18
Anne Street, Clonmel, 36, 43
Anner River, 45
Antrim County, 33, 57
Archdall, Captain Mervyn MP, 223, 274
Arima, 210
Arouca, 210
Askeaton, 3, 5
Assistant county surveyors, 63-65
Athy, 8
Atkins, William, 98
Australia, 278-279
Austria, 175, 182
Baghdad, 176
Bagwell, John, 41
Balaklava, 182, 221
Bald, William, 22, 27
Balkan Mountains, 185
Ballinadee, 80
Ballincollig, 76, 84
Ballingeary, 80
Ballinspittle, 80, 86
Ballydehob, 79
Ballydonegan Bay, 80
Ballydoyle, 8

Ballyhooley, 74
Ballymacquirk Bridge, 89
Ballyphehane, 129
Ballyvourney, 80
Baltimore, 98
Bandon Bridge, 87-88
Bandon Navigation, 118-121
Bandon River, 82-83, 118-121
Bandon, 73-74, 76, 80, 82-84, 99, 118, 129
Bandon, Earl of, 67, 133
Bann River, 12, 87
Bannister, John Henry, 218, 220 (note 79)
Banteer, 89
Bantry Bay, 147
Bantry, 68, 74, 79-80, 88, 98-99, 129, 147
Bantry, Earl of, 67
Baptist Grange, 45
Barleycove, 79
Barlow, W H, 242
Barrow River, 9-10
Basra, 176
Bazalgette, J W, 202
Beara Peninsula, 61, 80
Beatty, James, 182
Belfast, public buildings, 105
Belgooly, 86
Belgrade, 176, 179, 186, 189
Belgrade-Constantinople Railway, 176-180, 184-189
Benson, Sir John, 68, 97, 105, 152-153
Berehaven, Lord, 67
Berehaven, see Castletownbere
Bermuda, 166
Bernard, Hon H B, 120
Bernard, Viscount, 67, 121, 136
Besika Bay, 180
Betts, Edward, 182
Bianconi, Charles, 36, 41
Birkenhead, 41
Birmingham & Gloucester Railway, 137-138
Birmingham, 12, 39
Birr Castle, 45
Black Sea, 180
Blackpool, 127
Blackwater River, 74, 86, 88-89
Blarney, 84

Index

Blennerville, 113, 116
Blunt, Vice-Consul, 192
Board of Longitude, 14
Bogs Commissioners, 11-13
Bolton & Bury Canal, 38
Bombay, 160
Borlin Valley, 79
Bosporus, 176, 183
Boulogne, 284
Boundary Survey, 27
Branson, Charles, 65
Brassey, Thomas, 176, 180, 182, 184, 305
Bray, William B, 193
Brett, Henry, 55, 106-107, 224
Brett, John H, 224
Bride River bridges, 88, 89
Bride, Henry Nelson, 64
Bridges, Ireland, 86-90
Bridges, Jamaica, 228-230
Bright, Sir Charles Tilson, 277-278, 280 (note 24)
Bristol, 41
British Columbia, 222
Broughton, Lord, 177
Broughton, Lt W E D, 44
Brown, Captain Sir Samuel, 45
Brown, Henry, 64
Browne, Henry Philip, 65
Browne, John, 12-13
Bruin Lodge, Cork, 65, 159, 284
Brunett, Sandford, 64
Brussels, 279
Bucharest, 179
Buckingham & Chandos, Duke of, 254
Bulgaria, 178, 189
Bulwer, Sir Henry, 183, 187-188,
Bulwer-Lytton, Sir Edward, 223
Burgoyne, General John Fox, 47, 54-55, 57-59, 67, 90, 143, 163, 181-182, 208, 212, 223, 273-274
Cabra (Thurles), 65, 202
Cahir, 8-10, 26
Canal schemes, 7-10, 113-121
Canning, Sir Stratford, 177-178, 193 (note 13), 305
Cantwell, Dean Walter, 261

Cape Colony, 159-171, 199, 222-223, 296
Cape Town & Wellington Railway, 170
Cape Town, 162 -64, 166
Cappagh, 26
Carbery, Lord, 52
Cardwell, Edward, 248-249, 253
Carlow, 131, 133
Carnavon, Lord, 221-223
Carr's Hill, 84
Carrick-on-Suir, 8, 10
Carrigaline, 76, 84, 89
Carrigrohane Straight, 86, 108
Carter, Samson, 41
Cashel Election, 257-269
Cashel, 7-10, 26, 191, 197, 200-201, 257-267
Castlecomer Colliery, 11
Castlecomer, 8-10
Castlemore, 89
Castletownbere, 68, 74, 80-82, 98
Cathedral of the Assumption, Thurles, 261, 276, 285-286
Catholic University, 197, 283
Charleville, 74, 99, 131
Chelsea, 199, 202
Chetwynd Viaduct, 135
Church of Ireland Churches, 109-110
Church of Ireland, disestablishment, 259, 275-276
Civil Engineer's Society of Ireland, 47, 106
Clare County, 22
Clarendon, Frederick Villiers, 118
Clements, Hill, 17
Clonakilty, 73-74, 77, 99, 129, 146
Clonmel Borough, 42
Clonmel Commons, 42
Clonmel, 8-10, 15, 19, 22, 34, 36, 39, 41, 44-45, 65, 131
Cobh, 99, 126, 128-129
Colby, Colonel, 27
Colliers Quay, 82, 118
Commissioners of Public Works, 54-58, 61, 63-64, 67-68, 82, 100, 114-118, 120, 149
Commons Road, Cork, 84
Coneys, Joseph, 115-118
Connell, H K, 213
Connor, Daniel, 147, 151

Constantinople, 175, 177-181, 184, 186-190, 224, 265, 279
Coomhoola, 79
Coosane Gap, 79
Cork & Bandon Railway, 69, 83, 118, 127-128, 133-139, 146, 224, 249
Cork & Passage Railway, 126-127
Cork City Gaol, 100
Cork City, 57-58, 68, 74, 76, 84, 284
Cork City, liberties, 58, 70 (note 42), 83
Cork County Gaol, 100
Cork County, 22, 52, 74
Cork County, baronies, 60, 70 (note 53),
Cork County, bridewells, 98-99
Cork County, ridings, 60
Cork County, sessions houses, 98-99
Cork Courthouse, 63, 101
Cork Harbour, 121
Cork Lunatic Asylum, 97-98
Cork, Earl of, 67
Cork, Youghal & Cobh Railway, 128
Cork-Bandon Road, 83
Cork-Dublin Road, 84-86
County Surveyors, appointment, 20-21, 54-58
Courtmacsherry, 74, 98
Crimea, 182
Crimean War, 180-183
Croke, Archbishop Thomas, 286
Crookhaven, 74, 77, 79
Crookstown, 79
Crossbarry, 129
Cubitt, William, 46
Cullen, Cardinal Paul, 197, 201, 283, 286, 297
Curragh Chase, 3, 27 (note 4)
Custom House, Dublin, 47, 55, 64
Cyzicus, see Sizicus
Dalhousie, Earl of, 132
Danube Canal, 180, 194 (note 22)
Danube River, 179-180, 182
Dardanelles, 180
Darling, Governor Sir Charles Henry, 227, 230, 231 (note 38), 235, 244, 248
Dawson, Francis, 224, 229, 241-242
de Redcliffe, Viscount Stratford (see Canning, Sir Stratford)

de Vere, Aubrey, 3
Dea, Patrick, 64
Deane, Alexander Sharpe, 135, 141(note 38)
Deane, Sir Thomas, 65, 101, 109
Deane, William Henry, 213
Desart, Dowager Countess of, 250
Desart, Earl of, 250
Dickins & Jones, 218
Directors General of Inland Navigation, 7-10
Dolmabahce Palace, 184
Donegal County, 44, 273
Donemark Bridge, 88-89
Donoughmore, 84
Donovan, Daniel, 64
Douglas Bridge, 84
Douglas, 121
Downdaniel, 119
Doyne, William, 182
Drummond, Thomas, 62
Dry River Bridge, 228-230
Du Maurier, Daphne, 95 (note 63)
Dubbs, Henry, 134
Dublin & Kingstown Railway, 126
Dublin Castle, 13-14, 34, 55-56, 58, 61, 85, 144
Dublin Industrial Exhibition Building, 105
Dublin Tramways, 279
Dublin, 19, 21, 23, 66, 273, 284
Dublin-Cork mail coach road, 7
Dunboy, 80
Dundas, Vice-Admiral Sir James Whitley, 180-181
Dungarvan, 132
Dunkettle Bridge, 90, 95 (note 75)
Dunmanus Bay, 80
Dunmanway, 99
Dunsany, Baron, 250, 254
Durrow, 7
Durrus, 79
Dursey, 80
East India Railway, 176
Ecclesiastical Commissioners for Ireland, 109-110
Edgware Road Station, 287-288
Edwards, Osborne, 64
Engineering education, 34-35

Index

Engineering profession, 21-22, 46-47, 125
Ennis, 131
Estate surveys and maps, 6-7, 17-20
Eyre, Governor Edward John, 235, 238-239, 242-247, 250-253, 254 (note 9)
Family correspondence, 303-310
Famine, 153
Farrell, James Barry, 82, 213
Fenian Rising, 275-276, 300 (note 10)
Fennor, 5, 33
Fermoy, 86, 99, 129
Fethard, 44- 45, 261
Finsbury Park, London, 279-280, 287-288
Fitzgerald, Baron, 263-264
Fitzpatrick, Ulrica May, 218
Five-mile-bridge, 84
Fort Beaufort, 167
Foster, John Leslie, 13, 29 (note 74)
Fox Henderson & Co, 170, 186, 189,
Fox, Sir Charles, 170, 176-177, 180, 184, 305
France, 191
Franklin, Sir John, 222
Fraser, John, 105
Freshford, 10
Galgey, John Matthew, 284, 290
Galgey, William, 284
Galway, 131
Gardiner, Robert Barlow, 224-225
Gibbons, Barry Duncan, 35
Gilman, Edward, 83
Glandore, 68, 98
Glanmire River, 90
Glanmire, 65, 84, 85, 86, 90
Glenflesk, 74
Glengall, Earl of, 8
Glengariffe, 45, 68, 74, 79, 88
Glengariffe-Kenmare Road, 82
Glengoole, 3-4
Glenville, 65, 71 (note 96)
Goold's Cross, 260, 270 (note 20)
Gordon, Captain E R, 183
Gordon, George William, 251-253
Goresbridge, 10
Gortnahoe, 33, 47 (note 2)
Gosset, Sir William, 56
Goulburn, Henry, 13-14

Graham's Town, 164
Grand Canal, 8-10
Grand jury maps, 15-16
Grand jury reform, 52-54
Grand jury system, 51-52
Grant, Charles, 14
Grantham, Richard B, 212
Granville, Earl, 170, 277-278
Great Exhibition Building, Cork, 105
Great River Bridge, 229-230
Greece, 189
Grey, Earl, 53, 160-161, 165, 168-169, 171, 177, 298
Griffith, Dr Patrick OP, 167-168
Griffith, Richard, 20, 22, 27, 74, 114
Gunther, Edward, 25, 31 (note 133)
Harbour View, 80
Hartlepool Docks, 13
Hawkshaw, Sir John, 38, 125, 212, 241, 247-248
Holycross, 9
Holyhead Harbour, 13
Holyhead Road, 13
Hunt, Sir Vere, 3-5, 15
Innishannon, 82, 86, 119
Institution of Civil Engineers (London), 46-47, 223
Institution of Civil Engineers of Ireland, 47, 106
Irish Industrial Society, 274-275
Jamaica Railways, 233-234, 249
Jamaica, 221-230
Johnston, Francis, 19, 97
Johnston, William, 77
Johnstown, 11
Jones, Owen, 286
Joyce, George, 8
Kalamita Bay, 182
Kanturk Bridge, 89
Kanturk, 74, 84, 86, 88-89, 99
Keane, John B, 101
Kearney, Thomas, 213
Keate, Governor, 215-217
Keller, Charles McAuliffe, 159
Kempster, James Forth, 98
Kenmare Bridge, 45

Kenmare, 45, 74, 115
Kensington, 199, 283-288
Kerry County, 55, 74
Kettlewell, Samuel H, 147-149
Kilbrittain, 80
Kildare County, 55, 57
Kilgarvan, 79
Kilkenny County, 11, 17, 22
Kilkenny, 8, 131-133
Killaly, John, 9-10, 22-23, 27, 53
Killarney, 74, 79-80, 129, 131, 284
Killavullen, 86
Killenaule Estate, 19-20
Killenaule, 10, 17, 19, 25-26, 43, 257
Kilmacsimon Quay, 80, 83
King Fitzgerald, Catherine, 162, 199
King's County (Offaly), 55
Kingston, 229-230, 233-234
Kingston, Lord, 67
Kingwilliamstown, 74
Kinsale, 68, 74, 76, 84, 86, 99, 129
Klassen, PJ, 224
Knocklong, 34, 65
Knox, Major Laurence E, 268
Kustengee, 180
La Brea, 216
Laffan, Michael J, 259
Laffan, Patrick, 264
Lahy, see Leahy
Land surveyors, 6-7, 11, 25,
Lanyon, Sir Charles, 55, 57-58, 97, 105-107, 109
Laois County, 11
Larcom, Thomas, 55, 57, 68, 83, 273-274
Lardner, Rev Dionysius, 47
Larkin, William, 6
Layard, Sir Austen Henry, 186-187
Leahy, Albert William Denis, 213-214, 217-218, 219 (note 39), 284
Leahy, Alice, 33
Leahy, Anne, 33-34, 162, 198-201, 283-284, 286-290
Leahy, Catherine (see King Fitzgerald, Catherine)

Leahy, Denis:
- birth, education and training, 33-35
- early work, 36-37
- railway work in England, 37-38
- involvement in county surveyor work, 65-66, 80
- valuation of poor law unions, 110-111
- workhouse construction, 111-112
- contract for Tralee Ship Canal, 117-118
- railway promotion, 131-132
- marriage and family, 202, 213-214, 218
- work in London in the 1850s, 202-203
- Superintendent of Public Works in Trinidad, 211-217, 297-298
- death, 217
Leahy, Edmund:
- birth, education and training, 33-35
- early work, 36, 44-46
- railway work in England, 37-38
- employment on Ordnance Survey, 44, 273
- appointment as county surveyor, 54-58, 273-274
- activities as county surveyor, 59-68, 75-83, 86-89, 97-101, 143-150
- publication of his treatise on roads, 90-93
- private practice in 1834-46, 105-121
- railway work in Ireland, 126-140
- marriage and family, 162, 199-200, 204 (note 6), 283, 287
- at the Cape Colony, 159-171, 222-223
- activities in Greece and Turkey, 175-193
- report on visit to ruins of Sizicus, 191-192, 311-313
- correspondence with his brother, Patrick, 183-4, 190-193, 303-310
- financial transactions with family, 168, 198-203
- telegraph cable schemes, 221, 276-278
- steamship services, 160, 168-169, 222
- as Colonial Engineer in Jamaica, 223-230
- the Jamaica Tramway Scandal, 233-254
- involvement in Cashel Election, 257-269
- final years, 278-280, 283, 286
- death, 287-289
- assessment and attitudes, 294-299

Index

Leahy, Elizabeth, see Anderson, Elizabeth
Leahy, Gerald, 199, 261-262, 283, 287
Leahy, Helena (Ellen), 33, 47 (note 5), 162, 198-201, 283
Leahy, John, 3-4
Leahy, Juliet, 199-200, 204 (note 6), 252, 261, 283, 287
Leahy, Margaret (Cormack), 33, 65, 159
Leahy, Margaret (Mrs Ryan), 33, 162, 284, 286-287
Leahy, Matthew:
  - birth and education, 33-35
  - involvement in county surveyor work, 65-67, 146
  - contractor for Tralee Ship Canal, 114-118
  - railway promotion, 129-134, 138
  - at the Cape Colony, 162, 168, 170
  - in Turkey, 177-180, 189-191, 307-308
  - financial support from brothers, 198-203
  - Superintendent of Public Works in Trinidad, 207-210, 297-298
  - death, 210-211
Leahy, Mick (Lahy, Lahey), 4, 15
Leahy, Patrick, Archbishop of Cashel:
  - birth and education, 33-34
  - career, 1847-57, 197-198, 201
  - correspondence with brothers, 183-184, 190-193, 303-310
  - financial transactions with family, 168, 198-203
  - appointment as archbishop, 201
  - contacts with Colonial Office, 203, 246, 248-249
  - involvement in Cashel election, 258, 260-262, 265-267
  - contact sought by General Lord Strathnairn, 275-276
  - at Vatican Council, 279, 286
  - contacts with family in the 1860s, 283-286
  - construction of Cathedral in Thurles, 285-286
  - views on national issues, 296-297
  - death, 286
Leahy, Patrick (senior):
  - origins and early work, 3-7
  - family, 33-34
  - canal schemes, 7-10
  - work with the Bogs Commissioners, 11-13

- his surveying methods, 13-15
- estate surveys and manuscript maps, 17-20, 43
- applications for county surveyor post, 20-21, 54-58
- his Tables of Weights & Measures, 23-25
- railway work in Ireland, 25-26, 126-134
- work in England, 37-38
- maps of Waterford City and Clonmel, 38-42
- admission to Institution of Civil Engineers, 46-47
- as county surveyor, 59-68, 75-77, 83-90, 97-101, 143-145, 150-152
- private practice in 1834-46, 107-118
- at the Cape Colony, 162, 167-168, 200
- death, 167
- assessment and attitudes, 293-296
Leahy, Susan, 33-34, 162, 198-201, 209, 211, 283-284, 286-287
Leahy, Tom, 4
Leahy, Ulrica May, see Fitzpatrick, Ulrica May
Leahy, William Henry, 214, 217-218, 219 (note 40)
Lee, Joseph, x, 136-137, 295
Leeds, 38
Lennox, Lady Adelaide Constance, 199, 305
Lennox, Lord Arthur, 199, 204 (note 7)
Liberty Hill, 83
Limerick & Waterford Railway, 25, 35
Limerick City, 26, 131
Limerick County, 3, 74
Lincoln, 11
Lismore Workhouse, 112-113
Lismore, 86
Listowel, 74, 112-113
Littleton, 11
Liverpool & Leeds Railway, 38
Liverpool & Manchester Railway, 37-38
Liverpool, 37-39
London, 33, 160, 168, 184, 186, 192, 198, 201-203, 207, 211, 213-214, 217- 218, 221, 248, 250, 273, 279, 283-284
Longfield, John, 17-19,
Lough Neagh, 11-12
Lough Road, Cork, 84
Lower Glanmire Road, Cork, 65, 70 (note 51), 84, 90

MacCall, Mrs Eliza, 85-86
Macedonia, 189
MacNeill, Sir John, 35, 135-136
Macroom, 74, 80, 86, 99, 129
Mahmut II, Sultan, 178
Mail coach roads, 7, 73-74
Malcomson, David, 42
Mallet, J & R, 115-116
Mallow Workhouse, 112
Mallow, 73-74, 89-90, 99, 129
Mallow, elections, 267-269
Manchester, 37-38
Manuscript maps, 17-19, 38-42
Margate, 283
Market Hill, 45
Maryborough (Portlaoise), 10
Mauritius, 160
Mayo, Earl of (see Naas, Lord)
McCarthy, John C, 268-269
McClintock, F L, 222
McDowell, Daniel, 241
McNamara, James, OFM, 264
Mealagh River, 88
Meighan, Father Michael, 33
Mejid, Sultan Abdul, 184
Melbourne, 284
Menai Strait, 45
Meredith, Benjamin, 26
Merivale, Herman, 222
Metropolitan Railway, 287-288, 291 (note 43)
Michell, Lt Colonel, 165
Midleton, 99, 132
Millstreet, 80, 90, 99, 129
Mitchel, John, 166
Mitchell, James, 64
Mitchelstown, 99
Mizen Head, 79
Moldavia, 180
Monasterevan, 9-10
Montagu, John, 164-165
Montego Bay, 229
Moorsom, Captain William Scarth, 137-138
Morant Bay, 252
Moriarty, Bishop David, 276, 280 (note 18), 284
Morris, Mr Justice Michael, 268, 271 (note 47)

Morrison, Richard, 101
Morrison, William Vitruvius, 101
Mount Pelion, 190, 192
Mountcashel, Lord, 67
Mountrath, 10, 131
Moville, 44
Moyle River, 44-45
Mudge, General William, 12, 29 (note 70)
Mullinahone, 17
Munster, Henry, 257-269, 278-279, 286
Munster, William Felix, 268-269, 271 (note 54)
Murchison, Sir Roderick, 192, 311
Murphy, Denis B, 64
Murphy, Sergeant MP, 171
Murray, William Johnston, 97
Mushera Mountains, 80
Naas, Lord, 274-275
Natal, 168-169
Neale, Lt Colonel, 178-179
Nenagh Workhouse, 112
Nenagh, 101, 131
Neville, John, 5, 15,
New Bermingham, 4, 5, 8,10-11, 15, 17, 28 (note 15)
New Holland, 160
New York, 279
New Zealand, 160
Newcastle, Duke of, 207-208, 211-213, 225-226, 244-246, 248
Newcastlewest, 132
Newfoundland, 221
Newman, Cardinal John Henry, 197, 201, 204 (note 2)
Newmarket, 74
Nicholas I, Tsar of Russia, 180, 182
Nimmo, Alexander, 11, 22, 26-27, 35, 37-38, 58, 121, 212-213
Nish, 185-186
Nolan, David M, x
Norbury, Lord, 293, 296, 300 (notes 1 and 7)
North-east Passage, 222
North-west Passage, 222
O'Beirne, James Lyster, 258-263, 269 (note 4)
O'Brien, William Smith, 152
O'Carroll, Rev James, 201, 217, 294
O'Connell, Daniel, 52, 54, 114

Index

O'Connor, Feargus, 57
O'Connor, Rev Thomas, 37
O'Dwyer, Frederick, x
O'Leary, Edmund, 289
O'Leary, John, 289
Offaly County, 11
Old Harbour, 234, 249
Olivier, Lord Sydney, 249, 253
Ordnance Survey of Ireland, 21-22, 27, 41, 43-44, 83, 273
Orsova, 179
Ottoman Empire, 175-176, 184, 189
Overend, Gurney & Co, 258
Owen, Jacob, 54, 58, 117, 120
Paarl, 163-164
Pain, George Richard, 63, 100-101, 109
Pain, James, 63, 100, 109-110
Palmerston, Viscount, 175-177
Pandermo (Panorma, Bandirma), 191, 311
Parnell, Sir Henry, 90
Parry, John, 225, 246
Passage, 84, 126-129
Peel, Sir Robert, 147
Pelion Mines, 189-193, 202-203
Peto, Samuel Morton, 182
Petty, Sir William, 16
Phillipopoli, 179, 180, 185-186
Pilkington, Captain George, 163, 222-223
Pietermaritzburg, 169
Plymouth, 221
Pobal O'Keeffe, 74
Poor Law Commissioners, 110-113
Poor Law Valuation, 110-111
Porro, Ignazio, 15
Port Elizabeth, 164
Port of Spain, 210, 214-216
Portglenone, 33
Portlaoise, 10
Porus, 234, 249
Pouladuff Road, Cork, 84
Price, George, 237-240, 244, 250-253, 255 (notes 24-26)
Pruth River, 180
Puxley, John, 80-82, 95 (note 63)
Queen's College, Cork, 109
Queen's County (Laois), 9

Quirke, Archdeacon, 261
Radcliffe, John, 54-55, 70 (note 21)
Raglan, Field Marshal Lord, 181
Railway engineering, 125, 136-139
Railway schemes in Ireland, 25-26, 126-132
Railway surveys in England, 37-38
Ramsgate, 283, 289-290
Ranelagh Sewer, 202
Rastrick, John Urpeth, 39
Rathcoole Bridge, 89
Rathkeale, 132
Red Sea, 160
Red Strand, 77
Redesdale, Earl of, 130
Regan, James, 64
Rennie, John, 6, 13
Rennie, Sir John, 136
Reshid Pasha, Mustafa, 184, 305
Ring (Clonakilty), 74, 98
Rio Cobre Bridge, 228
Road construction, Cork East Riding, 83-86
Road construction, Cork West Riding, 77-83
Road contractors, 75-77
Road maintenance, 74-77
Roaring Water Bay, 80
Rock of Cashel, 262
Rome, 191, 265, 277, 279, 286
Roscrea, 9, 10
Rose, General Hugh Henry, 181, 273-276, 280 (note 13)
Rosscarbery, 74, 77, 99,
Roumelia, 178
Roy, General William, 12, 29 (note70)
Royal Canal, 7
Ruschuk, 179-180
Russell, Edward, 82
Russell, John Scott, 38, 208, 213, 228-229
Russell, Lord John (later Earl Russell), 187-188, 252, 298
Russell, W H, 182, 184-185
Russia, 176, 180
Ryan, George, 284
Ryan, James Patrick, 284
Ryan, Jane Mary (nee Tidmarsh) 284
Ryan, Louisa, 284

Ryan, Mary Elizabeth (Minnie) (Mrs Galgey), 284, 290
Ryan, Mrs Margaret, see Leahy, Margaret
Ryan, Patrick, 130, 284, 290 (note 10)
Rynne, Colin, x
Sadleir, John, 199, 204 (note 13), 295
Salonika, 192
Samuel, Lewis W, 207
Scutari, 183
Sea of Marmara, 19, 311-313
Sevastopol, 182
Shannon, Earl of, 67
Sheep's Head, 80
Sheerness Royal Dockyard, 13, 56
Shiel, R L, 163
Shrewsbury, Earl of, 250
Silistria, 180
Simon's Bay, 165
Sizicus (Cyzicus), 191-192, 195 (note 81), 311-313
Skibbereen, 74, 76-77, 80, 99, 129
Skibbereen-Crookhaven road, 106,
Skull, 99
Slattery, Archbishop Michael, 197-198, 201
Slieveardagh Barony, 3, 8
Slieveardagh Barony, map of, 10, 17-19, 22, 26, 29 (note 51)
Slieveardagh Collieries, 8, 10, 17, 19, 132
Smith, Adam, 91-92
Smith, David, 233-243, 247
Smith, Lieutenant-General Sir Harry 163, 165, 172 (note 24), 199-200, 304
Smith, William, 233
Smyrna (Izmir) 187
Snave, 79, 88
Sofia, 179, 185-186
South Africa, see Cape Colony
South Gate Bridge, 86
South Mall, Cork, 107
Spanish Town, 228, 233-234, 249
St Dorothy (Parish), 239-240
St Leger, Noblett, 41
St Mary's RC Cemetery, London, 289, 291 (note 46)
St Patrick's Bridge, Cork, 68
St Patrick's College, Maynooth, 34, 40, 65

St Patrick's College, Thurles, 34, 37, 197, 204 (note 2), 261
Stanford, Captain Robert, 166-167, 171
Stanley, Edward, (Earl of Derby), 53
Stanley, Lord, 277, 299
Stellenbosch, 163-164
Stephenson, Sir Rowland Macdonald, 176
Stokes, Charles, 180, 184, 305
Stokes, Henry, 55, 82, 107, 112
Storks, Brigadier-General Henry K, 183
Strathnairn, Baron (see Rose, Hugh Henry)
Suez, 160
Suir Navigation, 8, 19
Suir River, 8-10, 40
Sullivan, Edward QC, 267
Sullivan, Jeremiah, 64-65
Surveying instruments and methods, 13-14, 25
Suspension bridges, 44-45
Swansea, 80
Swellendam, 170
Synod of Thurles, 37, 197
Table Bay, 162
Tate, Alexander, 106
Taunton, John H, 212
Tayleur Engineering Works, 134
Taylor, Alexander, 19
Taylor, Thomas, 41
Telegraph cable schemes, 175-176, 221-222, 276-278
Telford, Thomas, 19, 45-46, 90, 120
Templemore, 8-9
Thessaly, 189-192
Thomas, John, 12-13, 46, 56
Thurles Poor Law Union, 111
Thurles Workhouse, 112, 123 (note 47)
Thurles, 9, 14-15, 17-19, 26, 33-34, 36, 48 (note 16), 56, 101, 191, 201, 217, 259, 264, 286
Tierney, Dom Mark, 286, 303
Timoleague, 74, 80
Tipperary Bank, 199, 204 (note 13), 295
Tipperary Baronies, mapping, 296
Tipperary County, 3, 7, 11, 15-19, 21-22, 26, 33, 36, 43, 56-57, 74
Tipperary County, elections, 265
Tipperary Poor Law Union, 111

Tipperary Town, 8, 26
Tipperary Workhouse, 112
Torrington, Viscount, 202
Townsend, Horace, 213
Townshend, Richard, 46
Townshend, Thomas, 7-13, 16, 29 (notes 59 and 64), 38-39, 46, 56
Tragumna, 80
Tralee Ship Canal 113-118
Tralee, 113, 129, 131, 284
Tramore, 132
Tramway schemes, 278-279
Treacy, William Augustus, 153, 155 (note 57)
Trinidad, 207-218
Trinity College, Dublin, 34-35, 297
Troughton, Edward, 13-15, 30 (note 83)
Tucker, John Scott, 223
Turkey, 202-203
Turkish baths, 279-280, 287
Two-Mile-Bridge, 19
Union Bridge, 45
Union Hall, 74
Urlingford, 5, 7, 9, 11, 17, 33
Valencia, 129, 221
Valentine, John, 134
Van Diemen's Land, 160, 167
Vancouver, 222
Varna, 178-181, 221
Vere Hunt Estate, 3
Vienna, 175, 186, 189, 264, 279
Vignoles, Charles, 126, 136, 137, 208
Volos, 190, 192
von Hirsh, Baron Maurice, 189

Waldron, Lawrence, 19
Walker, Lieutenant-Governor, 207-211
Wallachia, 180
Warrington, 132
Waterford City, 8, 26, 56, 132
Waterford City, maps of, 38-41
Waterford County, 22, 42
Waterford Harbour, 41
Waters, Captain M A, 44
Waters, George QC, 266-268, 271 (note 50)
Weale, John, 211
Webb, Sidney and Beatrice, 93
Weights and Measures, 23-24
Wellesley, Marquis, 11, 13, 57
Wellington, 164
Western District, Ireland, 38
Western Road, Cork, 86
Westmeath County, 11
White, John Davis, 259, 267
Wicklow Mountains, 21
Wicklow, 131
Wilkinson, George, 111-112
Williams, Col Edward, 12, 29 (note70)
Wilson, Edward, 14
Witham Navigation, 11
Woodhouse, Thomas Jackson, 55, 57, 62, 97, 109
Woodward, Benjamin, 109
Workhouse construction, 111-113
Youghal, 73, 90, 99-100, 128-129, 132, 151
Young, Arthur, 73
Zarify, George, 186, 189